Ecological Studies, Vol. 217

Analysis and Synthesis

Edited by

Ecological Studies

Further volumes can be found at springer.com

Volume 199
Hydrological and Biological Responses to Forest Practices: The Alsea Watershed Study (2008)
J.D. Stednick (Ed.)

Volume 200
Arid Dune Ecosystems: The NizzanaSands in the Negev Desert (2008)
S.-W. Breckle, A. Yair, and M. Veste (Eds.)

Volume 201
The Everglades Experiments: Lessons for Ecosystem Restoration (2008)
C. Richardson (Ed.)

Volume 202
Ecosystem Organization of a Complex Landscape: Long-Term Research in the Bornhöved Lake District, Germany (2008)
O. Fränzle, L. Kappen, H.-P. Blume, and K. Dierssen (Eds.)

Volume 203
The Continental-Scale Greenhouse Gas Balance of Europe (2008)
H. Dolman, R.Valentini, and A. Freibauer (Eds.)

Volume 204
Biological Invasions in Marine Ecosystems: Ecological, Management, and Geographic Perspectives (2009)
G. Rilov and J.A. Crooks (Eds.)

Volume 205
Coral Bleaching: Patterns, Processes, Causes and Consequences
M.J.H van Oppen and J.M. Lough (Eds.)

Volume 206
Marine Hard Bottom Communities: Patterns, Dynamics, Diversity, and Change (2009)
M. Wahl (Ed.)

Volume 207
Old-Growth Forests: Function, Fate and Value (2009)
C. Wirth, G. Gleixner, and M. Heimann (Eds.)

Volume 208
Functioning and Management of European Beech Ecosystems (2009)
R. Brumme and P.K. Khanna (Eds.)

Volume 209
Permafrost Ecosystems: Siberian Larch Forests (2010)
A. Osawa, O.A. Zyryanova, Y. Matsuura, T. Kajimoto, R.W. Wein (Eds.)

Volume 210
Amazonian Floodplain Forests: Ecophysiology, Biodiversity and Sustainable Management (2010)
W.J. Junk, M.T.F. Piedade, F. Wittmann, J. Schöngart, P. Parolin (Eds.)

Volume 211
Mangrove Dynamics and Management in North Brazil (2010)
U. Saint-Paul and H. Schneider (Eds.)

Volume 212
Forest Management and the Water Cycle: An Ecosystem-Based Approach (2011)
M. Bredemeier, S. Cohen, D.L. Godbold, E. Lode, V. Pichler, P. Schleppi (Eds.)

Volume 213
The Landscape Ecology of Fire (2011)
D. McKenzie, C.S. Miller, D.A. Donald (Eds.)

Volume 214
Human Population: Its Influences on Biological Diversity (2011)
R.P. Cincotta and L.J. Gorenflo (Eds.)

Volume 215
Plant Desiccation Tolerance (2011)
U. Lüttge, E. Beck, D. Bartels (Eds.)

Volume 216
Forest Hydrology and Biogeochemistry: Synthesis of Past Research and Future Directions (2011)
D.F. Levia, D. Carlyle-Moses, T. Tanaka (Eds.)

Volume 217
Peatlands of the Western Guayana Highlands, Venezuela (2011)
J.A. Zinck and O. Huber (Eds.)

Joseph Alfred Zinck • Otto Huber
Editors

Peatlands of the Western Guayana Highlands, Venezuela

Properties and Paleogeographic Significance of Peats

Springer

Editors
Professor (em.) Dr. Joseph Alfred Zinck
University of Twente
Faculty of Geo-Information Science and
Earth Observation (ITC)
PO Box 6
7500 AA Enschede, The Netherlands
zincka@itc.nl

Professor Dr. Otto Huber
Fundación Instituto Botánico de
Venezuela (FIBV)
Apartado 2156
Caracas 1010-A, Venezuela
ohuber@mac.com

ISSN 0070-8356
ISBN 978-3-642-20137-0 e-ISBN 978-3-642-20138-7
DOI 10.1007/978-3-642-20138-7
Springer Heidelberg Dordrecht London New York

Library of Congress Control Number: 2011934442

Cover design: deblik Berlin, Germany

Printed on acid-free paper

Springer is part of Springer Science+Business Media (www.springer.com)

In Memoriam

We dedicate this book to the memory of Richard Schargel, our mutual friend and contributor to the present volume, who passed away prematurely on March 24, 2011.

Foreword

When one stands somewhere on the Gran Sabana at about 1,400 m above sea level in southern Venezuela and looks around the vast open area, one recognizes the characteristic up to 1,400 m higher rock tables of the Tepuis limiting the horizon. By the sight the naturalist is caught by strong sentimental yearning for being there. He knows that it is an extreme environment up there continuously wet and cool. We already get a touch of it where we stand on the Gran Sabana with its palm swamps, the morichales of *Mauritia flexuosa* of which ALEXANDER VON HUMBOLDT wrote: "*Sie bildet an feuchten Orten herrliche Gruppen von frischem glänzendem Grün, das an das Grün unserer Ellergebüsche erinnert. Durch ihren Schatten erhalten die Bäume die Nässe des Bodens* ..." (Alexander von Humboldt 1849). The other face of the medal is that increasing comfort of (eco-) tourism lifts the mystery and provides access to almost everybody, increasingly endangering persistence of this extraordinary natural environment. What we need, of course, is neither yearning nor casual touristic encounter but solid scientific information, knowledge, and understanding. Comprehensive treatment so far does not exist, but this important gap is now filled by this book edited by ZINCK and HUBER.

Peatlands are more dominant in temperate and boreal than in subtropical and tropical zono-biomes of the Earth. The latter only comprise about 10% of all global peatlands. They are peatland ecosystems in tropical tidal flats, river deltas, coastal forest swamps, and flood plains of river basins with moor forests. Simply due to their low abundance they all are very unique. Most unique among all these tropical moor and peatlands are those of the tropical highlands and on top of the Tepuis in Venezuela, the major topic of this book. These peatlands provide the most vivid illustration of the general aspect that life on the Earth is working up the geological surface of the Earth. How life is shaping the geostructural ground to form peat and peatlands is unraveled in the book by interpretation of the chronological record in terms of past environmental changes in geological times. Morphological, physical, and chemical characteristics of the peats and their peculiar soils are described using modern methods of assessment including the use of carbon isotopes with ^{14}C dating of peat deposits and ^{13}C analyses of vegetation history. Coverage of geoecological developments and the structure of extant vegetation provide deep insights into life in these peat landscapes. Studies of physiological ecology of plants of the Tepui peatlands as far as I know are scarce or in fact virtually absent. One conspicuous exception is the pitcher plant genus *Heliamphora* which is endemic to the Tepuis and which comprises carnivorous plants. It has been studied in detail (Jaffe et al. 1992) and received great attention in the literature because as outlined in this book the peats are extremely poor in mineral nutrients especially phosphorus. Carnivory is a means of acquisition of nutrients by plants. The book shows us that other carnivorous plants are abundant in the Tepui peat vegetation, such as species of *Drosera* and *Utricularia*, and there are the species of the tank forming Bromeliaceae genus *Brocchinia* where – especially in *B. reducta* and *B. hechtioides* – we recognize kind of a "protocarnivory", i.e., indications toward evolution of full carnivory (Jolivet and Vasconcellos-Neto 1993). In addition to this subjectively picked example, the reader of the book will find a wealth of other suggestions provoking ecophysiological studies and stimulating further research.

In a changing world with largely unforeseeable global perturbations, the book is an invaluable documentation of an extraordinary natural heritage. The peats of the tropical highlands are a heritage like rainforests and savannas – but much less in the medial limelight – that global ethics should demand to protect and preserve not only because of their uniqueness and natural beauty but – due to farther reaching ecological and environmental implications – also in the very interest of keeping the planet inhabitable for mankind. This is forcefully advocated in this book, which provides a wealth of data, solid knowledge, and sympathetic understanding of the peatlands.

Darmstadt, Germany
March 2011

Ulrich Lüttge

References

Alexander von Humboldt (1849) *Mauritia flexuosa "at moist places forms magnificent groups of fresh and shiny greenery, which recalls the green of our alder bushes. With their shade these trees maintain the moisture in the soil ...".* Ansichten der Natur. Cotta, Stuttgart

Jaffe K, Michelangeli F, Gonzalez JM, Miras B, Ruiz MC (1992) Carnivory in pitcher plants of the genus *Heliamphora* (Sarraceniaceae). New Phytol 122:733–744

Jolivet P, Vasconcellos-Neto J (1993) Convergence chez les plantes carnivores. Recherche 24:456–458

Acknowledgments

This publication presents the results of fieldwork carried out in the western section of the Guayana Highlands between 1992 and 1996. Many persons and institutions have contributed to make our endeavor possible providing financial, administrative, and logistic support. To all of them we would like to extend our sincere thankfulness.

Fieldwork has been supported basically by the *Corporación Venezolana de Guayana – Electrificación del Caroní C.A.* (CVG-EDELCA in Puerto Ordaz), which has provided the overall logistic support to the numerous scientific expeditions planned and organized by O. Huber in the Venezuelan Guayana in the 1982–1996 period. Key persons for the successful realization of this program were H. Roo†, G. Febres, F. Barreat, A. Barreto, A. Lezama, and F. Díaz in EDELCA, as well as the intrepid helicopter pilots of AEROTECNICA (Charallave) and EDELCA. The expedition to the Duida-Marahuaka massif in 1992 has been organized and financed by the *Agencia Española de Cooperación Económica* (AECI, Caracas office) and led by J. Ayarzaguena, whereas the expedition to the Cuao-Sipapo massif in 1993 has been organized by O. Huber, *Centro Amazónico de Investigaciones Ambientales "Alejandro de Humboldt"* (CAIAH), and financed by the *Deutsche Gesellschaft für Technische Zusammenarbeit* (GTZ), together with the *Servicio Autónomo para el Desarrollo Ambiental del Estado Amazonas* (SADA-Amazonas) ascribed to the Venezuelan *Ministerio del Ambiente y de los Recursos Naturales Renovables* (MARNR). Both expeditions in the Estado Amazonas have been authorized by the *Instituto Nacional de Parques* (INPARQUES) with the active collaboration of R. García. E. Cuevas (IVIC) and C. Briceño (CVG) also participated in the fieldwork.

For the botanical collaboration, O. Huber acknowledges the support of the *Herbario "V.M. Ovalles"* (MYF) and its curator, S. Tillett of the Faculty of Pharmacy of the *Universidad Central de Venezuela* (UCV), and the *Herbario Nacional de Venezuela* (VEN) and its collaborator R. Riina, both in Caracas. Numerous specialists in plant taxonomy from American and European herbaria have offered their expertise in identifying and describing the voluminous plant collections deposited in VEN and MYF.

Analytical determinations of the peatsoils were carried out in the soil laboratory of the *Ministerio del Ambiente y Recursos Naturales Renovables* (MARNR) in Guanare, Venezuela. The radiocarbon age of selected peat samples was determined at the Center for Isotope Research (CIO) of the University of Groningen, the Netherlands, with the financial support of the International Institute for Geo-Information Science and Earth Observation (ITC), Enschede, the Netherlands, and CVG-EDELCA, Venezuela.

The editors and authors of this book thank U. Lüttge, Darmstadt, for accepting to write the preface to the book and gratefully acknowledge the active and helpful collaboration of O. Lange, Würzburg, and that of A. Schlitzberger, Springer-Verlag Heidelberg, during the final process of preparation and illustration of the manuscript for its publication in the Series *Ecological Studies*.

Enschede, The Netherlands
Meran, Italy
March 2011

Joseph Alfred Zinck
Otto Huber

Contents

Contributors

E. Cuevas College of Natural Sciences, University of Puerto Rico, PO Box 23360, San Juan PR 00931-3360, Puerto Rico, epcuevas@gmail.com

P. García Empresa EDELCA (Electrificación del Caroní), Urb. San Luis, Av. Principal, Residencias Torreana, El Cafetal, Caracas 1060, Venezuela, prgm2002@gmail.com

O. Huber Fundación Instituto Botánico de Venezuela (FIBV), Apartado 2156, Caracas 1010-A, Venezuela, ohuber@mac.com

D. Jiménez Barquisimeto, Estado Lara, Venezuela

E. Medina Centro de Ecología, Instituto Venezolano de Investigaciones Científicas (IVIC), Caracas 1020-A, Venezuela, medinage@gmail.com

R. Schargel Universidad Nacional Experimental de los Llanos "Ezequiel Zamora" (UNELLEZ), Urb. San Francisco Av. 4 No 54, Guanare, Estado Portuguesa, Venezuela, rschargel@cantv.net

J. van der Plicht Center for Isotope Research, Groningen University, Nijenborgh 4, 9747 AG Groningen, The Netherlands, J.van.der.Plicht@rug.nl

J.A. Zinck Faculty of Geo-Information Science and Earth Observation (ITC), University of Twente, PO Box 6, 7500 AA Enschede, The Netherlands, zincka@itc.nl

Chapter 1
Introduction

O. Huber and J.A. Zinck

The neotropical biogeographic region of Guayana lies mostly on the ancient Guayana Shield (or "Guiana Shield") in the northeast of South America, covering an area of approximately 2.5 million km^2. This large region extends from the shores of the Atlantic Ocean in the east (50°W) to the foothills of the Colombian Andes in the west (73°W), and from the lower Orinoco river in Venezuela in the north (8°N) to the lower Amazon river in Brazil in the south (3°S). It includes parts of Venezuela, Colombia, and Brazil, and the full extent of the three Guianas (i.e., Guyana, the former British Guiana; Suriname, the former Dutch Guiana; and the French Guiana or Guyane) (Fig. 1.1). This remote land, often called the "Lost World" after the title of a novel written by British author Conan Doyle (1912), is emplaced on one of the oldest continental nuclei of the western hemisphere, the Archean "Amazon craton" formed around 3,500 million years (3.5 Ga) ago.

The biogeographic Guayana region in Venezuela shows a variety of landscapes from sea level in the Orinoco river delta up to the summits of the Guayana Highlands, which are the second highest mountain massif in the Neotropics after the Andes. The densely forested lowland plains and peneplains are home to numerous indigenous groups among which the Yanomami and Ye'kwana Amerindians are best known. The intermediate uplands are formed by two large plateaus showing several planation surfaces: the Gran Sabana region, located in the southeast of the Bolívar state, and the extensive mid-elevation mountain range of Sierra Parima that forms the eastern boundary of the Amazonas state. More than 50 impressive table-mountains jut up from the forested lowlands and undulating savanna-covered uplands to form the Guayana Highlands. These mostly flat-topped mesetas, called "tepuis" by some local Indian groups, extend mainly between 2,000 and 2,500 m a.s.l. The highest elevations are Cerro de la Neblina (or Pico da Neblina), located near the Equator along the southernmost Brazilian–Venezuelan border, with 3,014 m a.s.l. (Brewer-Carías 1988), and Mts. Roraima and Marahuaka that reach elevations of about 2,800 m a.s.l.

Tepui mesetas harbor unique life forms and outstanding biological treasures. Spatially isolated since remote times from the surrounding biota by steep rocky walls of up to 1,000 m elevation, the high-mountain ecosystems of the Guayana

J.A. Zinck and O. Huber (eds.), *Peatlands of the Western Guayana Highlands, Venezuela*, Ecological Studies 217, DOI 10.1007/978-3-642-20138-7_1,

Fig. 1.1 Location of the Venezuelan Guayana (*bold contour line*), the biogeographic Guayana region (*hatched line*), and the underlying geologic Guayana Shield (*dotted line*) (modified from Huber and Foster 2003)

Highlands represent perhaps the only truly pristine environments still existing on earth. It was hardly 125 years ago that scientific exploration of the tepui summits started, yielding since then an astonishing amount of plants and animals never seen before elsewhere. Many of these unique plants and animals, biogeographically included in the term "Pantepui biota," are growing and living on thick, black, extremely acid, water-saturated organic soils that form high-mountain peatlands. Peat bogs lie directly atop the rocky surface of the table-mountains, which in most cases consists of quartz sandstone and, to a lesser degree, granitic and gneissic rocks.

The first paleoecological research on tepui peats was carried out scarcely 25 years ago in the eastern Guayana Highlands, Bolívar state (Schubert and Fritz 1985; Briceño and Schubert 1990; Briceño et al. 1990; Rull 1991; Huber 1992; Schubert et al. 1994). Following these pioneer studies, three multidisciplinary field missions including paleoecological research groups took place between 1991 and 1996 in the central and western parts of the Venezuelan Guayana Highlands, Amazonas state: the first (November 1991 and March 1992) to the tepuis Duida, Marahuaka and Huachamakari in the Duida-Marahuaka National Park; the second (February 1993) to the Cuao-Sipapo massif including Cerro Autana; and the third (March 1996) to the northern section of Sierra de Maigualida. One major objective of all three missions was the biophysical characterization of the tepui-summit environment. The present publication focusing on tepui peatlands is a product of these field missions.

For dealing with tropical highland landscapes, the subject of this book relates partly to two earlier volumes of Ecological Studies. Volume 146 on *Inselbergs:*

Biotic Diversity of Isolated Rock Outcrops in Tropical and Temperate Regions (edited by Porembski and Barthlott 2000) describes the variety of plant communities and vegetation types that develop on granitic and gneissic domes. Individual tepui mesetas could be etymologically assimilated to inselbergs (from German Insel = island and Berg = mountain) as they appear in the landscape in the form of rocky table-shaped islands overhanging the surrounding lowlands. Although inselberg and tepui bear some similarity in the sense that both are partly to largely devoid of vegetation cover, the concept of inselberg is usually restricted to dome-like reliefs developed on igneous–metamorphic rocks in the lowlands, while tepuis are assemblages of mesetas, rather than isolated relief units, constituted by siliceous sedimentary rocks and circumscribed by high vertical walls reaching the highlands. Igneous–metamorphic mountain massifs are much more extensive in the western Guayana Highlands than siliceous sedimentary plateaus, but bare inselberg-type rock outcrops are less frequent than rocky mesetas. Therefore, peatlands are mainly found on the large, flat to slightly concave tepui summits, while they are less common and less extensive on the narrow, convex dome-backbones. This is the reason why our work concentrates on tepui peats and secondarily refers to peats developed on igneous–metamorphic mountains. The second volume bearing some relationship with our study topic is volume 198 on *Gradients in a Tropical Mountain Ecosystem of Ecuador* (edited by Beck et al. 2008). The tepui landscape with vertical cliffs several hundreds meter high offers limited possibilities for vegetation gradients to develop, although vertical vegetation shifts caused by past climate changes are still matter of debate. By contrast, vegetation gradients with elevation from lowlands to uplands to highlands are characteristic of the igneous–metamorphic massifs. Thus our study is to a given extent complementary to these two former volumes.

The book is composed of nine chapters covering a variety of features including geology, geomorphology, vegetation, and soils that characterize the biophysical environment of tepui summits in the western Guayana Highlands, southern Venezuela. The core subject concentrates on the properties, age record, and paleogeographic significance of the peats that have deposited during the Holocene in (pseudo-)karstic depressions on siliceous sedimentary and igneous–metamorphic rocks. As tropical peatlands, especially tropical highland peats, are much less known than temperate and boreal peats, the book starts providing in Chap. 2 an overview on tropical and subtropical peats. Chapter 3 gives a description of the biophysical environment of the Guayana region as a whole, with emphasis on geology, climate, and vegetation, and then zooms into the highland study areas focusing on the vegetation cover and the floristic composition of the formations found atop the tepuis and specifically on peatlands. Types of vegetation cover, hydrologic regime, and organic materials make the tepui peatlands a very special ecosystem. However, what really distinguishes tepui and dome peatlands from others is their geomorphic setting, as peat formation and development are intimately related to specific (pseudo-)karstic landscape features that are described in Chap. 4. The physical and chemical characterization of organic soils in general and peats in particular requires the application of special laboratory techniques and procedures. These are described in Chap. 5, highlighting the ad hoc adaptations that were needed to carry out the determinations. Using laboratory and field data, the morphological, physical, and chemical properties of the peats are described, analyzed, and interpreted in Chap. 6, followed by a discussion on the taxonomic classification of the peat materials. The following Chap. 7 deals with the description of the peat sampling sites, the carbon-14 dating of selected peat layers, and the interpretation of the peat age record with respect to peat formation and environmental changes during the Holocene. The peat age record of the western Guayana Highlands documented here is unique, as no new data collection has taken place in these areas since our exploration missions in the early 1990s. Chapter 8 is an extra-regional contribution focusing on the origin of organic matter that leads to peat formation based on data from the southeastern Guayana Highlands. A synthesis compiled from the relevant conclusions of the individual contributions is presented in Chap. 9. The book closes with an Appendix which provides the field description of profiles and sites and the corresponding laboratory data. Included are also some mineral soils that are usually associated with the organic soils on the tepui summits as well as some typical lowland soils that are derived from the weathering of the tepui rocks.

References

Beck E, Bendix J, Kottke I, Makeschin F, Mosandl R (eds) (2008) Gradients in a tropical mountain ecosystem of Ecuador, vol 198, Ecological Studies, Springer, Berlin

Brewer-Carías C (ed) (1988) Cerro de la Neblina: resultados de la expedición 1983–1987. FUDECI, Caracas

Briceño H, Schubert C (1990) Geomorphology of the Gran Sabana, Guayana Shield, southeastern Venezuela. Geomorphology 3:125–141

Briceño H, Schubert C, Paolini J (1990) Table-mountain geology and surficial geochemistry: Chimantá Massif, Venezuelan Guayana Shield. J S Am Earth Sci 3(4):179–194

Doyle AC (1912) The lost world. Hodder & Stoughton, New York

Huber O (ed) (1992) El macizo del Chimantá, Escudo de Guayana, Venezuela. Un ensayo ecológico tepuyano. Oscar Todtmann Editores, Caracas

Huber O, Foster M (eds) (2003) Conservation priorities for the Guayana Shield: 2002 consensus. Conservation International – CABS, Guiana Shield Initiative, UNDP, IUCN, Washington DC

Porembski S, Barthlott W (eds) (2000) Inselbergs: biotic diversity of isolated rock outcrops in tropical and temperate regions, vol 146, Ecological Studies. Springer, Berlin

Rull V (1991) Contribución a la paleoecología de Pantepui y la Gran Sabana (Guayana Venezolana): clima, biogeografía y ecología. Scientia Guaianae 2, CVG-EDELCA, Caracas

Schubert C, Fritz P (1985) Radiocarbon ages of peat, Guayana Highlands (Venezuela). Some paleoclimatic implications. Naturwissenschaften 72:427–429

Schubert C, Fritz P, Aravena R (1994) Late quaternary paleoenvironmental studies in the Gran Sabana (Venezuelan Guayana Shield). Quatern Int 21:81–90

Chapter 2
Tropical and Subtropical Peats: An Overview

J.A. Zinck

2.1 Introduction

Peats are frequent in cool temperate and boreal regions, but occur also in tropical and subtropical areas. While the distribution and characteristics of peats at higher latitudes have been well documented, peats at lower latitudes are less known. Unrecorded tropical peat reclamation efforts date back several centuries, as, for instance, the reclamation of coastal soils undertaken by the Dutch in the seventeenth century north of Colombo in Sri Lanka. However, it was not until the late 1890s when Koorders provided the first formal description of tropical peats from the forests of Sumatra (Andriesse 1988). Since this early reference, knowledge on tropical peatlands has made progress but by far not at the same pace as the knowledge on temperate and boreal peatlands. The lack of systematic surveys limits the scope of the updating reviews (Shier 1985). One of the most comprehensive documents on tropical and subtropical peats is from the late 1980s (Andriesse 1988).

This chapter provides an overview of tropical and subtropical peats. After describing worldwide peat extent and distribution, the factors controlling peat formation and development, peat features and peat classification are analyzed. The chapter also addresses issues related with peats and peatlands as resources.

2.2 Peat Extent and Distribution

Our knowledge about the areal extent of nontropical peats is supported by reliable statistics, mainly because of the early importance of peat as a source of energy. For tropical peats, by contrast, it was not until the mid-1900s that their extent became better known (Polak 1950). Statistics are not always directly comparable because they may reflect different acceptations and definitions of peat and peatland. The concept of peat varies from that of true peat that contains 100% organic material to that of organic soil defined on the basis of a combination between percent organic

J.A. Zinck and O. Huber (eds.), *Peatlands of the Western Guayana Highlands, Venezuela*, Ecological Studies 217, DOI 10.1007/978-3-642-20138-7_2,

matter (organic matter = organic carbon × 1.724) and thickness of the organic horizons. According to Andriesse (1988), organic soils have more than 50% organic matter in the upper 80 cm. For Rieley and Page (2005), organic soils are at least 50 cm thick and contain at least 65% organic matter, while Joosten and Clarke (2002) fix these thresholds at 30 cm thickness and 30% organic material. In Histosols, as defined in the USDA Soil Taxonomy (USDA 1999), the amount of organic matter is at least 20–30% in more than half of the upper 80 cm of the soil. Similarly, the term peatland is frequently used as a generic proxy of concepts which are not strictly synonymous, such as wetland, peatswamp, moor, muskeg, pocosin, bog, marsh, mire, and fen (Andriesse 1988; Joosten and Clarke 2002). Martini et al. (2006b) recognize four basic classes of peatland in nontropical environments: bogs, fens, swamps, and marshes, the first one being ombrotrophic and the others being minerotrophic with additional distinctive attributes based on acidity, type of vegetation cover, and water regime. All types of peatland are wetlands, but not all wetlands are peatlands. The global wetland area is estimated at 5.3–6.4 M km^2 (Matthews and Fung 1987; Lappalainen 1996), while only about 60–75% of this surface is actually covered by peat (Armentano and Menges 1986; Andriesse 1988; Wikipedia 2008).

2.2.1 *Peats in Temperate and Boreal Regions*

Worldwide, peatlands cover an estimated 4.26 M km^2 (Bord na Mona 1984; Andriesse 1988; Chimner and Ewel 2005; Chimner and Karberg 2008). This represents about 3% of the global land mass. The largest proportions concentrate in the temperate and boreal regions of North America (49%) and Eurasia (42%) (Table 2.1). Russia has the world's largest contiguous peat bog, while Canada has the largest total area of peatland, estimated at 1.7 M km^2 (Wheeler 2003).

A recent update of the areas covered by peat and peat-topped soils in Europe, derived from the 1:1,000,000 European Soil Database covering roughly the EU territory, provides a surface area of 291,600 km^2 (Montanarella et al. 2006).

Table 2.1 Worldwide distribution of peatlands

Continent	km^2	%
North America	2,096,400	49.19
Eastern Europe	1,519,578	35.65
Western Europe	259,862	6.10
Asia	248,865	5.84
South America	61,730	1.45
Africa	48,565	1.14
Central America	25,240	0.59
The Pacific	1,650	0.04
Global peatlands	4,261,890	100.00

Source: Data summarized from Bord na Mona (1984) and Andriesse (1988)

This extent is not substantially higher than the total peat area of 279,440 km^2 estimated by Bord na Mona (1984) for Western and Eastern Europe together (ex-Soviet Union excluded). In fact, these two figures are not directly comparable because of, among other reasons, recent changes in the territorial configuration of Europe and the inclusion of peat-topped soils (0–30 cm) in the concept of peatland. In spite of that, the updated figure reflects progress made and accuracy achieved in peat mapping using modern information technology. In some cases, peatlands may cover more than 10% of the surface area of individual countries, such as in Finland (30%), Estonia (22%), Ireland (17%), Netherlands (16%), Sweden (16%), Latvia (12%), and United Kingdom (11%). Finland alone concentrates almost one-third of the peat and peat-topped soils in Europe, and Sweden more than a quarter (Montanarella et al. 2006).

2.2.2 Peats in Tropical and Subtropical Regions

From a practical point of view, based on peatland similarities for reclamation and management purposes, Andriesse (1988) sets the boundaries between tropical–subtropical peats and temperate–boreal peats at latitudes 35° North and South. This territorial belt includes Southeast Asia, most of Africa, and a large stretch of the Americas from Florida and North Carolina to southern Brazil and Uruguay.

The most relevant features that distinguish intertropical lowland peats from temperate–boreal ones are surplus rainfall and high temperatures (Andriesse 1988). High temperature, with little diurnal and seasonal variations, accelerates the rate of peat oxidation. Rainfall controls peat hydrology and also has an effect on vegetation type and composition. Peat initiation is mainly a response to substrate waterlogging because of abundant rainfall or flooding by river overflow in areas with impeded drainage. Many tropical peatlands in low-elevation areas are forest-covered peat swamps, and that represents a striking difference with temperate peatlands commonly covered by sedges and moss. In the subtropical areas of the mid-latitudes, surplus rainfall remains important, but the annual temperature regime is more contrasted. Peatlands and peats occurring on highlands within the tropical and subtropical belts, such as, for instance, in Central Africa above 2,000 m elevation, are closer to those of the temperate regions. Worldwide, peat development has taken place over thousands of years converging at the end into the formation of ombrotrophic peat bogs in both temperate and tropical regions.

Compared with the areal importance of peat in temperate and especially in boreal regions, peatlands are much less extensive in tropical and subtropical environments. Only 0.36 M km^2 peatland, or 8.5% of the global 4.26 M km^2, occur in the warm and moist regions of the world, especially in Southeast Asia in the areas surrounding the South China Sea and areas in Papua-New Guinea that together concentrate 68% of the known tropical peats (Immirzi et al. 1992). Other areas with peatlands of some extent are the Caribbean and the basins bordering the Gulf of Mexico, the Amazon basin, and the wet equatorial belt of Africa, especially

Table 2.2 Extent of tropical and subtropical peatlands

Region	km^2	Global %	Tropical %
Southeast Asia	202,600	4.65	56.6
Caribbean	56,700	1.30	15.8
Africa	48,600	1.11	13.6
Amazonia	15,000	0.34	4.2
South China	14,000	0.32	3.9
Other regions	21,100	0.49	5.9
Tropical and subtropical peatlands	358,000	8.21	100.0

Source: Andriesse (1988)

Table 2.3 Areal ranges of tropical peatlands

Region	Minimum km^2	Maximum km^2
Southeast Asia	196,404	332,152
South America	37,136	96,380
Africa	26,607	88,657
Central America	14,465	25,935
Asia (Mainland)	622	6,245
Pacific	190	21,240
Total areas	275,424	570,609

Source: Page et al. (2007)

the depressional areas around the Gulf of Guinea (Table 2.2). Data on peats and peatlands in the tropics and subtropics are much less accurate than those concerning the higher latitudes, mainly because they are derived from small-scale soil maps in which organic soils are frequently mapped in association with poorly drained mineral soils. Andriesse (1988) considers that the extent of organic soils in the Amazon basin and in the wet equatorial belt of Africa is underestimated and that peat deposits in the tropics and subtropics might occupy areas larger than those so far reported.

As soil and land inventories proceed, data on peat extent are becoming more accurate. However, there are still large data gaps and data discrepancies between sources. According to Page et al. (2006), the total area of undeveloped tropical peatland is in the range of 310,000–460,000 km^2, which is about 10–12% of the global peatland extent. Page et al. (2007) computed data from different sources and found that the range between low and high estimates can be considerable (Table 2.3). In the case of Indonesia, for instance, estimates range from 168,000 km^2 (Bord na Mona 1984) to 270,000 km^2 (Jansen et al. 1985) for the same reference period. For the tropics as a whole, peatland area estimates vary roughly from simple to double, between a minimum of 275,424 km^2 and a maximum of 570,609 km^2 (Page et al. 2007) (Table 2.3). Other estimates set the tropical peat surface area closer to 0.5 M km^2 (Immirzi et al. 1992; Lappalainen 1996; Maltby and Proctor 1996). While the knowledge about tropical lowland peats is steadily increasing, tropical highland peats still remain poorly documented. For instance, all papers on tropical peatlands presented at the most recent (2007) international peat congress held in Tullamore, Ireland, deal exclusively with lowland peats (Farrell and Feehan 2008).

In Central and South America as a whole, peatlands cover about 87,000 km^2, representing 2% of the worldwide peat distribution (Table 2.1). The largest part of this extent is in the Caribbean (56,700 km^2) and in Amazonia (15,000 km^2) that together account for 20% of the tropical peatland areas (Table 2.2). A recent estimate (2007) sets the extent of tropical peatlands in South America between a minimum of 37,000 km^2 and a maximum of 96,000 km^2 (Table 2.3). Peats in tropical and subtropical South America are found in a variety of landscapes, including coastal plains (e.g., deltas of the Amazon and Orinoco rivers), inland sedimentary basins (e.g., Llanos in Colombia and Venezuela, Pantanal in Brazil), and highlands (e.g., filled glacial lakes in the Andes, (pseudo-)karstic depressions and other kinds of pond on the Guayana sandstone plateaus and mesetas).

2.3 Peat Formation and Development

Peat formation results from an unbalance between accumulation and decomposition of organic materials. In places where the speed of deposition exceeds the rate of decay, there will be a surplus of organic matter. Deficit of decomposition is caused by insufficient or low biological activity as a consequence of adverse environmental factors, basically excessive acidity and/or prolonged waterlogging causing anaerobic conditions. In tropical lowlands, the fluctuation of the groundwater level, controlled by rainfall and evapotranspiration, has an important effect on peat formation, especially in forest swamps (Ludang et al. 2007). In tropical highlands, lower temperatures slow down the rate of biomass decomposition in contrast to what occurs in the warm-to-hot lowland areas.

2.3.1 Topography and Water Regime

Topography plays an important role in water concentration and in situ retention. Concave, basin-shaped sites favor water accumulation, especially when coupled with rock or soil substrata of low permeability. Such relief types occur in a variety of landscapes, including coastal tidal marshes, inland depressional plains, undulating peneplains, karstic plateaus, volcanic reliefs, and glacial and periglacial mountains, all present in the tropics and subtropics. In temperate and boreal regions, glacial till plains offer the best conditions for peat formation.

Waterlogging is the main factor controlling peat initiation because it allows the colonization by adapted pioneer vegetation and the preservation of at least part of the decay residues. Water concentrates and stagnates in depressional sites where water outflow is less than water inflow so that excess water remains in situ. Water inflow is runoff, ground- and rainfall water, while water outflow includes water exiting through surface outlets, underground flow, and evapotranspiration. The presence of free water over longer periods favors specialized plants to establish, while water stagnation in more or less closed depressions creates an anaerobic environment that prevents the dead vegetation from decomposition or retards it. The activity of decomposing organisms is suppressed in waterlogged conditions (Lappalainen 1996). As a consequence, vegetation debris accumulates as partly decomposed biomass that constitutes the initial stage of primary peat formation. The process of peat inception and histosol formation in waterlogged conditions is termed paludification (Andriesse 1988) or paludization (Buol et al. 1997). As the initial peat mass continues growing, the peat surface rises above the water level retained in the pool, causing secondary peat formation. The top layers tend then to expand beyond the physical limits of the original depression, and the peat mantle may encroach onto the surrounding slopes, leading to tertiary peat formation. In this enlarged peat reservoir, a perched water table forms that is fed only by rainwater, resulting in the formation of ombrogenous peat.

Gore (1983) and Martini et al. (2006b) clearly contrast paludification, the process of colonization of poorly drained landforms by plant communities, with the process of terrestrialization that consists in a gradual overgrowth and infilling of water bodies by the litter of moss and aquatic plant remains. Comparing the shape of pollen isochrones in kettle hole-shaped basins in northeastern Germany, Gaudig et al. (2006) suggest that the peat-forming-upwards mechanism (i.e., paludification) better explains rapid peat formation in the studied mires than the commonly assumed peat-forming-downwards mechanism (i.e., terrestrialization).

When the water balance is largely positive, peat grows vertically and horizontally, covering with peat blankets the surrounding terrain surfaces. This occurs in the cool wet climates of North America and Northern Europe as well as in tropical highlands. It happens also in tropical and subtropical coastal lowlands, with excess rainfall and poor drainage conditions. Otherwise, peat formation remains approximately confined to the configuration and topographic limits of the original basins.

In the tropics, peat is always associated with waterlogged conditions where oxygen is deficient and the underlying substratum is poor in nutrients (Sieffermann et al. 1988). The temperature regime plays a minor role, in contrast to what happens in temperate and boreal regions. However, tropical peat is not exclusively topogenous, and ombrogenous peat in the tropics is not restricted to high elevations. There are large coastal swamps in Indonesia that are covered by ombrogenous peat with its typical dome-shaped relief (Notohadiprawiro 1997).

2.3.2 Source and Quality of Water

Water pH controls the types of plant that can adapt and the nature of the organic residues that contribute to peat formation. Low pH water creates oligotrophic conditions, poor in minerals. Neutral pH water favors eutrophic conditions, rich in nutrients. Intermediate conditions are termed mesotrophic. The rate of peat decomposition decreases from eutrophic to oligotrophic environments as biological activity is increasingly inhibited. Often site conditions tend to change over time, as peat grows vertically and gets out of reach of the nutrient-providing substratum and groundwater. As a consequence, conditions switch from originally eutrophic species-diverse to oligotrophic species-poor, with mainly acidity-tolerant plants, frequently endemic. Peats forming on bare rock exposure might show, in their initial stage of development, substantial differences in plant residues according to the mineralogy of the rocks (for instance, siliceous sedimentary versus igneous-metamorphic substrata).

Water reaction depends on the origin of the water that feeds the peat ecosystem. Kulczynski, quoted by Moore and Bellamy (1974), distinguishes several swamp types according to the water source. The rheophilous type is a swamp fed by cation-rich surface and ground flow that collects water running on or permeating the surrounding landscapes. This is frequently the case of the eutrophic peats that develop in tropical and subtropical lowlands exposed to river flooding and avulsion of mineral sediments. When the water source is mainly rainwater, with some contribution of local surface runoff, the swamp type is called ombrophilous. Rainwater has generally low pH and is poor in nutrients, providing the conditions for oligotrophic peat formation. In highlands, peat sites are commonly of the ombrophilous type. Also lowland peatlands become oligotrophic ombrophilous in their later stage of development, when the peat deposit rises above the groundwater level. In transitional swamp types, fed both by some lateral water seepage and by rainwater, peat is mesotrophic.

Peat formation systems that mainly depend on the inflow of nutrient-rich surface runoff and groundwater from upland soils and surrounding landscapes are called topogenous (Andriesse 1988) or minerotrophic (Anderson et al. 2003). Such peats are frequent in tropical and subtropical lowlands, especially in their initial stages of formation. When the peat system relies mainly on rainwater and the recycling of minerals, peat is called ombrogenous (i.e., bog peat). Lowland peats in advanced stages of development and highland peats belong primarily to this type. Peat deposits are generally topogenous in the early stages of formation and become ombrogenous in later stages. This evolution has been documented in the Sarawak Lowlands, Malaysia, by Anderson (1964) and Andriesse (1974).

2.3.3 Geodynamics

Geomorphogenic processes contribute to creating conditions favorable to peat formation, while sometimes also inhibiting it. In tropical and subtropical regions receiving heavy rainfall, flooding, slope instability, and disruption of the drainage network are common features that have influence on peat formation.

Torrential floods carry large amounts of mineral sediments that are trapped in coastal and inland depressions. The seasonal entry of sediments raises the floor of the depressions and favors the outflow of excess water, while the inflow of freshwater allows oxygenation of the organic residues at a rate higher than that of accumulation. Thus, this process inhibits peat formation when it operates repeatedly.

In contrast to the former, peat accumulation is favored when incoming mineral sediments block the drainage system by clogging the natural water outlets or creating barriers. This occurs in landscape units such as delta depressions surrounded by levees, coastal lagoons with plugged outlets, cut-off meander lakes, abandoned stream beds, small tributary valleys blocked by debris, large basins in coastal plains, river valleys without organized drainage network and lacking drainage outlets, moraine lakes in mountains, karstic and pseudokarstic depressions, among others.

Blanket peats on mountain slopes are exposed to scarring, fragmentation or even large dismantlement because of sliding. Sliding is frequent in soligenous peats that become unstable because of sustained water flow along the interface between the rock substratum and the base of the peat mantle and/or because of oversaturation of the peat body upon heavy rainfall. Failures of blanket peats in temperate regions are frequently due to anthropogenic causes, while they are mainly related to natural instability in tropical and subtropical regions where slope peat exploitation is uncommon. It can be argued whether peat failures are likely to become more frequent in response to climate change effects. A study carried out in Northwest Ireland shows that the high frequency of large rainfall events since 1961 did not trigger landslides, because the latter are in fact controlled by slowly changing internal thresholds (Dykes and Kirk 2006; Dykes et al. 2008). Peat flows on hillslopes under forest cover have been reported from Tierra del Fuego, southernmost Argentina, where the triggering mechanisms are heavy snowfalls or earthquakes (Gallart et al. 1994).

2.3.4 Time and Rate of Formation

In temperate and boreal regions, peat started forming after the last glaciation, as glaciers receded and left exposed poorly drained terrains with irregular moraine and glacial till topography that favored the accumulation of water and organic materials. These peat deposits are thus mainly younger than 10,000 years. Similarly in the tropics, change from glacial-induced aridity to a moister and warmer climate triggered peat formation at the beginning of the Holocene. Some early studies on lowland peats report maximum ^{14}C ages of 4,300 BP in Sarawak (Anderson 1964), 4,400 BP in the Everglades of Florida (Lucas 1982), 6,000 BP in the coastal areas of Southeast Asia (Andriesse 1974; Driessen 1977), and 8,000 BP in central Kalimantan, Indonesia (Sieffermann et al. 1988). In general, inland basal peats are older than coastal basal peats, the age of which is worldwide in the range of 5,500–4,000 calBP, reflecting the time at which rising sea levels stabilized. In contrast, peat started forming already in the Late Pleistocene, around 26,000 calBP in an inland peatland of Kalimantan (Page et al. 2004, 2006). In this place, after a period of rapid organic accumulation in the early Holocene (11,000–8,000 calBP), peat formation continued at a slow rate until nowadays (Neuzil 1997). In the Nilgiri hills, a mountain region at more than 2,000 m a.s.l. in southern India, peat formation was dated back to 40,000 ^{14}C BP (Rajagopalan et al. 1997).

Peat studies carried out on intertropical highlands provide a range of peat initiation dates. Peat inception ^{14}C dates go back to ca 3,000 BP in the Ecuadorean Andes (Chimner and Karberg 2008), 4,300 BP on the high-plateau of the Gran Sabana, Venezuela (Rull 1991), 6,000 BP on table-mountain summits in the eastern Guayana Highlands, Venezuela (Schubert and Fritz 1985; Schubert et al. 1994), 7,500 BP on table-mountain summits of the western Guayana Highlands, Venezuela (Zinck et al. 2011), and 10,630 BP at one site in the Chimantá massif, eastern Guayana Highlands (Nogué et al. 2009).

Peat formation during the Holocene has been neither continuous nor linear. Climate change, sliding in slope conditions, disruption of river networks, and changes in sedimentation are some of the factors that usually affect peat deposition and influence the peat accumulation rate. According to Lucas (1982), it takes between 600 and 2,400 years for 1 m peat to form in temperate and boreal conditions, thus at a rate of 0.42–1.67 mm yr^{-1}. Abundant rainfall and high temperature in most tropical and subtropical lowlands stimulate biomass production, although high temperatures also accelerate oxidation and decomposition of the organic matter. According to Anderson (1964), forest peat formation in Sarawak varies from 4.67 mm yr^{-1} in the lower layers to 2.2 mm yr^{-1} in the upper ones. Similarly, Page et al. (2004) provide data on differential rates of peat formation in inland Kalimantan, with rates of >2 mm yr^{-1} at the beginning of the Holocene but decreasing to 0.15–0.38 mm yr^{-1} in the mid- and late Holocene.

In the cooler tropical highland conditions, average peat formation rates are in general lower than in the warm moist lowlands: for instance, 0.2–0.3 mm yr^{-1} in the Guayana Highlands at ca 1,000–2,600 m elevation (Rull 1991; Zinck et al. 2011). These figures are close to the average accumulation rates common in subarctic and boreal peatlands, with often less than 0.5 mm yr^{-1} (Gorham 1991). However, higher deposition rates have been reported from some specific places: 1.3 mm yr^{-1} in the Ecuadorean Andes at 3,968 m elevation (Chimner and Karberg 2008), and 1.84 mm yr^{-1} in the Chimantá massif, eastern Guayana Highlands, at 2,700 m elevation (Nogué et al. 2009).

2.3.5 Vegetation

As peat is made of decayed plant residues, vegetation obviously is a main peat formation factor. Distinction between aerated forest peat and anaerobic swamp peat, as proposed by Kurbatov (1968) for cool moist regions, does not hold in the tropics and subtropics where swamps have frequently forest cover, as, for instance, in the Orinoco river delta (Rodriguez 1999; White et al. 2002) or in the coastal lowlands of Southeast Asia (Anderson 1964). In tropical lowland forest swamps, the build-up of peat is often due essentially to the intrinsic properties of the leaves that accumulate and inhibit decomposition rather than to the properties of the swamp site itself, i.e., the acidic, anaerobic, high tannin conditions (Yule and Gomez 2009). In tropical highland peats, usually covered by meadow and shrub vegetation, roots are mainly responsible for biomass accumulation and peat formation (Chimner and Ewel 2005).

Peat profiles are archives recording vegetation change or stability during the time of peat formation. Andriesse (1988) suggests a classical succession of plant associations in a lowland peat swamp, starting at the bottom with floating aquatic plants, algae, and plankton, followed by perennials emerging from shallow water, and then grassy perennials, shrubs, and trees. As peat systems tend to become ombrogenous in later stages of development, thus nutrient-poor and oligotrophic, vegetation is increasingly dominated by acidophilic plants. In coastal swamps of Southeast Asia, Anderson (1964) recognized a succession starting with pioneering mangrove plants, followed by brackish-water communities, and finally replaced by freshwater swamp communities. In highlands, trees are infrequent in peat profile records. The current vegetation types are mainly shrubland on peat forming in drying lakes (glacial or others) and meadows with sedges and reeds on peat forming in lithogenous depressions.

2.3.6 Peat and Climate Change

Peatlands are natural archives that register the paleoenvironmental conditions associated with peat formation, and for that reason they have been used to reconstruct Holocene climate change (Charman and Warner 2002; Martini et al. 2006a). In a recent paper on peat humification and climate change in southeast Alaska, Payne and Blackford (2008) refer to a set of publications that have attempted to infer climate change from the stratigraphy of peat deposits. For that purpose, a variety of climate indicators has been used, including plant macrofossils, testate amoebae, fossil lipids, and isotope ratios. Peat humification analysis, based on estimating the degree of decomposition using alkali extraction and colorimetry, is a widely used technique for paleoclimatic inference from peatlands. This proxy approach has been successfully applied in northern Europe to infer climate change from the degree of peat humification. By contrast, the study carried out by Payne and Blackford (2008), comparing low-resolution peat humification records from several mire sites in southeast Alaska, did not provide strong evidence of climatic forcing of humification in their area. Such studies deducing climate change from peat stratigraphy and degree of humification are less common in tropical and subtropical areas.

Stable isotope ratios of carbon and oxygen can provide valuable paleoclimate records, as they are able to reflect past moisture variations. Based on a $\delta^{13}C$ record spanning the past 20 kyr from peats in the Nilgiri hills at 2,000 m a.s.l., southern India, Sukumar et al. (1993) identified climate shifts corresponding to the last glacial maximum at 18 kyr BP, an arid phase at 6–3.5 kyr BP, and a short wet phase at 0.6 kyr BP. Rajagopalan et al. (1999) measured the stable carbon and oxygen isotope ratios in cellulose of C3 and C4 plants growing on a montane peat bog in the same region of southern India. The mean monthly $\delta^{13}C$ values were found to be significantly related to rainfall, while the $\delta^{18}O$ values revealed to be sensitive to changes in maximum temperature and relative humidity. The authors suggest that plant isotope–climate correlations could be used for reconstructing past temperature and rainfall conditions of the tropics from the isotopic ratios of peat deposits. Using $\delta^{13}C$ ratios, Medina et al. (2011) and Zinck et al. (2011) reached the conclusion that peat vegetation in the Guayana Highlands, southern Venezuela, did not change significantly during the mid- and late Holocene, a fact that could be interpreted as reflecting climate stability during the same period.

Pollen records from peat deposits allow reconstructing past plant communities and inferring, from their succession in time, correlative climate changes. Using peat pollen assemblages from Guayana Highlands, southern Venezuela, Rull (1991) recognized Holocene climate shifts including a dry phase before about 5,000 BP, followed by a moist phase between 4,000 and 2,700 BP. Working in the same region, Nogué et al. (2009) refined the former finding using the pollen record of the Eruoda-tepui in the Chimantá massif. After a long period of about 9,000 years, between 12.7 and 4.3 kyr calBP, showing little peat formation probably because of limited rainfall, perhumid conditions established between 4.3 and 4.0 kyr calBP with high peat accumulation rates. From 4.0 kyr calBP until nowadays, the vegetation at the tepui summit did not change substantially, suggesting the permanence of climatic conditions similar to the present ones.

The use of peat records for inferring Holocene climate changes in the tropics and subtropics is still lagging behind when compared to the development of such studies in temperate and boreal areas. In a recently published volume on the evolution and records of environmental and climate changes in peatlands (Martini et al. 2006a), only one out of 23 contributing papers addresses specifically the tropical peats, restricted to lowlands.

2.4 Features of Peats and Peatlands

Peat material is strongly related to the characteristics of the landscape and landform in which it forms. To some respect, peat and peatland features in tropical and subtropical regions are not fundamentally different from those of temperate–boreal peats. Peat properties and peatland characteristics are commonly used as diagnostic criteria to classify peat and peatlands.

2.4.1 Physical Peat Properties

Differences in organic matter content of organic soils are primarily related to the position of the water table at each site and, in a lesser extent, to the cooler

Table 2.4 Moisture relationships in three kinds of organic material

Moisture parameters (% oven-dry weight)	Fibric	Mesic	Sapric
Maximum moisture holding capacity	1,057	374	289
Water retention 10 kPa	570	193	163
Water retention 33 kPa	378	150	144
Water retention 1,500 kPa	67	84	100

Source: Data summarized from Farnham and Finney (1965)

temperatures in the lower layers of the peat (Moore et al. 2007). Moisture content is an essential peat property as it controls the accumulation and evolution of organic materials. When expressed on an oven-dry weight basis, both the water holding capacity against gravity and the water retention at lower pressure decrease as a function of the stage of peat decomposition, while water retention at higher pressure increases. Farnham and Finney (1965) report on earlier figures that illustrate the difference in moisture relationships between little decomposed fibric, moderately decomposed mesic (i.e., hemic), and well-decomposed sapric materials (Table 2.4). When the same moisture parameters are expressed on a volume basis, water retention figures increase as the degree of decomposition increases. Microporosity that controls water retention is higher in sapric material than in fibric material.

Hydraulic conductivity is influenced by the degree of decomposition and the bulk density of the organic material. Because of the presence of large pores, water moves faster in fibrous material than in more decomposed material. Hydraulic conductivity is a dynamic property that decreases as peat evolves from fibric to hemic to sapric because of decreasing pore space and increasing water retention. Sometimes, horizontal hydraulic conductivity rates are faster than vertical rates because of peat stratification and the flattened way organic residues often settle in peat profiles. This may result in the formation of perched water tables, especially in the later stages of peat development.

Bulk density is a good proxy of several other peat features, in particular the level of compaction and the degree of decomposition. In general, bulk density values are much lower in organic soils than in mineral soils. Bulk density can vary from 0.05 Mg m^{-3} in non or poorly decomposed material up to 0.5 Mg m^{-3} in well-decomposed material (Andriesse 1988). Bulk density values of less than 0.1 Mg m^{-3} in fibric materials and 0.13–0.2 Mg m^{-3} in sapric materials have been reported from lowland peats of Malaysia and Indonesia (Driessen and Rochimah 1976; Tie and Kueh 1979). In most tropical peats, surface layers are more decomposed than the underlying ones, and have thus higher bulk density values. Specific density of solid peat material ranges from 1.26 Mg m^{-3} to 1.80 Mg m^{-3} (Driessen and Rochimah 1976).

Organic soils are sensitive to irreversible drying after exposure to the sun and/or upon drainage and fire. Many peat soils, especially those with low bulk density, are difficult to rewet. The hydrophobic nature of dried peat has been related to different factors that would prevent the reabsorption of water, including the formation of resinous coating upon drying (Coulter 1957), high lignin content in acid peats (Lucas 1982), and the presence of iron coating around the organic particles. Organic soils that are able to rewet after drying experience changes in volume resulting from swelling–shrinking. Volume changes range from 40% in fibric peats to 90% in aquatic peats (Andriesse 1988).

2.4.2 Chemical Peat Properties

In young peats, the water-soluble compounds, including polysaccharides, monosugars, and some tannin, are the first to be washed away upon decomposition. Cellulose and hemicelluloses also decompose easily. As a result, lignin and lignin-derived substances that are fairly resistant to microbial activity constitute the largest part of older peats. Tropical lowland forest peats have frequently reached an advanced stage of development and, for that reason, may contain up to 75% of lignin and lignin-derivates (Andriesse 1988). In lowland peats, especially those of coastal areas, iron, aluminum, sodium, and sulfur can be present in high levels, inherited from the early paludification stage.

Acidity in organic materials varies widely according to the environmental conditions and the stage of peat development. Reaction is neutral in eutrophic peats (pH 6–7), alkaline in brackish-water peats (pH 7–8), and very strongly acid in tidal peats upon oxidation of pyritic materials (pH 2–3). In general, tropical lowland peats are acid to extremely acid (pH 3–4.5). The acidity of peats forming directly on bedrock reflects commonly, at least in the early stage of development, the mineralogical composition of the substratum, with pH being higher on igneous-metamorphic rocks than on sedimentary siliceous rocks. Strongly developed ombrogenous peats are usually oligotrophic, with extremely acid top layers.

Cation exchange capacity (CEC) of organic soils is strongly pH-dependent, but is in general high because of the abundance of hydrophilic colloids, in particular humic acids and hemicelluloses. CEC usually increases as peat develops, because decomposition of the organic material generates increasing amount of lignin-derivates rich in exchange sites. At pH 7, CEC of fibric material is around 100 cmol(+) kg^{-1}, while it is about 200 cmol(+) kg^{-1} in sapric material (Andriesse 1988). Eutrophic peat is usually fully saturated by divalent cations, while in oligotrophic peat CEC is dominated by hydrogen and base saturation values are low.

Organic carbon content increases with increasing decomposition of the organic materials. EKONO (1981) reports organic carbon values of 48–50% in fibric peat, 53–54% in mesic (i.e., hemic) peat, and 58–60% in sapric peat. Since decomposition of the organic material in tropical lowland peats decreases with depth, top layers show usually higher organic carbon values than deeper layers. The loss on ignition is related to the percentage of organic carbon. According to Tie and Lim (1976) working in Sarawak, the ratios of loss on ignition to carbon content were around 2 on average, but they were usually higher in shallow peats (ratio of 4) than in deep peats (ratio of 2.5). Nitrogen levels are higher in surface layers of deep peats than in those of shallow peats, and this is assumed to be related to the concentration of residual nitrogen in lignin as peat continues decomposing (Hardon and Polak 1941). In general, tropical lowland peat derived from rainforest trees has higher lignin and nitrogen contents, and lower ash, carbohydrate, and water-soluble protein contents than those of temperate peat; the organic matter is less biodegradable leading to faster peat accumulation (Notohadiprawiro 1997).

Other chemical substances present in peat material include phosphorus, free lime, and sulfur. Phosphorus content in oligotrophic tropical peats is generally very low, less than 0.04% on average (Andriesse 1988). Phosphorus release may be significant in completely humified peat (Jordan et al. 2007), but limited to the surface layers under flooded conditions (Stêpniewska et al. 2006). In general, tropical peats are also very poor in lime (<0.3%). Eutrophic peats occur only in exceptional situations, provided with calcium carbonate sources (shell deposits, coral reefs, marl, limestone). High sulfur contents, often in the form of pyrite, are comparatively more frequent in the tropics, especially in coastal lowlands (e.g., the Orinoco river delta in Venezuela). Pyrite forms in peatswamps in the presence of brackish water, commonly associated with potential acid sulfate soils. Pyrite oxidation upon exposure to air in drained peatlands causes extreme acidification (pH 2–3).

2.4.3 Biological Activity in Peats

The aerobic conditions that prevail during the initial stage of peat development allow micro-organisms, including actinomycetes, fungi, and bacteria, to rapidly decompose freshly accumulated organic material. As the peat deposit grows, mostly under fairly permanent water table, micro-organisms adapted to anaerobic conditions take over the decomposition process, using oxygen derived from organic material. High temperatures in the tropical lowlands increase significantly microbial activity. Yule and Gomez (2009) suggest that bacteria and fungi must be responsible for leaf breakdown in a peatswamp they studied in Malaysia, because no aquatic invertebrates to ingest leaf material were found. Usually, biological activity decreases with depth. However, acidity-resistant anaerobic bacteria have been found to increase with depth in oligotrophic peats (Andriesse 1988). Microbial decomposition of organic materials contributes to peat subsidence.

The intensity and type of biological activity depend on a variety of factors including the peatland setting, the physical and chemical properties of the peat material, the water regime and the water quality, among others. The acrotelm–catotelm model is an ecohydrological model that aims to account for the multiple interactions between these factors and understand their effect on the functioning of peatlands (Holden 2006). A distinction is made between an upper acrotelm tier, periodically aerated and partly living, and a lower catotelm tier, always water-saturated, impermeable, anaerobic/anoxic, and mostly biologically dead. This diplotelmic system has still to be tested in tropical environment.

2.4.4 Characteristics of Peatlands

Peat forms in sites with physical features that favor the accumulation and preservation of organic material. In this respect, two factors are important, namely, the geomorphic setting and the water regime. Andriesse (1988) provides a list of landscape units that are most likely to promote the accumulation of organic materials and subsequent peat formation in swampy sites, including deltas, coastal basins, lagoons, narrow inland valleys, depressions in major valleys, meander bends, isolated small bottomlands, and atolls.

Peat thickness varies from a few centimeters to several meters, even tens of meters, sometimes over short distance as in the case of soligenous slope peats. Thick peat mantles are usually found in forest-covered low-lying tropical peatlands (i.e., peatswamp forests). For instance, on the west coast of peninsular Malaysia, the thickness of the organic deposits that formed in marine clay depressions varies from less than 3 m to more than 5 m (Hashim and Islam 2008). In Kalimantan, peat thickness varies from 0.3 to 20 m (Anderson 1983). In highland landscapes, deepest peats are usually found in filled lakes.

In initial stages of formation, peat deposits are usually enclosed within the boundaries of a depression. As peat goes on growing, it overlaps the topographic limits of the original site and tends to expand horizontally, forming peat blankets over the surrounding slopes and hillocks. Vertical peat growth results often in dome-shaped peatswamps that are several meters higher than adjacent river floodplains. The micro-topography of the peat cover can be irregular because of the formation of tussocks on the surface of grassy swamps and the presence of aerial roots and shoots in forest peat areas. The relief under the peat mantle is in general concave, but meso- and micro-topography can vary from flat in the case the buried surface is a depositional one, to irregular when the substratum is fractured bedrock.

Due to the presence of organic substances and iron compounds, water in peatswamps is usually brown to black, but clear. Rivers feeding from overflowing peatlands are sometimes called after their dark-colored water, such as, for instance, the Rio Negro which crosses large oligotrophic lowland peats in Venezuela and Brazil. Oligotrophic drainage water is very poor in nutrients and basic cations (<5 ppm) (Andriesse 1988).

2.5 Peat Classification

Peat and peatland have been classified using a variety of criteria, including the topographic setting, current vegetation cover, peat-originating plant communities, physical and chemical peat properties, genetic processes, and soil taxonomy (Andriesse 1988). These classifications have been mainly applied to peats from the boreal and temperate regions, but much less to tropical and subtropical peats.

2.5.1 Classifications Based on Landscape Features

Topography is a key factor for peat initiation and development. A concave depressional setting favors the accumulation of organic material and the concentration of surface runoff and groundwater flow. As the peat deposit grows vertically, it tends to expand from the initial core area to the surrounding landscape. This topographic basis was used to distinguish between raised bogs in bottom areas and blanket bogs on slopes (Hammond 1981).

The current vegetation cover allows separating, at a first level, forest peatswamps that are more frequent in lowlands, especially in coastal basins, from shrubby and grassy peatswamps more frequent in highlands. Nonwoody peats are further subdivided according to the dominant vegetation type, including moss peat (e.g., *Sphagnum* peat), sedge peat (e.g., Cyperaceae peat), reed peat, heath peat, saw-grass peat, among others. Vegetation may have changed during peat development time, and such changes are recorded in the botanical composition of the successive peat layers. In coastal swamps, for instance, organic residues may be derived from mangrove vegetation at the swamp bottom, brackish-water species in intermediate layers, and freshwater plants in the top layers. This can lead to more complex classifications to account for vegetation successions.

2.5.2 Classifications Based on Peat Properties

Chemical properties, basically nutrient availability, help distinguish between nutrient-rich eutrophic peat, moderately rich mesotrophic peat, and nutrient-poor oligotrophic peat. The provision of nutrients is more related to the source and quality of water than to the nature of the peat material itself. Eutrophic peats are enriched in nutrients by overland flow or river flooding, while oligotrophic peats receive only or mainly nutrient-poor rainwater. In the tropics, raised peats in the lowlands but also less developed peats in highlands are dominantly oligotrophic. The concepts of low moor and high moor reflect, in a somehow similar manner, differences in nutrient content as a function of the peat development stage. Low moor is young moor fed by nutrient-rich groundwater, while high moor refers to older raised peat fed mainly by nutrient-poor rainwater.

Physical characteristics implemented for peat classification are essentially based on the degree of decomposition of the organic material. An early classification proposed by Von Post (1924), still used in some countries, recognizes ten decomposition steps, from little decomposed light-colored fibrous peat to well-decomposed dark-colored, colloidal peat. Modern classification systems of organic soils are based on the same principle but use a much lower number of decomposition degrees. For instance, the USDA Soil Taxonomy (USDA 1999, 2006) distinguishes three types of organic material on the basis of fiber content (fc) expressed in volume after rubbing, wet bulk density (bd) in Mg m^{-3}, saturated water content (wc) as percent of oven-dry material, and color. Fibric materials have fc > 2/3, bd < 0.1, and wc 850–3,000$^+$. Sapric materials have fc < 1/6, bd > 0.2, and wc < 450. In hemic materials, the selected parameters have intermediate values. Using these criteria, organic soils are classified in the order of the Histosols into Folists, Fibrists, Hemists, and Saprists. These classes are further subdivided to account for specific characteristics that may occur or not in tropical and subtropical environments (e.g., Sulfohemists, Sulfihemists, Luvihemists, Cryohemists, and Haplohemists, in the case of the Hemist suborder). In the World Reference Base for Soil Resources (FAO 2006), Histosols are also recognized as a class at the highest level of the classification system. They are further subdivided into a large number of subclasses that reflect not only decompositional ranges of the organic material but also other soil properties and environmental features such as the presence of salt or sulfur, drainage conditions or base saturation, among others.

2.5.3 Multicriteria Classifications

Integrated multicriteria classifications have also been proposed. For instance, the International Peat Society (Kivinen 1980) developed a system which combines some of the properties above analyzed, specifically the botanical composition (distinguishing between moss, sedge, and wood peats), the degree of decomposition (distinguishing between weakly, moderately, and strongly decomposed), and the trophic status (distinguishing between oligotrophic, mesotrophic, and eutrophic). Unfortunately, this kind of classification accounts only for the peat material itself and ignores the features of the physical environment, in particular the geomorphology and hydrology of the peatlands.

Dammon and French (1987) propose a scheme that combines hydrology and geomorphology as basic criteria for classifying bogs. Three categories of bog, each with several classes, are distinguished according to (1) the source of water (ombrotrophic and minerotrophic bogs), (2) the landforms where the bog development occurs (ombrogenous, topogenous, limnogenous, and soligenous bogs), and (3) the landforms produced by the bog (peat bog lake systems, perched water peatland systems, peat bog stream systems, and ombrogenous peatland systems).

Referring specifically to tropical and subtropical peats, Pfadenhauer (1990) makes a basic distinction between peats forming at low and mid-elevations (up to 1,000–1,500 m a.s.l.) and mountain peats (above 1,000 m a.s.l. in the subtropics and above 1,500 m a.s.l. in the tropics). In each of these two categories, several peat classes are recognized on the basis of the current vegetation cover and floristic composition, the nutrient status, and the water regime. Main peat classes occurring in the lowlands and uplands include floating meadows, swampy oligotrophic low peats covered by grasses and sedges, swampy eutrophic low peats covered by grasses and sedges, and forest peats. In the highlands, distinction is made between sedge and grass low peats, *Sphagnum* peats, and cushion bogs.

Most peat classification schemes were developed in boreal and humid temperate regions, and therefore fail recognizing the distinctive features of tropical peats, such as peat texture and ash content. Based on their studies in the Tasek Bera Basin, Malaysia, Wüst et al. (2003) propose a classification framework that combines the morphological fibric, hemic, and sapric classes of the USDA Soil Taxonomy (USDA 1999) with ash and C contents. Organic deposits are defined as having less than 65% ash (i.e., >35% loss on ignition). Four main groups are distinguished on the basis of ash content: peat (0–55%), muck (55–65%), organic-rich soil/sediment (65–80%), and mineral soil/sediment (80–100%). The peat class is further divided into five subclasses spanning from very low to very high ash content.

2.6 Peat and Peatland as Resources

2.6.1 Reclamation and Multipurpose Uses

Indigenous people developed early experience in managing peatlands in a variety of tropical and subtropical areas (e.g., chinampas in Xochimilco and the Patzcuara lake basin in Mexico; raised fields in poorly drained tropical lowlands; abandoned riverbeds filled with organic deposits). For a long time, large-scale reclamation and use of peats were constrained by factors such as poor drainage, low fertility, risk of disease, and inaccessibility (Andriesse 1988). However, because of the need to meet increasing food demands, tropical peatlands are becoming new agricultural frontier areas, increasingly settled by newcomers, small farmers, and entrepreneurs as well, who frequently lack the required experience to manage such problem-soils.

Reclamation requires taking into account not only the peat material itself, but also the topographic, geomorphic, and hydrologic characteristics of the landscape units (peatland or peatswamp) in which peat forms and develops. When using peat for industrial or energy purposes, it might be enough to know and assess the properties of the material. But when peat is meant to be used for farming or to provide environmental services, then the physiographic setting of the peatland plays a role probably not less important than that of the peat material proper.

Peatland is used in agricultural production and peat is used as a source of energy. The vegetation cover of peatlands provides raw materials that are transformed by local people into crafts, artifacts, and goods, such as raffia palm and papyrus from African swamps and timber from coastal forest swamps in Southeast Asia. From an environmental point of view, peats and peatlands perform regulating functions that play a role in the carbon cycle as carbon stocks, in hydrology as water reservoirs and catchment areas for flood mitigation, in the adsorption of heavy metals and organic pollutants, and in the buffering between salt- and freshwater systems in coastal marshes.

The preservation of pristine peatlands and the management of reclaimed peatlands strongly depend on the hydropedological conditions of the site. Ideally, groundwater levels should be maintained between 40 cm below and 100 cm above the peat surface to prevent subsidence and fire. Working in central Kalimantan, Indonesia, Wösten et al. (2008) noticed that the above-mentioned water-level thresholds were altered during dry years. Relatively intact peatland was able to recover from the disturbance, but degraded peatland became more susceptible to fire. Using a hydropedological modeling approach, they produced groundwater-level prediction maps that can be used for fire hazard warning and coordinated land use planning.

So far, peat in its natural state was considered unsuitable for supporting foundations owing to low bearing capacity. However, because of increasing population pressure, especially in Southeast Asia, new settlements are encroaching on peatlands. The reclamation and use of peat soils for construction require special treatments. Based on laboratory experimentation on peat material, Hashim and Islam (2008) have shown that unconfined compressive strength increased significantly after stabilization of peat soil with a mixture of cement–bentonite–sand and that the strength of stabilized columns was reinforced by enlarging the curing time.

2.6.2 Carbon Storage and Potential Release of Greenhouse Gases

Since the inception of peat formation after the last glacial period, about 10–12 kyr ago, considerable amounts of organic material have slowly built up in peatlands. With the current relevance of climate change on the global agenda, the role of peats as carbon stores and the impact of the potential release of greenhouse gases from peatlands are given increasing importance (ClimSoil 2008). Climate warming may cause changes in the balance and annual distribution of rainfall and evapotranspiration that, in turn, will affect peatland productivity and peat decomposition (Alm et al. 1999). However, peatlands are not yet formally included in global climate models and predictions of future climate change (Limpens et al. 2008).

According to early estimates, the world content of carbon in peat deposits ranges from 150 Gt (Moore and Bellamy 1974) to 300 Gt (Sjörs 1980). Maltby and Immirzi (1993) considered that peatlands could store up to 525 Gt of carbon globally. Uncertainty of the estimates depends on the basic assumptions adopted for peat surface area, peat thickness, and peat density (Shimada et al. 2001; Page and Banks 2007). In a recent attempt to estimate the total carbon stored in tropical peatlands, Page et al. (2007) obtained a range of 16.5–68.5 Gt, based on a minimum area of 275,424 km^2 with 1 m peat thickness and a maximum area of 570,609 km^2 with 2 m peat thickness, respectively, and considering in both cases a volumetric carbon density of 60 kg m^{-3}. Jaenicke et al. (2008), taking into account the biconvex shape of the tropical peatlands to determine peat volume, estimate that Indonesian peatlands alone may store at least 55 $\pm$ 10 Gt of carbon. Tropical forest swamp peats in particular may store substantial amounts of carbon and function as carbon sink (Buringh 1984; Satrio et al. 2009).

Although carbon sequestration in peats is considerable, it is generally believed that the contribution of peat burning to the global rise in atmospheric CO_2 will remain subordinate to that of other fossil fuels. However, increasing deforestation of tropical lowland peats will generate large carbon dioxide emissions that may negatively influence the global climate. During the 1997 El Niño-driven dry season, peat and forest fires in Indonesia have released to the atmosphere 0.81–2.57 Gt of carbon, a range equivalent to 13–40% of the average annual carbon emissions caused by the burning of fossil fuels worldwide (Page et al. 2002). Once ignited, smoldering peat fires can continue burning for long periods of time, even centuries, propagating through underground peat layers, especially in peatlands that have been artificially drained. In Central Kalimantan, peatland drainage and the change in land cover from forest to secondary vegetation, which is more flammable, have contributed to increase the frequency of peatland fires during the last decade (Hoscilo et al. 2008).

Jauhiainen et al. (2005), studying the carbon fluxes in a tropical ombrotrophic peatland ecosystem in central Kalimatan, Indonesia, show that CO_2 emissions are highest during the dry season when the water table is low. High water table is thus a relevant condition to restrict carbon emissions from the peatland to the atmosphere. Peatland cultivation not only leads to lowering the water table, but also causes subsidence. Hairiah et al. (2001) estimated a subsidence rate of 2.5 cm per year in oil palm plantations on peat soils in Malaysia. Half of the former was attributed to the process of decomposition/respiration of the organic matter that resulted in a C loss to the atmosphere of 10–20 Mg C ha^{-1} yr^{-1}, an amount ten times greater than the losses from upland soils after forest conversion.

2.7 Conclusion

Worldwide, peats and peatlands are increasingly used as agricultural resource, source of energy, water regulation body, biodiversity reservoir, carbon pool, and provider of other environmental services. However, knowledge on tropical peats is still lagging behind when compared to the development of peat and peatland studies in temperate and boreal areas, although considerable progress has been made in mapping tropical peats, identifying their specific characteristics, assessing their use potentials, and calling the attention on their vulnerability. The online journal "Mires and Peat." jointly published by the International Mire Conservation Group and the International Peat Society, produces yearly volumes since 2006 that so far include one paper concerned with peat in tropical environment. In a recently published volume on the evolution and records of environmental and climate changes in peatlands (Martini et al. 2006a), only one out of 23 contributing papers addresses specifically the tropical lowlands. Tropical lowland peats, extensive in Southeast Asia, have been relatively well documented, while tropical highland and mountain peats have been so far less studied. The present volume on "Peatlands of the Western Guayana Highlands, Venezuela," is a contribution to the latter.

References

Alm J, Schulman L, Silvola J, Walden J, Nykänen H, Martikainen PJ (1999) Carbon balance of a boreal bog during a year with exceptionally dry summer. Ecology 80:161–174
Anderson JAR (1964) The structure and development of peat swamps of Sarawak and Brunei. J Trop Geogr 18:7–16
Anderson JAR (1983) The tropical peat swamps of western Malaysia. In: Gore AJP (ed) Mires: swamp, bog, fen and moor, B-regional studies. Elsevier, Amsterdam, pp 181–199
Anderson RL, Foster DR, Motzkin G (2003) Integrating lateral expansion into models of peatland development in temperate New England. J Ecol 91:68–76
Andriesse JP (1974) Tropical lowland peats in South-East Asia. Royal Tropical Institute, Department of Agricultural Research, Communication 63, Amsterdam
Andriesse JP (1988) Nature and management of tropical peat soils. FAO Soils Bulletin 59, Food and Agriculture Organization of the United Nations, Rome
Armentano TV, Menges ES (1986) Patterns of change in the carbon balance of organic soil wetlands of the temperate zone. J Ecol 74:755–774
Bord na Mona (1984) Fuel peat in developing countries. Study Report for the World Bank, Dublin
Buol SW, Hole FD, McCracken RJ, Southard RJ (1997) Soil genesis and classification, 4th edn. Iowa State University Press, Ames
Buringh P (1984) Organic carbon in soils of the world. In: Woodwell GM (ed) The role of terrestrial vegetation in the global carbon cycle: measurement by remote sensing. Wiley, New York, pp 91–109
Charman DJ, Warner BG (2002) Peatlands and environmental change. Wiley, London
Chimner RA, Ewel KC (2005) A tropical freshwater wetland. II: production, decomposition and peat formation. Wetland Ecol Manage 13:671–684
Chimner RA, Karberg JM (2008) Long-term carbon accumulation in two tropical mountain peatlands, Andes Mountains, Ecuador. Mires and Peat 3: Art. 4. http://www.mires-and-peat.net/
ClimSoil (2008) Review of existing information on the interrelations between soil and climate change. Final Report. Alterra, Wageningen UR, The Netherlands
Coulter JK (1957) Development of the peat soils in Malaya. Malays Agric J 40:188–199
Dammon AWH, French TW (1987) The ecology of peat bogs of the glaciated northeastern United States. U.S. Fish and Wildlife Service, Biological Report 85 (7.16). Supt. of Documents, Washington DC
Driessen PM (1977) Peat soils. In: Soils and rice. International Rice Research Institute, Philippines
Driessen PM, Rochimah L (1976) The physical properties of lowland peats from Kalimantan and their significance for land suitability appraisal. In: Research on peat and podzolic soils in Indonesia and their potential for agriculture. Soil Research Institute, Bogor
Dykes AP, Kirk KJ (2006) Slope instability and mass movements in peat deposits. In: Martini IP, Martinez-Cortizas A, Chesworth W (eds) Peatlands: evolution and records of environmental and climate changes, vol 9, Developments in earth surface processes. Elsevier, Amsterdam, pp 377–406
Dykes AP, Gunn J, Convery (née Kirk) KJ (2008) Lanslides in blanket peat at Cuilcagh Mountain, northwest Ireland. Geomorphology 102(3–4):325–340
EKONO (1981) Report on energy use of peat. Contribution to U.N. conference on new and renewable sources of energy. Nairobi
FAO (2006) World reference base for soil resources 2006. A framework for international classification, correlation and communication. World Soil Resources Reports 103. FAO, ISRIC and IUSS, Rome
Farnham RS, Finney HR (1965) Classification and properties of organic soils. Adv Agron 17:115–162
Farrell C, Feehan J (eds) (2008) Proceedings of the 13th international peat congress: "After wise use – the future of peatlands". International Peat Society, Jyväskylä

Gallart F, Clotet-Perarnau N, Bianciotto O, Puigdefàbregas J (1994) Peat soil flows in Bahía del Buen Suceso, Tierra del Fuego (Argentina). Geomorphology 9:235–241

Gaudig G, Couwenberg J, Joosten H (2006) Peat accumulation in kettle holes: bottom up or top down? Mires and Peat 1: Art. 6. http://www.mires-and-peat.net/

Gore AJP (1983) Introduction. In: Gore AJP (ed) Ecosystems of the world – 4A mires: swamp, bog, fen, moor. Elsevier, New York, pp 1–34

Gorham E (1991) Northern peatlands: role in the carbon cycle and probable responses to climatic warming. Ecol Appl 1:182–195

Hairiah K, Sitompul SM, van Noordwijk M, Palm C (2001) Carbon stocks of tropical land use systems as part of the global C balance: effects of forest conversion and options for 'clean development' activities. International Centre for Research in Agroforestry (ICRAF), Southeast Asian Regional Research Programme. Bogor

Hammond RF (1981) The peatlands of Ireland. Soil Survey Bulletin 35. An Foras Taluntais, Dublin

Hardon HJ, Polak B (1941) De chemische samenstelling van enkele venen in Nederlandsch Indië. Landbouw 17:1081–1093

Hashim R, Islam S (2008) A model study to determine engineering properties of peat soil and effect on strength after stabilization. Eur J Sci Res 22(2):205–215

Holden J (2006) Peatland hydrology. In: Martini IP, Martinez-Cortizas A, Chesworth W (eds) Peatlands: evolution and records of environmental and climate changes, vol 9, Developments in earth surface processes. Elsevier, Amsterdam, pp 319–346

Hoscilo A, Page SE, Tansey K (2008) Repeated and extensive fire as the main driver of land cover change in Block C of the former Mega Rice Project. In: Farrell C, Feehan J (eds) Proceedings of the 13th international peat congress: "After wise use – the future of peatlands", vol 1. International Peat Society, Jyväskylä, pp 206–208

Immirzi CP, Maltby E, Clymo RS (1992) The global status of peatlands and their role in carbon cycling. Report 11, Friends of the Earth, London

Jaenicke J, Rieley JO, Mott C, Kimman P, Siegert F (2008) Determination of the amount of carbon stored in Indonesian peatlands. Geoderma 147(3–4):151–158

Jansen JC, Diemont WH, Koenders N (1985) Peat development for power generation in West Kalimantan – an ecological appraisal. The Netherlands Economic Institute, Rotterdam

Jauhiainen J, Takahashi H, Heikkinen JEP, Martikainen PJ, Vasanders H (2005) Carbon fluxes from a tropical peat swamp forest floor. Glob Change Biol 11(10):1788–1797

Joosten H, Clarke D (2002) Wise use of mires and peatlands. International Mire Conservation Group and International Peat Society, Jyväskylä

Jordan S, Velty S, Zeitz J (2007) The influence of degree of peat decomposition on phosphorus binding forms in fens. Mires and Peat 2: Art. 7. http://www.mires-and-peat.net/

Kivinen E (1980) Proposal for general classification of virgin peat. Proceedings of the 6th international peat congress, Duluth, pp 47–51

Kurbatov IM (1968) The question of the genesis of peat and its humic acids. In: Robertson RA (ed) Transactions of the 2nd international peat congress, Leningrad, vol 1. HMSO, Edinburgh, pp 133–137

Lappalainen E (ed) (1996) Global peat resources. International Peat Society, Jyväskylä

Limpens J, Berendse F, Blodau C, Canadell JG, Freeman C, Holden J, Roulet N, Rydin H, Schaepman-Strub G (2008) Peatlands and the carbon cycle: from local processes to global implications – a synthesis. Biogeosciences 5:1475–1491

Lucas RE (1982) Organic soils (Histosols). Formation, distribution, physical and chemical properties and management for crop production. Research Report 435 (Farm Science). Michigan State University, East Lansing, MI

Ludang Y, Jaya A, Inoue T (2007) Geohydrological conditions of the developed peatland in Central Kalimantan. World Appl Sci J 2(3):198–203

Maltby E, Immirzi P (1993) Carbon dynamics in peatlands and other wetland soils: regional and global perspectives. Chemosphere 27:999–1023

Maltby E, Proctor MCF (1996) Peatlands: their nature and role in the biosphere. In: Lappalainen E (ed) Global peat resources. International Peat Society, Jyväskylä, pp 11–19

Martini IP, Martinez-Cortizas A, Chesworth W (eds) (2006a) Peatlands: evolution and records of environmental and climate changes, vol 9, Developments in earth surface processes. Elsevier, Amsterdam

Martini IP, Martinez-Cortizas A, Chesworth W (2006b) Peatlands: a concise guide to the volume. In: Martini IP, Martinez-Cortizas A, Chesworth W (eds) Peatlands: evolution and records of environmental and climate changes, vol 9, Developments in earth surface processes. Elsevier, Amsterdam, pp 1–13

Matthews E, Fung I (1987) Methane emission from natural wetlands: global distribution, area, and environmental characteristics of sources. Glob Biogeochem Cycles 1:61–86

Medina E, Cuevas E, Huber O (2011) Origin of organic matter leading to peat formation in the southeastern Guayana uplands and highlands. In: Zinck JA, Huber O (eds) Peatlands of the Western Guayana Highlands, Venezuela, Chap. 8. Springer, Heidelberg, doi: 10.1007/978-3-642-20138-7_8

Montanarella L, Jones RJA, Hiederer R (2006) The distribution of peatland in Europe. Mires and Peat 1: Art. 1. http://www.mires-and-peat.net/

Moore PD, Bellamy DJ (1974) Peatlands. Elek Science, London

Moore TR, Bubier JL, Bledzki L (2007) Litter decomposition in temperate peatland ecosystems: the effect of substrate and site. Ecosystems 10:949–963

Neuzil SG (1997) Onset and rate of peat and carbon accumulation in four domed ombrogenous peat deposits, Indonesia. In: Rieley JO, Page SE (eds) Biodiversity and sustainability of tropical peatlands. Samara, Cardigan, pp 55–72

Nogué S, Rull V, Montoya E, Huber O, Vegas-Vilarrúbia T (2009) Paleoecology of the Guayana Highlands (northern South America): Holocene pollen record from the Eruoda-tepui in the Chimantá massif. Palaeogeogr Palaeoclimatol Palaeoecol 281:165–173

Notohadiprawiro T (1997) Twenty-five years experience in peatland development for agriculture in Indonesia. In: Rieley JO, Page SE (eds) Biodiversity and sustainability of tropical peatlands. Samara, Cardigan, pp 301–310

Page SE, Banks C (2007) Tropical peatlands: distribution, extent and carbon storage – uncertainties and knowledge gaps. Peatlands Int 2:26–27

Page SE, Siegert F, Rieley JO, Boehm H-DV, Jaya A, Limin S (2002) The amount of carbon released from peat and forest fires in Indonesia during 1997. Nature 420:61–65

Page SE, Wüst RAJ, Weiss D, Rieley JO, Shotyk W, Limin SH (2004) A record of Late Pleistocene and Holocene carbon accumulation and climate change from an equatorial peat bog (Kalimantan, Indonesia): implications for past, present and future carbon dynamics. J Quatern Sci 19(7):625–635

Page SE, Rieley JO, Wüst R (2006) Lowland tropical peatlands of Southeast Asia. In: Martini IP, Martinez-Cortizas A, Chesworth W (eds) Peatlands: evolution and records of environmental and climate changes, vol 9, Developments in earth surface processes. Elsevier, Amsterdam, pp 145–172

Page SE, Banks CJ, Rieley JO (2007) Tropical peatlands: distribution, extent and carbon storage – uncertainties and knowledge gaps. http://www.geog.le.ac.uk/carbopeat/media/pdf/yogyapapers/p1.pdf

Payne RJ, Blackford JJ (2008) Peat humification and climate change: a multi-site comparison from mires in south-east Alaska. Mires and Peat 3: Art. 9. http://www.mires-and-peat.net/

Pfadenhauer J (1990) Tropische und subtropische Moore. In: Göttlich K (ed) Moor- und Torfkunde. 3. Auflage. E. Schweizerbart'sche Verlagsbuchhandlung (Nägele und Obermiller), Stuttgart, pp 102–113

Polak B (1950) Occurrence and fertility of tropical peat soils in Indonesia. In: Proceedings of the 4th international congress of soil science vol 2, pp 183–185

Rajagopalan G, Sukumar R, Ramesh R, Pant RK, Rajagopalan G (1997) Late Quaternary vegetational and climatic changes from tropical peats in southern India – an extended record up to 40,000 years BP. Curr Sci 73(1):60–63

Rajagopalan G, Ramesh R, Sukumar R (1999) Climatic implications of $\delta^{13}C$ and $\delta^{18}O$ ratios from C3 and C4 plants growing in a tropical montane habitat in southern India. J Biosci 24(4):491–498
Rieley JO, Page SE (2005) Wise use of tropical peatlands: focus on Southeast Asia. Alterra-Wageningen University and Research Centre and the EU INCO-Strapeat and Restorpeat Partnerships, Wageningen
Rodríguez AR (1999) Conservación de humedales en Venezuela: inventario, diagnóstico ambiental y estrategia. Comité Venezolano de la UICN, Caracas
Rull V (1991) Contribución a la paleoecología de Pantepui y la Gran Sabana (Guayana Venezolana): clima, biogeografía y ecología. Scientia Guaianae 2, CVG-EDELCA, Caracas
Satrio AE, Gandaseca S, Ahmed OH, Ab Majid NM (2009) Effect of precipitation fluctuation on soil carbon storage of a tropical peat swamp forest. Am J Appl Sci 6(8):1484–1488
Schubert C, Fritz P (1985) Radiocarbon ages of peat, Guayana Highlands (Venezuela). Some paleoclimatic implications. Naturwissenschaften 72:427–429
Schubert C, Fritz P, Aravena R (1994) Late Quaternary paleoenvironmental studies in the Gran Sabana (Venezuelan Guayana Shield). Quatern Int 21:81–90
Shier CW (1985) Tropical peat resources – an overview. In: Proceedings symposium tropical peat resources: prospects and potential. International Peat Society, Kingston, pp 29–46
Shimada S, Takahashi H, Haraguchi A, Kaneko M (2001) The carbon content characteristics of tropical peats in Central Kalimantan, Indonesia: estimating their spatial variability in density. Biogeochemistry 53(3):249–267
Sieffermann RG, Fournier M, Triutomo S, Sadelman MT, Semah AM (1988) Velocity of tropical forest peat accumulation in Central Kalimantan Province, Indonesia (Borneo). Proceedings of the 8th international peat congress, Leningrad, pp 90–98
Sjörs H (1980) Peat on earth: multiple use or conservation? Ambio 9:303–308
Stêpniewska Z, Borkowska A, Kotowska U (2006) Phosphorus release from peat soils under flooded conditions of the Leczynsko-Wlodawskie Lake District. Int Agrophysics 20:237–243
Sukumar R, Ramesh R, Pant RK, Rajagopalan G (1993) A $\delta^{13}C$ record of late Quaternary climate change from tropical peats in southern India. Nature 364(6439):703–706
Tie YL, Kueh HS (1979) A review of lowland organic soils of Sarawak. Department of Agriculture, Technical Paper 4, Research Branch, Sarawak
Tie YL, Lim CP (1976) Lowland peat soils for sago-growing in Sarawak. In: Tan K (ed) Sago-76. Papers of the 1st international sago symposium, Kuching, pp 187–189
USDA (1999) Soil taxonomy. A basic system of soil classification for making and interpreting soil surveys. Agricultural Handbook 436, 2nd edn. US Department of Agriculture, Natural Resources Conservation Service, Soil Survey Staff. Washington, DC
USDA (2006) Keys to Soil Taxonomy. US Department of Agriculture, Natural Resources Conservation Service, National Cooperative Soil Survey. Washington, DC
Von Post L (1924) Das genetische System der organogenen Bildungen Schwedens. Mémoires sur la nomenclature et la classification des sols. International Committee of Soil Science, Helsinki, pp 287–304
Wheeler R (2003) Forest ecosystems and bogs in arctic, temperate and tropical locations. http://www.uaf.edu/ces/forestry/pdfs/Ecosystems%20and%20Bogs.pdf
White WA, Warne AG, Guevara EH, Aslan A, Tremblay TA, Raney JA (2002) Geo-environments of the northwest Orinoco delta, Venezuela. Interciencia 27(10):521–528
Wikipedia (2008) Peat. http://en.wikipedia.org/wiki/Peat
Wösten JHM, Clymans E, Page SE, Rieley JO, Limin SH (2008) Peat-water interrelationships in a tropical peatland ecosystem in Southeast Asia. Catena 73(2):212–224
Wüst RAJ, Bustin RM, Lavkulich LM (2003) New classification systems for tropical organic-rich deposits based on studies of the Tasek Bera Basin, Malaysia. Catena 53(2):133–163
Yule CM, Gomez LN (2009) Leaf litter decomposition in a tropical peat swamp forest in Peninsular Malaysia. Wetlands Ecol Manage 17:231–241
Zinck JA, García P, Van der Plicht J (2011) Tepui peatlands: age record and environmental changes. In: Zinck JA, Huber O (eds) Peatlands of the Western Guayana Highlands, Venezuela, Chap. 7. Springer, Heidelberg, doi: 10.1007/978-3-642-20138-7_7

Chapter 3
The Venezuelan Guayana Region and the Study Areas: Geo-ecological Characteristics

O. Huber and P. García

3.1 Introduction

The Venezuelan Guayana located in northeastern South America covers approximately 454,000 km^2. It includes three states, namely the Delta Amacuro state corresponding to the Orinoco river delta to the east, the Bolívar state in the north and to the west of the Delta Amacuro state, and the Amazonas state that is the southernmost political entity of the country (Fig. 3.1). The area is very sparsely populated with a regional average of 2.7 inhabitants per km^2 (INE 2001). More than 90% of the population concentrates in a few cities and towns along the northern and northwestern periphery of the region, mainly in the Bolívar state where industrial and commercial activities continue to attract people from other parts of the country and from abroad.

The largest part of the Venezuelan Guayana (about 83%) is covered by a variety of tropical and montane forests. The rest of the area shows savannas and other herbaceous vegetation types, shrublands, and pioneer vegetation growing either on local rock outcrops or extensive whitesand deposits. The flora is very rich, including around 10,000 different species of vascular plants with more than one third being endemic. The degree of endemism increases significantly with elevation, while the total number of species decreases in the same direction.

Within this regional context, the Guayana Highlands stretch like an archipelago over the states of Bolívar and Amazonas in southern Venezuela, with spurs extending into the Pakaraima region of northwestern Guyana and the Roraima and Amazonas states of northern Brazil. Some lower outliers are found in Surinam (Wilhelmina Gebeergte, ca 1,280 m a.s.l.) and in central-eastern Colombia (Serranía de Chiribiquete, ca 840 m a.s.l.). Typically, the Guayana Highlands include mid-elevation plateaus, high mesetas, and dissected mountain ranges with interspersed dome-shaped hillands. Individual rocky mesetas are called "tepuis" by the Pemón people, an Arawak-speaking ethnic group of about 20,000 members dwelling mainly in the southeastern uplands of the Venezuelan Guayana (Gran Sabana). Nowadays, the term "tepui" is commonly used in geologic and geomorphic

J.A. Zinck and O. Huber (eds.), *Peatlands of the Western Guayana Highlands, Venezuela*, Ecological Studies 217, DOI 10.1007/978-3-642-20138-7_3,

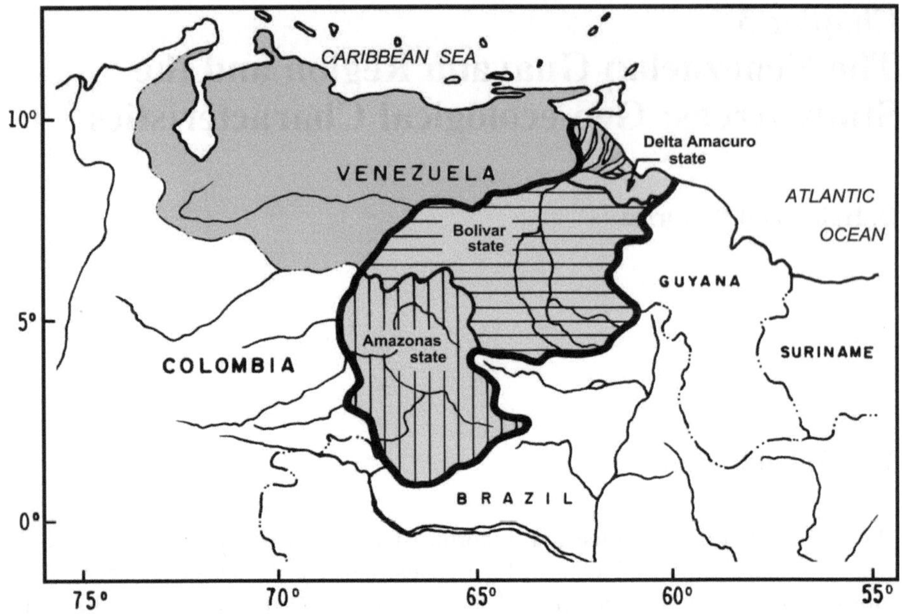

Fig. 3.1 Administrative entities included in the Venezuelan Guayana: Amazonas state, Bolívar state, and Delta Amacuro state

descriptions of the Guayana Highland landscape. A classical tepui example is Cerro Roraima on the border between Venezuela, Brazil, and Guyana. This is a large rocky massif subdivided in mesetas that emerges from forest- and savanna-covered lowlands and rises to 2,800 m elevation. In his novel "*The Lost World,*" Conan Doyle imaginarily describes the Cerro Roraima landscape as an archaic, dinosaur-infested, inaccessible, and inhospitable land of the dark American tropics (Doyle 1912).

Scientific exploration of the Guayana Highlands started in the mid-1830s with the first travels of Richard and Robert Schomburgk to the former colony of British Guiana. The Schomburgk brothers explored the geography and biology of a large region extending between Georgetown, Mount Roraima, and the southern Gran Sabana uplands (Schomburgk 1840, 1848). They penetrated the upper Orinoco watershed after crossing the Sierra Pakaraima from east to west, and from there they navigated on the Casiquiare channel and the Río Negro down to its confluence with the Río Branco in the Brazilian Amazonas. They returned by land to Boa Vista, the Rupununi, and finally back to the mouth of the Essequibo river. The Schomburgks tried to ascend to the summit of Mt. Roraima, but it was only in 1884 when Everard Im Thurn and Henry Perkins were able to reach the tepui summit and gathered the first plant and animal collections from this completely unknown environment. Early exploration expeditions were not supported by indigenous people because tepuis and in particular tepui summits are sacred sites in their cosmogony. It was only around 1930 when younger tribe members accepted to participate in the first modern explorations carried out by scientists from USA and Venezuela.

After intensive exploration and research during the last 70 years, using helicopter logistics since the early 1960s, all tepuis and surrounding landscapes in the Venezuelan and Brazilian Guayana were included in strictly protected areas, mainly national parks and natural monuments. Since 1991, some of these tepuis such as Parú, Duida-Marahuaka-Huachamakari, and Sierra La Neblina in the Venezuelan Amazonas state are part of the large Biosphere Reserve "Alto Orinoco-Casiquiare." The well-known tepui massifs of Auyantepui, Chimantá, Roraima, and others located in the Canaima National Park in southeastern Bolívar state belong since 1995 to the World Natural Heritage Site "Canaima," the only WNH site existing in Venezuela. Also nontepui landscapes on igneous–metamorphic rocks, including for instance Sierra de Maigualida, Cerros Moriche, Morrocoy, Aratityope and Tamacuari, and Serranías Tapirapecó and Unturán, are protected as natural monuments.

The Guayana Highlands are not only of incredible scenic beauty, they are also one of the most interesting biological and geologic areas in the Neotropics. The tepuis, more than 50 in southern Venezuela, and their surroundings harbor an extraordinarily rich, specialized, and diversified assemblage of biota and ecosystems not easily found in other tropical mountains. Together, they constitute the biogeographic province of Pantepui, one of the four provinces forming the biogeographic region of Guayana (Huber 1994a; Berry et al. 1995; Berry and Riina 2005). Because of their sheer high walls, most tepui summits were practically inaccessible until recently except by helicopter. This has largely contributed to prevent human intrusion in the past and helped preserve these landscapes in a truly pristine state. Thus, tepui summits and many other highland landscapes above ca 1,500 m elevation can be considered to be true benchmark ecosystems for comparative ecological studies in tropical high mountains, such as for instance the study on peat ecology presented in this volume. Peatlands are a characteristic feature throughout these areas; their preservation is relevant because they control the regime of most lowland rivers.

This chapter provides an overview of the biophysical conditions of the Guayana region that constitute the environmental context where peatlands have originated and evolved. First, the geo-ecological structure of the region as a whole is described. Then, the characteristics of the study areas are analyzed.

3.2 Geology and Paleoecology

3.2.1 Geologic History

The Guayana Highlands consist of a mosaic of plateaus on siliceous sedimentary rocks and mountain ranges and hillands on igneous–metamorphic rocks at elevations of 600–3,000 m a.s.l. They are part of a continental area that has probably one of the oldest geologic histories on earth. The first terrestrial nuclei started consolidating presumably already between 3.7 and 3.4 Ga ago leading eventually to the creation of the Amazon Craton. This early basement underwent further expansion and accretion along a northeast–southwest gradient for another 2 Ga, until approximately 1.8–1.5 Ga ago when it reached its present configuration consisting of a northern Guayana Shield and a southern Guaporé (or Brazilian) Shield (Gibbs and Barron 1993; Sidder and Mendoza 1995) (Fig. 3.2).

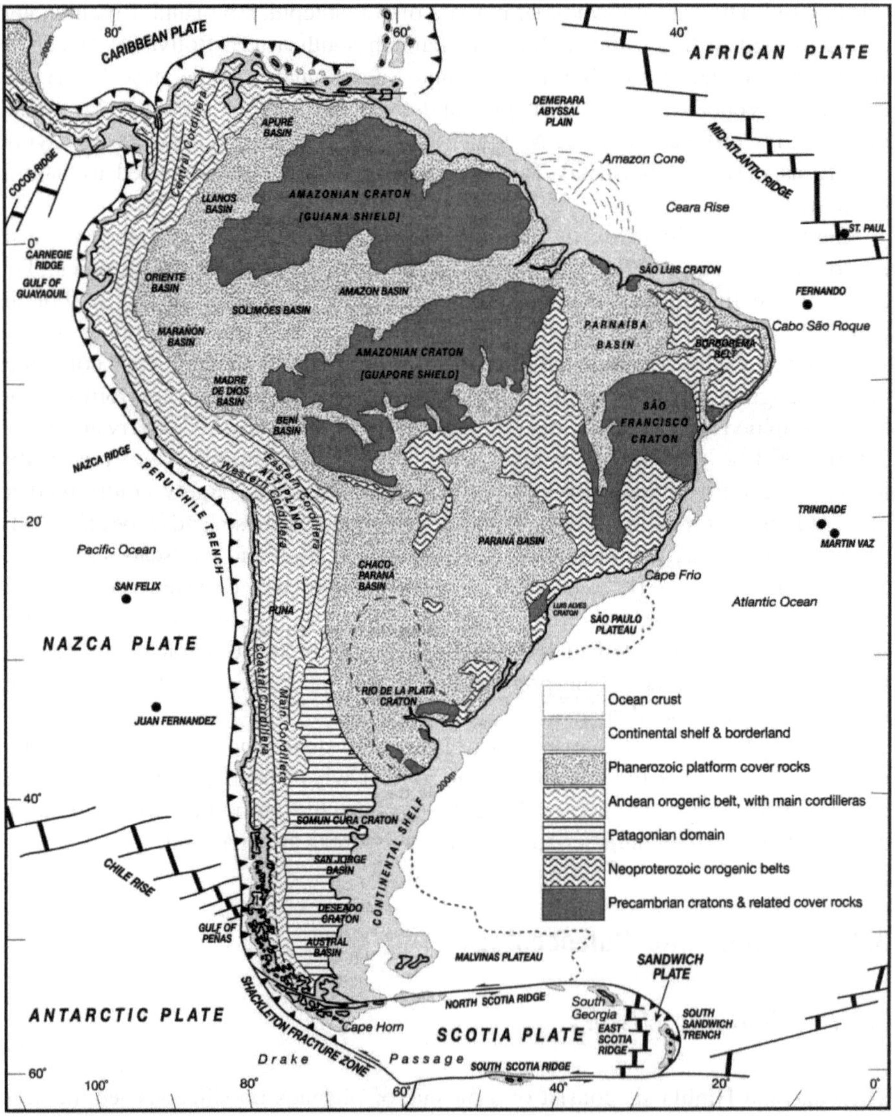

Fig. 3.2 Tectonic Provinces of South America, (from Orme 2007 by permission of Oxford University Press, Inc)

Geotectonic processes taking place in the northeastern and northern parts of the Guayana Shield resulted in the formation of several geologic provinces including (1) the Imataca Complex consisting of ortho- and paragneisses older than 2.8 Ga, (2) the Greenstone Belt of submarine mafic volcanic rocks of Early Proterozoic (2.25–2.1 Ga), and (3) the Supamo Complex consisting of intrusive granites of Early Proterozoic (2.23–2.05 Ga). According to Sidder and Mendoza (1995), a major cycle of metamorphism, deformation, and magmatic activity, corresponding to the Trans-Amazonian Orogeny, took place during the Early Proterozoic (ca 2.15 to 1.96 Ga ago, possibly continued to 1.73 Ga). At the end of this regional orogenic event, a large area of the northern and central Guayana Shield was covered by volcanic, subvolcanic, and plutonic rocks of Early to Middle Proterozoic (1.93–1.79 Ga) belonging to the Cuchivero Group that forms a fourth geologic province. Granitic mountains appeared then in several parts of southern Venezuela, in particular the Sierra de Maigualida range. The territory created during these primeval geotectonic events constitutes the igneous–metamorphic basement of the Guayana Shield.

During the final period of the craton accretion, a long and persistent, though irregular process of sedimentation took place atop the basement mainly in shallow fluvial and marine environments, lasting about 400–700 million years between 1.9 and 1.5 (possibly 1.3) Ga ago. This has resulted in an extensive cover of sandstones, quartzites, and other siliceous sedimentary rocks included in the Roraima Group and forming the Roraima province, approximately one million km^2 wide and up to 3,000 m thick (Gosh 1985; Sidder and Mendoza 1995; Mendoza 2000). The tepui mesetas, the most typical features of the Guayana Shield, have developed from this sedimentary mass.

The Roraima sedimentary cover, deposited and shaped in the Proterozoic over the western portion of land masses that later developed into the Paleozoic Gondwana continent, is spatially heterogeneous because the long-lasting sedimentation processes on the northern Amazon Craton have shifted over time along a geographic gradient running roughly from northeast to southwest. Resulting from the former, several distinct sedimentary basins have been recognized, especially in the Venezuelan Guayana south of the Orinoco river (Gosh 1985). These regional basins are split into an eastern sector (mainly Gran Sabana) and a southwestern sector (mainly Amazonas state) by the north–south oriented granitic Maigualida-Parima mountain system of the Cuchivero geologic province. The northern part of the Maigualida mountains is the only place in the entire Guayana Highlands where igneous–metamorphic rocks belonging to the Santa Rosalía granite (Urbani 2008) reach elevations similar to those of many neighboring, typically flat-topped sandstone tepuis (2,200–2,400 m a.s.l.).

During the Late Proterozoic, the Guayana Shield has undergone tectonic events including regional uplifting, folding, and upbreaking processes. Several magmatic events intruded the sedimentary cover of the Roraima Group between 1,850–1,650 Ma (Early Proterozoic) and 200 Ma (Mesozoic), producing numerous sills and dikes of ultramafic diabase rocks of variable extent over the entire Guayana region (USGS and CVG-TECMIN 1993). Metasedimentary rocks show textural and mineralogical evidence of low-to-medium grade metamorphism with strong recrystallizations (Urbani et al. 1977; Cordani et al. 2010).

Since the Paleozoic at least, the Guayana Shield has been a relatively stable and continuously emerged land mass, a fact that is unique in the otherwise so eventful geologic history of the tropical South American continent. Long-lasting erosion operating differentially on the sedimentary and igneous–metamorphic rocks has resulted in the present landscape structure and features. Granitic batholiths have been exhumed upon partial removal of the sedimentary cover, while tepui mesetas have been sculptured upon relief inversion between anticlines and synclines. About 50 larger plateaus and smaller isolated mesetas with summits between 2,000 and 3,000 m a.s.l. are today scattered over the area south of the Orinoco river.

Figure 3.3 shows a generalized distribution of the geologic provinces of the Guayana Shield. The litho-stratigraphic units composing these provinces are

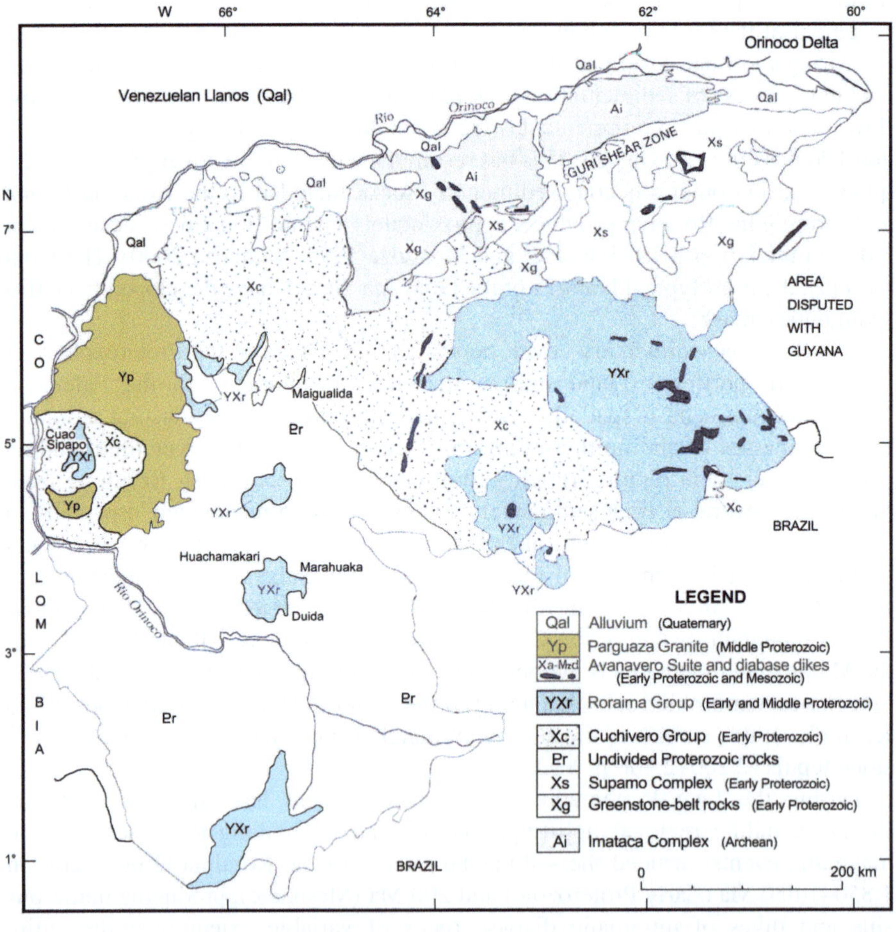

Fig. 3.3 Geologic provinces of the Guayana Shield (modified from Sidder and Mendoza 1995)

reported in two geologic maps of the Venezuelan Guayana based on recent satellite imagery and extensive fieldwork carried out during the last two decades (Wynn et al. 1993; Hackley et al. 2005).

3.2.2 Regional Geology

The study areas treated in this volume belong to two of the geotectonic spatial entities that form the Guayana Shield, namely the Cuchivero and the Roraima provinces.

3.2.2.1 The Cuchivero Geologic Province

In the Amazonas state and adjacent areas of the Bolívar state, the Cuchivero geologic province is subdivided into five petrotectonic subprovinces or dominions (Mendoza 2000).

- The Ayacucho subprovince consists of younger anorogenic and posttectonic granites (mainly less than 1.5 Ga) such as the Parguaza rapakivi granite (1.53 Ga), the Cuao massif, the Atabapo granites, and the Sipapo granodiorite. All these rocks intrude volcanic ignimbrites 1.76–1.97 Ga old, and biotite granites such as the Santa Rosalía granite (1.88 Ga) that is covered by unconformable sediments of the Roraima Group (1.5 Ga).
- The Manapiare subprovince includes intrusions of volcanic rocks and biotite granites into gneisses and migmatites. The Santa Rosalía granite is a fine- to coarse-grained biotite rock with 35% quartz and 5% biotite. Ages between 1.1 and 1.9 Ga have been reported by several authors (Olmeta 1968; Gaudette et al. 1977). According to Mendoza (2000), the Santa Rosalía granite belongs to the group of potassium-rich granites of the Trans-Amazonian event that took place about 1.9 Ga ago. Pre-Roraima metasedimentary rocks consisting of an unmetamorphosed sequence of fine- to very fine-grained, clay-rich sandstones, overly with apparent unconformity the volcanic substratum (Caicara Formation).
- The Casiquiare subprovince, corresponding to the low-lying Casiquiare peneplains, consists of rocks with different degrees of metamorphism including gneisses, migmatites, gneiss-granites, and intrusive granites. Volcanic and plutonic rocks, as well as Roraima sediments and metasediments do not occur in this subprovince.
- The upper Orinoco subprovince extends from the confluence between the Ventuari and Orinoco rivers to the Orinoco headwaters in the southern Sierra Parima at ca 1,200 m elevation. It consists of volcanic and igneous rocks (e.g., Santa Rosalía granite) intruding a substratum of gneiss, augengneiss, granite, and tonalite.
- The Siapa subprovince covers the very little studied southern part of the Amazonas state, and includes metasediments of quartz-micaceous schists and quartz-feldspatic gneisses atop tonalitic gneisses and migmatites.

In several parts of the Amazonas state, Middle Proterozoic granites commonly penetrate the sedimentary strata of the Roraima Group. Such is the case of the Parguaza granite, a massive biotite-containing, coarse-grained, porphyritic granite with rapakivi texture, rich in feldspars and hornblende. Circular structures of about 5–10 km diameter have been identified and are believed to represent alkaline intrusive bodies of probable Middle Proterozoic age (1.3 Ga by Rb/Sr). Such a circular body is located in the eastern foothills of Cerro Duida (USGS and CVG-TECMIN 1993).

3.2.2.2 The Roraima Geologic Province

The Roraima Group consists of a relatively undeformed continental clastic sedimentary sequence of Early to Middle Proterozoic (1.9–1.5 Ga) that covers large areas in the Venezuelan Guayana. Total thickness has been estimated at more than 3,000 m by Reid and Bisque (1975). It is composed by quartzarenite, arkose, silty arenite, conglomerate arenite, and shale. Roraima rocks do not show regional metamorphism, but post-1.5 Ga intrusive granites have caused contact metamorphism. The low degree of metamorphism of the Roraima rocks suggests that a large cover of sedimentary materials has been removed (Ghosh 1977; Urbani et al. 1977; Grupo Científico Chimantá 1986). Locally, the sedimentary beds have been penetrated by diabase dikes and sills. Roraima strata show low-grade regional folding but, since Cretaceous times, have been tectonically stable (Briceño and Schubert 1990, 1992; Briceño et al. 1990). However, local tectonics has caused rock dip to vary from slight to strong, as expressed in the landscape by the presence of monoclinal reliefs from cuestas to hogbacks. The main structural elements are fields of fractures, mainly vertical, which cut the plateaus into prisms of quadrangular shape.

It was on Mt. Roraima, located south of the Gran Sabana plateau at the borders between Venezuela, Brazil, and Guyana, where the first formal geologic description of the Roraima sandstones has been made (Dalton 1912) and where four main sedimentation phases were identified along a stratigraphic column of more than 2,000 m (Reid 1972; Briceño et al. 1990). The sequence includes the following formations from the most recent at the top to the oldest at the base: the Matauí Formation (600–900 m thick), the Uaimapué Formation (up to 650 m thick), the Kukenán Formation (50–400 m thick), and the Uairén Formation (>850 m thick). Each of these formations is made of many strata of sandstones, quartzites, lutites, cherts, and jaspers with highly variable thickness, color, hardness, and acidity.

The Roraima Group shows regional variations from east to west in terms of composition, age, and postdepositional evolution. Conglomerates, jaspers, and arkoses, which are very common in Bolívar state, Brazil and Guyana, are infrequent to the west, in Amazonas state. Exposed strata of similar lithology are often uncorrelated as observed at sites of Cerro Parú, Cerro Sipapo, and Cerro Yapacana

(Ghosh 1977). In Amazonas state, Roraima rocks lie on the Parguaza granite (1.5 Ga) and are younger than those in the east. The tepui erosion surface in Amazonas state does not correlate with the Gran Sabana surface located in the northeastern part of the Guayana Shield, in Bolívar state.

The most emblematic geoforms of the Roraima Group are the tepuis, isolated mesetas or groups thereof whose summits are mostly between 1,300 m (Cerro Yapacana) and 2,992 m a.s.l. (Pico Phelps near Cerro Neblina). The best known tepui massifs in Amazonas state are the Cuao (including the spectacular tower-like Cerro Autana), Guanay, Coro-coro, Yutajé, and Yaví in the north, the Parú, Duida, Marahuaka, Huachamakari, and Yapacana in the center, and the Aracamuni, Avispa, and Neblina in the south, at the border with Brazil.

3.2.3 Paleoecological Considerations

The Guayana Shield is one of the oldest continental nuclei of the western hemisphere. The lack of fossils in any of the lithological entities is evidence that this area is much older than the surrounding mountain systems in northern South America. Since Precambrian times, the mountains and mesetas of the Guayana have undergone long periods of geologic evolution and adaptation to continuously changing environmental conditions.

The sharp separation of the different altitudinal levels observed in many sandstone and quartzite table-mountains (tepuis) of the Guayana has led to the idea of strong isolation processes acting both horizontally and vertically. The main 50 tepui massifs are distant from each other by often more than hundred kilometers and their summits are separated by 1,000–2,000 or more meters from their basements. Yet, the biota found on these summits are remarkably similar and have in common a basic floristic matrix at the genus level, so much that they are now grouped together in a special biogeographic province called Pantepui (Fig. 3.4).

Recently, new lines of field research have been designed, after an initial period of mainly biological collecting carried out on a number of tepuis during the last century. Systematic vegetation studies started in the 1980s, together with palynological and paleoecological research. In 1984, Carlos Schubert initiated peat surveys on several tepui summits in the eastern Guayana Highlands, mainly the Chimantá massif, Auyantepui, and Guaiquinima (Schubert and Fritz 1985; Schubert et al. 1986). The thickness of the peat deposits under different geographic and climatic conditions was measured and their age determined using modern dating methods (Schubert et al. 1992). At the same time, María Lea Salgado-Labouriau began collecting modern pollen samples from several tepui summit floras, with the purpose of building a reference database of pollen types, a basic instrument for further paleoecological research on tepui flora and vegetation (Salgado-Labouriau and Villar 1992). This research line was continued by Valentí Rull who has produced a significant amount of paleoecological papers together with his research group based in Barcelona, Spain (Rull 1991, 2004; Rull et al.

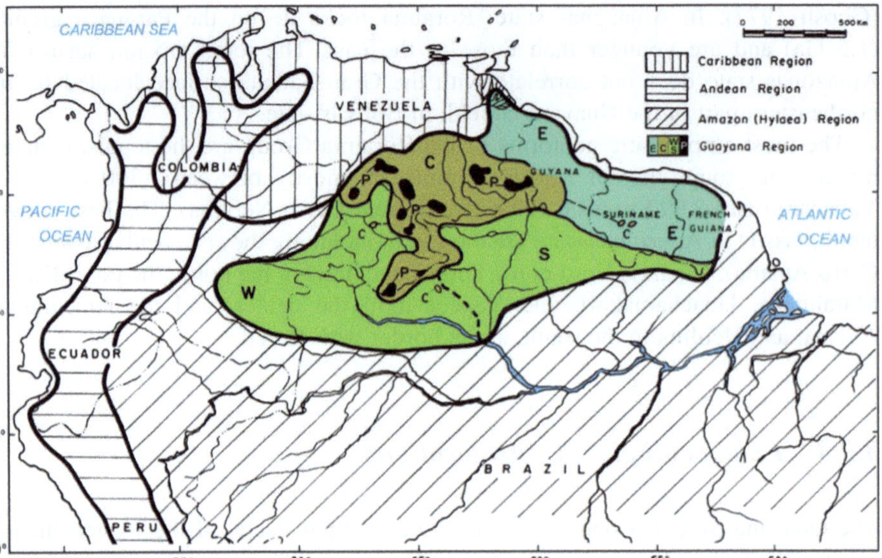

Fig. 3.4 Phytogeographic divisions of northern South America, including the Venezuelan Guayana Region with its four provinces: *E* Eastern Guayana province, *C* Central Guayana province, *W* Western and *S* Southern Guayana province, *P* Pantepui province (modified from Huber 1994a)

2009; Nogué et al. 2009). Following these first studies in highland environment (Chimantá massif and Auyantepui, 1,860–2,250 m a.s.l.), palynological research was expanded to the eastern Guayana uplands, specifically on the Guaiquinima-tepui at 1,350 m a.s.l. (Rull 2005) and the plateau of the southern Gran Sabana at 800–940 m a.s.l. (Rinaldi et al. 1990; Rull 1992, 1999, 2007, 2009; for the location of the sites, see Fig. 8.1). An important outcome of these palynological studies was the initiation of a pollen catalog of the high-tepui flora published by Salgado-Labouriau and Villar (1992) and by Rull (2003).

The peat studies made by Schubert and collaborators on the high-tepui summits of the Chimantá massif and Auyantepui, in the eastern Guayana Highlands, revealed that peat accumulations were not deeper than 2–2.5 m and all younger than 10,000 years. Similar results were obtained for the western Guayana Highlands (see Zinck et al. 2011). Peat accumulations occur on all large table-mountain summits above 1,600 m a.s.l. in the entire Guayana Highland region, from the Chimantá massif in the northeast to the Cuao-Sipapo massif in the northwest and to the Neblina massif in the extreme south of the Venezuelan Amazonas state (personal observations of the first author). Only the summits of the eastern tepui chain from Roraima to Ilú-tepui, along the Venezuela-Guyana demarcation line, show little peat formation probably because of excessive leaching on the flat meseta surfaces under heavy, almost permanent rainfall and strong eastern trade winds, which prevent the accumulation of organic materials on top of the bare, windswept rock surfaces. Likewise, peat deposits are scarce on the tepui range of Ptari-Los Testigos-Agparamán that extends north of the Gran Sabana region, probably because of the same weather conditions mentioned earlier (personal observation by the first author).

A new research project was started by V. Rull in 2006 focusing on issues that relate to the evolution and speciation of the unique Pantepui flora, with intensive pollen sampling on several high-tepui summits of the Chimantá massif and Uei-tepui south of the Roraima massif. It was postulated that the distribution of the present tepui flora could be explained either "...as the result of a long history of evolution in isolation (Lost World hypothesis or LW) or by alternating upward and downward displacements during the Quaternary glacial–interglacial cycles (Vertical Displacement hypothesis or VD)..." (Rull 2004). At present, it seems that both hypotheses must be considered for a meaningful explanation of the floristic diversity and endemism on high-tepui summits. Nogué et al. (2009) found that modern vegetation (broadleaved meadows) in high-tepui environment (>2,200 m a.s.l.) of the Chimantá massif started ca 4,300 years ago and has since then remained mainly unchanged. Vegetation constancy during the upper Holocene can be explained either by environmental stability or site insensitivity. Shedding light on the various hypotheses geared towards explaining the evolution of the Pantepui flora can help assess the impact of potential climate change on the Guayana Highland environment.

Paleoecological and palynological research on the southern Gran Sabana plateau also raised the question of how far human intervention around early settlements in the uplands may have altered the natural vegetation cover. The main debate centers on the role of natural vs anthropogenic fires in forest and savanna vegetation during the Holocene. A relatively large body of literature on this controversial subject is now available, without providing clear-cut answers on the past and present vegetation dynamics and floristic degradation (e.g., Rull 1991, 1992, 1999, 2007, 2009; Leal 2010).

Further research should also focus on the extensive peatlands that occur in the northeast of the Gran Sabana uplands between approximately 1,000 and 1,500 m a.s.l., especially above 1,200 m a.s.l. These peatlands cover slightly inclined slopes and harbor an impoverished flora originated from the high-tepui level and mixed with a few local endemic species. The vegetation is a broadleaved meadow with sparse shrubs that differs notably from the contiguous homogeneous, species-poor grass savannas. Because of their strong affinity with the high-tepui flora, these upland meadows have sometimes been called "sub-tepui meadows" (e.g., in Huber et al. 2001). So far no paleoecological or palynological research has been made on this vegetation type of the Guayana upland region.

Although access to most tepui summits is easy by helicopter, replacing the early long-lasting expeditions with porters and climbers, nowadays scientific research in the Guayana Highlands is restricted mainly because of bureaucratic regulations (Rull et al. 2008) and fewer funding opportunities. However, these high-mountain ecosystems are fragile and sensitive to upcoming climate changes. Therefore, reliable field data and careful interpretation thereof are needed to support the ecological monitoring of this unique biota. Comparative paleoecological studies in the different areas of Pantepui and adjacent upland ecosystems should also be undertaken.

3.3 Physiography, Climate, and Vegetation

In general, the physiography of the Guayana region is controlled by the mainly horizontally lying sandstone and quartzite beds of the Roraima Group. Resulting from long-lasting erosion, the current landscape consists of a series of stepwise topographic levels ("Stufenlandschaft") from plains and peneplains in the lowlands, to large plateaus at intermediate elevation, to the high-elevation tepui mesetas. This table-shaped pattern alternates with rugged mountain ranges and hillands of dome-shaped relief developed from igneous–metamorphic Guayana Shield basement.

Pouyllau (1989) distinguishes two topographic levels in Amazonas state: highlands ("hautes terres") above 200/300 m a.s.l., and lowlands ("basses terres") between 35 and 250 m a.s.l. The highlands include the sandstone mesetas of the Roraima Group at 800–3,000 m a.s.l. (e.g., Sierra Avispa-Pico Neblina, Duida, Parú, and Guanay-Yutajé-Yaví), as well as the mainly granitic undulating plateaus and mountain ranges of Unturán-Tapirapecó (1,000–2,000 m a.s.l.), Sierra Parima (1,000–1,800 m a.s.l.), Serranía de Maigualida, and the Parguaza massif (1,000–1,900 m a.s.l.). The lowlands comprise four subunits called depressions: (1) the Río Ventuari plains in the north, (2) the Casiquiare depression in the center, (3) the upper Orinoco depression in the east, and (4) the transitional Orinoco-Guaviare-Atabapo system along the western border with Colombia. This is one of the first physiographic–geomorphic classifications published on the Amazonas state. It reflects the limited and sometimes inaccurate (e.g., altitudes) knowledge available only 30 years ago on this remote area.

In a more recent attempt of landscape classification, Huber (1995a, b) proposes a geo-ecological division of the Venezuelan Guayana into three altitudinal belts: lowlands, uplands, and highlands (Figs. 3.5 and 3.6). The main geomorphic, climatic, and ecological characteristics of each zone are hereafter described. Climatic information is derived from data published in MARNR (1979) and MARNR-ORSTOM (1987). The general description of flora and vegetation is based on Steyermark et al. (1995–2005) and Huber (1995b, 2005), whereas more detailed information sources are quoted in the text. An explanatory list of plant taxa mentioned in the text is given in Table 3.1 at the end of this chapter.

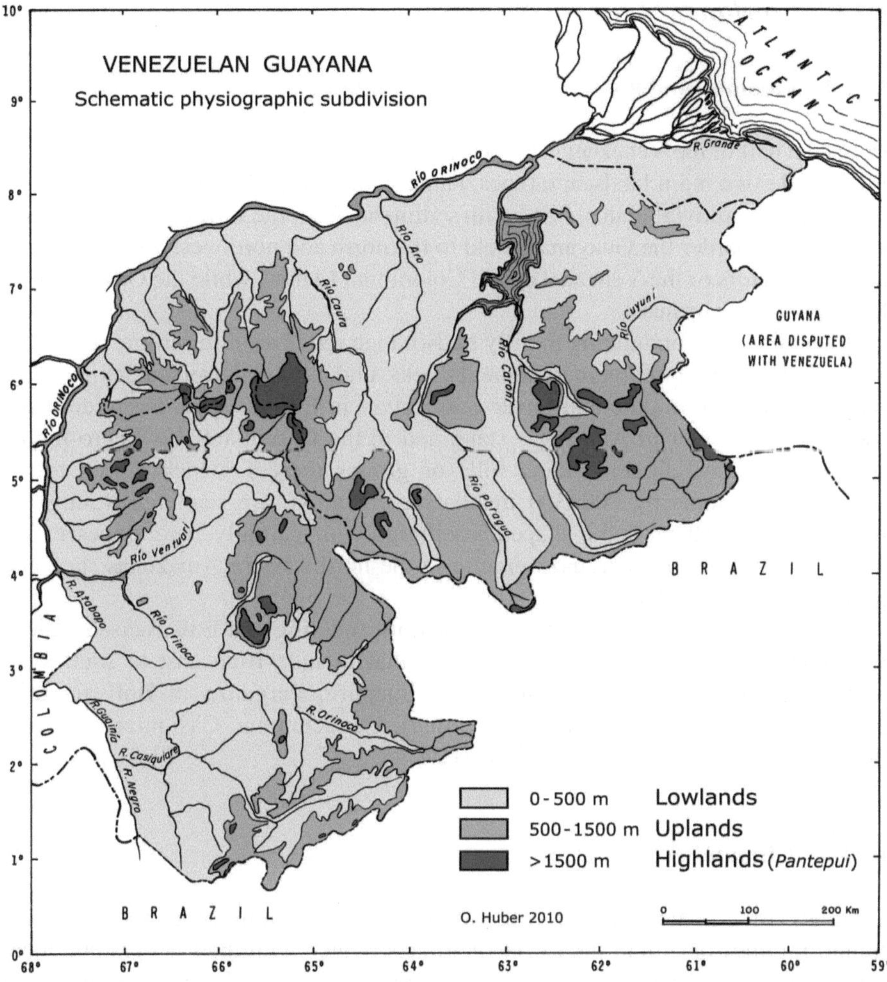

Fig. 3.5 Physiographic landscape distribution of the Venezuelan Guayana

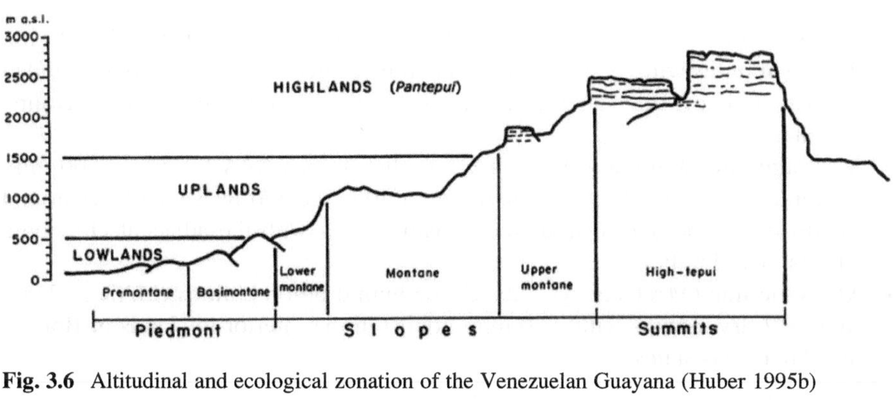

Fig. 3.6 Altitudinal and ecological zonation of the Venezuelan Guayana (Huber 1995b)

3.3.1 *Lowlands*

3.3.1.1 Geomorphology

More than half of the Venezuelan Guayana consists of lowland areas (0–500 m a.s.l.) that include two main landscape types, namely alluvial plains and peneplains.

Quaternary alluvial plains, filled with sediments from the surrounding mountains and mesetas, border the Guayana Shield to the north and northwest, in contact with the distal sectors of the Venezuelan and Colombian Llanos, while the Orinoco delta develops to the northeast.

Peneplains with undulating to hilly half-orange relief resulting from weathering of the igneous–metamorphic basement rocks under humid tropical climate are frequent along the middle and lower Caura river in Bolívar state, the middle and lower Ventuari river in Amazonas state, and in the Casiquiare-Río Negro fluvial system. Typically, alone-standing hills or groups thereof, which dominate the surrounding terrains by 200–300 m relative elevation, are scattered throughout the peneplains. These dome-shaped inselbergs with partially bare slopes called "lajas" are frequent on Parguaza granite in the northwestern Amazonas state and the western Bolívar state.

Plains and peneplains are crossed by flat, meandering and little incised valleys corresponding to the main Guayana rivers that include from east to southwest: Caroní-Paragua, Aro, Caura, Cuchivero, Suapure, Parguaza in Bolívar state; Cataniapo, Cuao-Sipapo, Atabapo, Ventuari, Cunucunuma, Casiquiare channel, Guainía, and Río Negro in Amazonas state.

3.3.1.2 Climate

A variety of climate patterns with different combinations of temperature and rainfall regimes occurs in the lowlands in accordance with their geographic location. They include the following types differentiated on the basis of mean annual temperature (MAT), mean annual rainfall (MAR), and the average number of dry months per year.

- Macrothermic (MAT always >24°C), tropophilous (MAR 1,000–2,000 mm, 2–5 dry months with <50 mm/month) in the northeastern plains around Ciudad Bolívar and towards the transition area with the adjacent Llanos savanna climate.
- Macrothermic (MAT always >24°C), ombrophilous (MAR 1,500–2,500 mm, maximum 3 dry months with <50 mm/month) in the northwestern plains around Puerto Ayacucho and towards the transition area with the adjacent Orinoquia Llanos of Colombia.
- Macrothermic (MAT always >24°C), perhumid ombrophilous (MAR > 2,000 mm, <2 dry months with <50 mm/month) in the interior lowlands of Bolívar and Amazonas states.

3.3.1.3 Vegetation

Lowland vegetation types range from dense forests to shrublands and savannas (Fig. 3.7). The mosaic of plant formations is determined firstly by macroclimatic parameters and secondly by soil factors, including critical soil surface water regime. Most of the vegetation cover is still pristine because population pressure is very low, as the area is inhabited by sparsely distributed indigenous people.

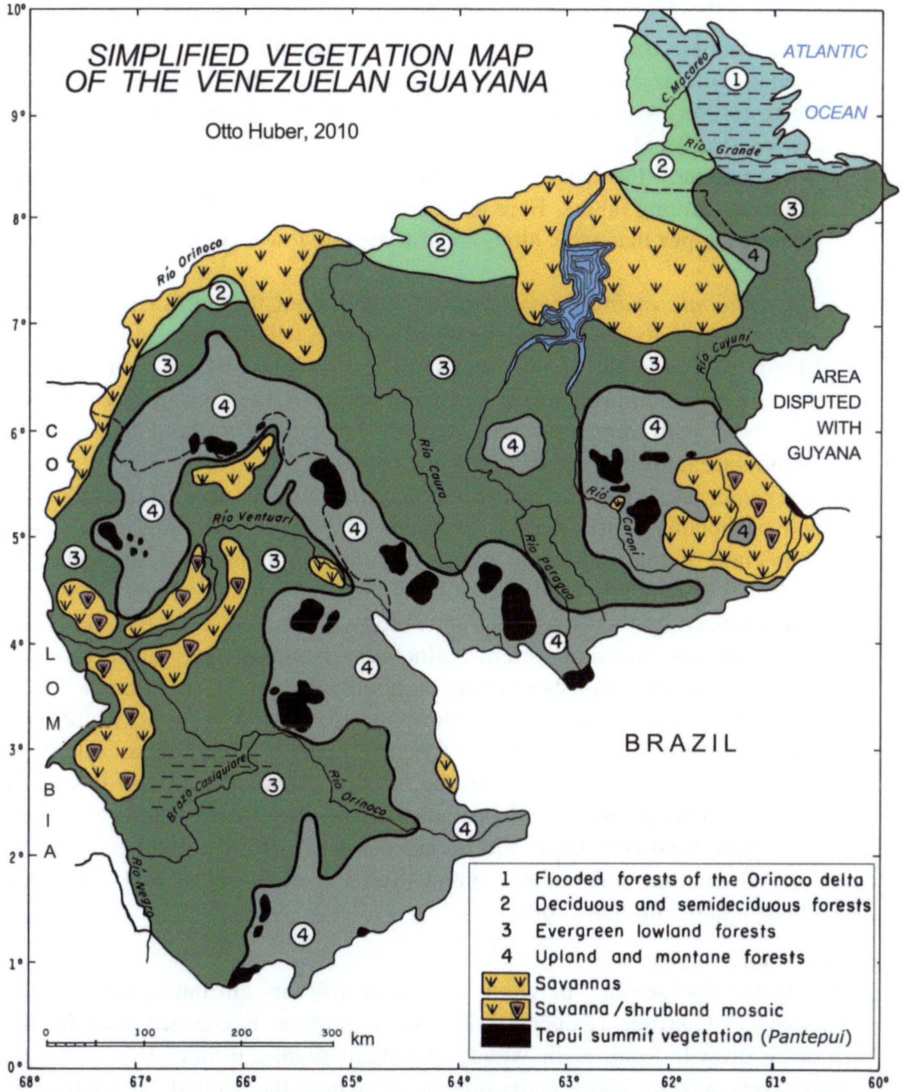

Fig. 3.7 Simplified vegetation map of the Venezuelan Guayana (based on Huber 1995c)

However, vegetation degradation can be observed in the neighborhood of larger towns and cities such as Puerto Ordaz and Ciudad Bolívar in the lower Orinoco valley, or Puerto Ayacucho in the north of the Amazonas state.

(a) Forests

Lowland forests cover large parts of the Venezuelan Guayana belonging to the tropical macrothermic life zone, where mean annual air temperature never drops below 24°C. Under such favorable thermic conditions, forest diversification is mainly controlled by available soil moisture and soil nutrient levels. The following three major lowland forest types can be distinguished.

- *Macrothermic tropophilous forests* occur in the northeastern plains around Ciudad Bolívar and towards the transition area with the adjacent Llanos. These forests are moderately tall (mostly <20 m) and dominated by deciduous or semideciduous tree families, such as Bignoniaceae, Bombacaceae, Boraginaceae, Fabaceae s.l., Anacardiaceae, Chrysobalanaceae, and Tiliaceae, among others. Here are also included some semideciduous riparian forest types that grow along the middle and lower course of the Orinoco river.
- *Macrothermic ombrophilous forests* extend from the northwestern plains around Puerto Ayacucho and the transition area with the adjacent Orinoquia Llanos of Colombia in the west until the middle Cuyuní river basin in the east. The dominant forest is a semievergreen, medium-to-low premontane type with numerous tree species belonging to the Fabaceae s.l., Bignoniaceae, Burseraceae, and a few palms.
- *Macrothermic perhumid ombrophilous forests* cover the extensive plains and peneplains of the interior lowlands of Bolívar and Amazonas states. This class of true rainforest, so far little known, contains a large number of physiognomically and floristically different forest types. They are characterized by their evergreen phenology and the presence of usually tall upper forest trees. An important distinction is made between forests growing on floodplains and forests growing on nonflooded land ("tierra firme"). The water quality is also used as a criterion to distinguish between whitewater forests ("várzea") and blackwater forests ("igapó"). The peculiar, relatively low-stemmed forests growing in the flooded blackwater area of the upper Río Negro basin are called "Amazonian or Río Negro caatinga." Their evergreen, relatively thick and coriaceous leaves form a litter mat on the forest floor that accumulates directly on top of bleached and extremely oligotrophic whitesand soils.

(b) Shrublands

Shrublands (or scrubs) play an important role in all landscapes of the Venezuelan Guayana (Huber 1989). The vegetation is characterized by the shrubby growth form, with woody perennial plants, usually 0.5–5 m tall, which have several short stems arising at or near the ground. In contrast to the forest cover that is generally continuous over large extents, shrublands occur in mosaics of small glades interspersed within the woodland. Most shrublands are related to poor soil conditions (e.g., rocky substratum, whitesands) or limited soil–water regimes (e.g., surface water stress). The main lowland shrub formations and the areas where they occur are as follows.

- Dense sclerophyllous shrub colonies, 1–5/8 m tall, associated with patches of bare whitesand, form islands scattered throughout the forest cover in the Sipapo, Atabapo, and Guainía river plains (50–200 m a.s.l.). The most important shrub families are Humiriaceae, Malpighiaceae, Aquifoliaceae, Icacinaceae, and Sapotaceae. Similar shrublands up to 3 m tall, with several endemic shrubs and herbs, exposed to short-duration surficial flooding are called "bana" in the Río Negro region.
- Small saxicolous bush islands develop in depressions and along grooves on bare slopes ("lajas") of the granitic rock outcrops (inselbergs) that typically occur at 50–400 m elevation along the northwestern border of the Guayana Shield (northwestern Amazonas and Bolívar states). These pioneer communities include a variety of endemic shrub species belonging to the families Apocynaceae, Bombacaceae, Burseraceae, Melastomataceae, and Erythroxylaceae (Gröger and Huber 2007). Also the herbaceous compartment of these ecosystems contains several endemics, mainly from the Bromeliaceae and Cyperaceae families.
- Open shrublands are also found on whitesands (Quartzipsamments and deep Spodosols). Loosely spaced shrubs 1–3 m tall, with usually very thin and open crowns and fairly dense herbaceous layer, grow in open formations that give the tiger-pattern landscape (when seen from the air) characteristic of the peneplains and alluvial plains at 50–200 m a.s.l. in the middle watersheds of the Ventuari and Casiquiare rivers. Most of the species are endemic and belong to such families as Sapotaceae, Malpighiaceae, Ixonanthaceae, Ochnaceae, Combretaceae, and Tepuianthaceae.

(c) Herbaceous plant communities

Herbaceous vegetation is widespread in the American Tropics from sea level up to almost 5,000 m elevation in the Andes. The largest extents of herbaceous ecosystems in South America are in the lowland plains of Colombia and Venezuela (Llanos del Orinoco), lowlands and uplands of central Brazil, northern lowland plains of Bolivia (Llanos de Mojos), and subtropical lowland plains of Paraguay and northern Argentina. The predominant vegetation type consists of a fairly continuous cover of low-to-tall grasses, with or without woody elements sparsely distributed through the herbaceous layer. Such vegetation is generally called savanna ("sabana" in tropical Central and South America), field ("campo" in Brazil), or grassland ("pastizal" in subtropical and temperate South America).

Field exploration and research have revealed the presence of distinct herbaceous communities in the Venezuelan Guayana that needed a new classification approach (Huber 1995b, 2006). Herbaceous plant communities, generally called meadows, were divided into two classes: one dominated by tropical grasses and another dominated by nongramineous herbs. In the first case, the gramineous meadows or grasslands are called savannas, typically formed by tropical grasses with C4 photosynthetic pathway. Other herbs and subfruticose plants are usually present in much lower proportions, as well as woody shrubs, treelets and trees. The woody elements growing widely spaced never form a continuous canopy in tropical savanna. In the second case, instead of Poaceae (grasses) with C4 photosynthetic pathway, herbs belonging to entirely different families such as Rapateaceae, Bromeliaceae, Xyridaceae, Eriocaulaceae, and others, form the main compartment of the ecosystem. This is then called nongramineous meadow or simply meadow with further specification based on the predominant growth form.

In the Guayana lowlands, savannas and meadows are present, sometimes growing side by side as for instance in the upper Orinoco plains.

- Typical neotropical savannas dominated by *Trachypogon spicatus*, together with many other grasses belonging to the genera *Axonopus, Panicum*, and *Andropogon*, among others, are found in several Guayana lowland areas. A fairly continuous belt of savannas with treelets of *Curatella* and *Byrsonima* extends over areas with strong rainfall seasonality, such as in the lower and middle Orinoco rightbank plains between Puerto Ayacucho and Puerto Ordaz. These savannas are clearly related to the large body of savannas of the adjacent Llanos region, which stretches north of the Orinoco over central Venezuela and represents one of the major neotropical savanna regions.
- Nongramineous meadows are widespread in central Amazonas state as herblands often quite dense and 20–50 cm tall. They grow on deep whitesand soils (Quartzipsamments) that are subject to strong soil–water fluctuations from temporary flooding to extreme desiccation of the uppermost soil horizons. The diversified and highly specialized herbaceous and sometimes shrubby vegetation living on this extremely oligotrophic substrate has been documented in the Venezuelan (Huber 1982), Colombian (Cárdenas 2007), and Brazilian Guayanas (Projeto RADAMBRASIL 1975, 1976). Nongramineous meadows form together one of the principal Guayana endemicity centers, comparable to similar ecosystems found on the summits of the sandstone and quartzite mesetas (Pantepui) where the whitesands deposited in the lowlands are coming from. Main floristic constituents belong to the Rapateaceae (several species of *Schoenocephalium, Duckea,* and *Monotrema*), together with numerous species of Xyridaceae, Eriocaulaceae, Bromeliaceae, and Cyperaceae. Endemic shrublets scattered throughout the herbaceous stratum belong to Fabaceae, Simaroubaceae, Ochnaceae, Olacaceae, and several other families.

(d) Pioneer plant communities

Pioneer plant communities typically grow in open habitats such as bare rock surfaces, sandy beaches, riverbed bars, active river margins, or anthropically degraded areas, depleted of their original plant cover. In the Guayana region, a highly diversified spectrum of natural pioneer plant communities develops on rocky slopes from sea level to 3,000 m a.s.l. in accordance with altitudinal variation in climate and lithology. Open rock surfaces cover relatively small total areas in the lowlands and uplands. But they are significantly extensive in the highlands where tepui walls and summits provide an ideal habitat for the development of the saxicolous biome (or lithobiome) of Pantepui.

Rock outcrops are frequent all over the northwestern lowlands of the Guayana Shield, especially in central and northwestern Amazonas state and in the adjacent northwestern Bolívar state, where the underlying Parguaza batholith reaches the terrain surface in many places. Granitic hills, often dome-shaped, 50–500 m high, occur as extensive hillands or isolated inselbergs (Barthlott et al. 1993; Gröger 2000). Some are covered by various types of vegetation from shrub to forest, while others are totally or partially bare. Bare hillslopes, typical of isolated inselbergs as in the vicinity of Puerto Ayacucho, are locally called "lajas." Their convex surface is spotted with depressions and incisions occupied by small shrub or forest islands. The black color of the open rock surface was believed by Humboldt in 1800 to be caused by manganese coating. The surface temperature of the dark rock slabs reaches frequently 55°C or more, while water supply is scarce and extremely irregular both per day and per year. However, a well developed and partly endemic flora grows on these sites. Thus, the recognition of a Guayanan lithobiome with several ecosystem types appears fully justified.

Plant life on rocky surfaces is commonly considered under the aspect of evolutionary progression in time and space (lithosere). Microscopic organisms belonging to Bacteria (especially *Cyanobacteria*) and Algae initiate the early colonizing phase of the rocky substratum through chemical erosion and dissolution of the mineral components (Büdel et al. 1994); these are usually followed by other groups of nonvascular plants occupying larger areas, such as lichens and bryophytes. After some time, the ongoing erosion of the rock surface leads to the formation of small depressions and fissures between the rock strata. These habitats are suitable for a new set of colonizers belonging to vascular plant groups. The latter have specialized root systems that penetrate into the fissures and expand in the shallow ponds filled with dust and other organic particles favoring their permanent installation. Initially, small ephemeral meadows dominated by delicate herbs of Lentibulariaceae (*Utricularia, Genlisea*) and Mayacaceae (*Mayaca*) develop in the rainwater captured in the depressions on the open rock surface. These pioneer colonizers are followed by perennial grass and sedge communities mixed with colonies of Eriocaulaceae, Xyridaceae, and Bromeliaceae (especially *Pitcairnia*) that expand over increasingly larger areas. In a later seral stage, specialized shrub communities (with *Clusia, Acanthella, Pseudobombax*, among others) take over and eventually evolve towards deciduous forest patches with palms (*Syagrus*) and herbaceous understory (e.g., *Ananas, Tillandsia*), once a sufficiently stable process of soil formation has succeeded to operate on the site.

3.3.2 *Uplands*

3.3.2.1 Geomorphology

Uplands consist mainly of mid-elevation plateaus and undulating mountain ranges (500–1,400/1,500 m a.s.l.), dissected by deep river incisions. They include the following regional units.

- The Gran Sabana plateau in southeast Bolívar state between 700 m a.s.l. in the south and 1,350 m a.s.l. in the north is a complex mosaic of Roraima sandstones, ultramafic diabase intrusions, and Cuchivero granites.
- The Sierra Parima is a multilevel plateau and mountain range along the eastern border of Amazonas state with Brazil, between 700 and 1,450 m a.s.l., consisting mostly of granitic rocks.
- The Cuao-Sipapo plateau in northwestern Amazonas state is a large erosional surface between 800 and 1,200 m a.s.l., showing a mosaic of mesetas on Roraima sandstones and dome-shaped hillands on Parguaza granite.
- The Unturán plateau and mountain range in southern Amazonas state is an unexplored rugged area probably not higher than 1,000 m a.s.l., drained by the upper Orinoco, Mavaca, and Siapa rivers.

The main Guayana river watersheds cross these upland areas. They are from east to southwest: Caroní-Paragua, Caura, Cuchivero, Suapure, and Parguaza in Bolívar state; Cataniapo, Cuao-Sipapo, Ventuari, Cunucunuma, Padamo, Matacuni, and Mavaca in Amazonas state.

3.3.2.2 Climate

The uplands belong to two climatic zones:

- Submesothermic (MAT 18–24°C), tropophilous (MAR 1,000–2,000 mm, 2–4 dry months with <50 mm/month) in the southern Gran Sabana and northern Parima uplands.
- Submesothermic (MAT 18–24°C), ombrophilous (MAR > 2,000 mm, <2 dry months with <50 mm/month) in the northern Gran Sabana, Cuao-Sipapo, and Unturán uplands.

3.3.2.3 Vegetation

The upland vegetation of the Venezuelan Guayana is transitional between the macrothermic lowland life zone and the mesothermic upper montane life zone (see Fig. 3.5). The lower, submontane belt extends here mainly between 400/500 and 800 m a.s.l., followed by a montane belt between 800/1,000 and 1,500/1,600 m a.s.l. These belts differ considerably from the altitudinal divisions found in the tropical Andes because of (1) the lower elevation of the Guayana Highlands in general, and (2) the orographic arrangement of the Guayana Highlands into discrete, stepped regional relief units. The upper limit of the slope forests and scrubs in the montane life zone is controlled by the abrupt knick at the foot of the tepui scarps. Many tepui mesetas are large enough to allow watersheds to develop on their usually concave summits. Rivers and creeks reach the basimontane life zone in the piedmont through gorges incised in the tepui rims and walls.

Due to the relative proximity of the Atlantic Ocean, moisture-laden trade winds from the northeast supply abundant and evenly distributed rainfall that favors the growth of evergreen vegetation types in almost all uplands and highlands, with local exceptions in the southern Gran Sabana and in some parts of the Parima range.

(a) Forests

The upland forests grow predominantly on low to mid-elevation slopes, thus covering most of the basal life zones of the tepuis between 400/500 and 1,500 m a.s.l. Two well recognizable belts can be distinguished, one belonging to the submontane and the other to the montane life zones. Unfortunately, most of these slope forests are still unexplored. They include the following types.

- Submontane forests on lower slopes are evergreen, multilayered rainforests with moderately high to tall trees and a dense understory. Dominant families are Myristicaceae, Burseraceae, Annonaceae, and Fabaceae. Some of the evergreen lowland tree species reach their upper limit in these forests.
- Submontane forests grow also atop the large plateaus of the Gran Sabana, in southeastern Bolívar state, and Parima in eastern Amazonas state. In the first case, medium-sized, evergreen forests with Myrtaceae, Vochysiaceae, and Fabaceae predominate, whereas tall, dense, evergreen forests with Lecythidaceae, Meliaceae, Anacardiaceae, and Burseraceae have been recorded in the second case.
- Montane forests, also called "cloud forests," grow preferably on the upper slopes of many of the larger and higher tepui massifs. Epiphytes are frequent because of high air moisture. Trees are rather low, with dense crowns formed by sclerophyllous, evergreen leaves. The main arboreal families are Clusiaceae, Elaeocarpaceae, Podocarpaceae, Caryocaraceae, Malpighiaceae, while Bonnetiaceae become more abundant at higher elevation.

(b) Shrublands

Several shrubland types associated with poor soil conditions (e.g., rocky or sandy substratum) and unfavorable soil–water regime have been described in the highly diversified landscape mosaic of the submontane and montane life zones (Huber 1989, 1994b; Riina and Huber 2003). They include the following types.

- A bauxite-specific shrubland type is found on hilltops between 400 and 800 m a.s.l. in the lower and middle Suapure and Parguaza watersheds. Various members of the Humiriaceae, Olacaceae, Apocynaceae,

Chrysobalanaceae, and Annonaceae families form low, stunted woodlands with relatively small and open crowns, few lianas, and sparse understory.

- Shrublands are extensive in the Gran Sabana uplands of southeastern Bolívar state at 800–1,500 m a.s.l., especially on sandstone layers but also on large whitesand accumulations. These shrublands vary in height, density, and floristic composition. They include several endemics belonging to the families Bonnetiaceae, Melastomataceae, Humiriaceae, Rutaceae, Clusiaceae, and Euphroniaceae.
- Fairly extensive shrublands, with variable density and height but in general below 8 m, cover many of the lower plateaus and mesetas (800–1,500 m a.s.l.) such as Guaiquinima, Marutaní, Ichún, Sarisariñama, Parú, Yapacana, Autana, and Vinilla. These communities show often strong floristic relationships with neighboring high-tepui shrublands, but they always contain a significant number of local endemic species distinguishing them clearly from the true highland scrub types. Endemic upland species belong mainly to the Bonnetiaceae, Ochnaceae, Malpighiaceae, Rubiaceae, Clusiaceae, and Asteraceae families which play a relevant role in the Pantepui biogeography, the core area of the Guayana region.

(c) Herbaceous plant communities

Upland meadows are widespread only in the Gran Sabana region of southeastern Bolívar state. Elsewhere, herbaceous ecosystems are restricted to small areas on some of the lower tepui summits. The main types of herbaceous plant communities and the areas where they occur are as follows.

- Large grassland areas alternate with dense forests and shrublands on the undulating plateau of the Gran Sabana that gently descends from ca 1,350 m a.s.l. in the north to ca 700 m a.s.l. in the south. Most of this region is covered by acid Ultisols and Inceptisols, with open, treeless grass savanna dominated by a few bunchgrasses (e.g., *Trachypogon, Axonopus*). In the southern part of the plateau, around 800 m a.s.l., extensive palm swamps ("morichales") and flooded savannas with small palm ponds are found on the flat valley bottom of the Kukenan river. Palynological and paleoecological studies have been carried out in this environment by Schubert and collaborators in the late 1980s (Schubert and Fritz 1985; Schubert and Huber 1990; Schubert et al. 1986, 1994; Rinaldi et al. 1990; Rull 1991). Ongoing palynological research is being carried out by Rull and Montoya on samples from small ponds of the palm savannas, and by Leal in the wider Gran Sabana context.
- Grass savannas are also found around 750–1,000 m a.s.l. on the Parima plateau in eastern Amazonas state along the border with Brazil. In the central-northern Parima uplands, savannas form small islands within a forested landscape. These grasslands seem to be natural plant communities or at least stabilized since long time, because their floristic spectrum is surprisingly rich and diversified. In contrast, unvaried, sometimes monospecific anthropic savannas are steadily expanding in southern Parima as a

consequence of increasing demographic pressure (Huber et al. 1984; Guerra et al. 1998).

- Peatlands have developed in large depressions and on slightly inclined slopes above 1,000 m elevation in the central and northern parts of the Gran Sabana. Peat bog vegetation is a nongramineous meadow type that is physiognomically and floristically related to similar broadleaved meadows found on high-tepui summits throughout the Pantepui region. Several herbaceous species otherwise recorded only at much higher elevations are present including, for instance, two species of *Stegolepis* (Rapateaceae), several Xyridaceae (*Orectanthe sceptrum, Xyris* sp.), and Eriocaulaceae. This formation can thus be interpreted as being relict highland vegetation found at current upland level as a result of Pleistocene climatic oscillations.
- Patches of nongramineous, dense broadleaved meadows have also been recorded on lower sandstone mesetas as, for instance, on the summits (1,400–1,600 m a.s.l.) of Cerro Guaiquinima, Cerro Parú, and Cerro Aracamuni. These plant communities are physiognomically similar but dominated by different species of *Stegolepis* (Rapateaceae) according to the places considered.
- A special case of upland meadow can be observed on the summit of the mythical tower-like Cerro Autana (1,300 m a.s.l.), near the northwestern border of the Guayana Shield in Amazonas state. A dense tubiform meadow, dominated by the peculiar tubular terrestrial bromeliad *Brocchinia hechtioides,* covers the largest part of the small meseta summit (Steyermark 1974). This plant occurs on all tepui summits of the Venezuelan Guayana and is particularly well developed on the plateau of Cerro Jaua, between ca 1,500 and 2,200 m a.s.l. (Steyermark and Brewer-Carías 1976). Its massive presence on the isolated, small, low-elevation Cerro Autana tepui calls the attention.

3.3.3 Highlands

3.3.3.1 Geomorphology

Highlands correspond to the elevation belt from 1,400/1,500 to 3,000 m a.s.l. They include plateaus cut into individual mesetas (tepuis) on sandstones/quartzites, and dissected mountains and hillands on granites, with fairly even summit areas. Because most highland vegetation in the Venezuelan Guayana grows above the tree line, this orographic belt can be considered analogous to the tropical subalpine and alpine belts of other mountain systems (Smith 1994).

The highest elevations of the Guayana Shield are reached at Pico da Neblina (3,014 m a.s.l.; Brewer-Carías 1988) along the southernmost Brazilian–Venezuelan border in Amazonas state, Cerro Marahuaka (ca 2,800 m a.s.l.) in central Amazonas state, and Mt. Roraima (ca 2,750 m a.s.l.) on the tri-national checkpoint between Venezuela, Brazil, and Guyana. Most tepuis with well developed Pantepui biota on their summits and upper slopes reach 1,500–2,500 m a.s.l. The largest tepuis (600–1,200 km^2) are Auyantepui and the Chimantá massif in the Canaima National Park, southeastern Venezuela, Jaua and Sarisariñama in central-southern Bolívar state, and the Duida-Marahuaka massif in central Amazonas state. The greatest number of tepuis (about 50) occurs in southern Venezuela. Narrow floodplains and creek incisions, controlled by joints and fractures in the sedimentary rocks, are common on tepui summits. Deep gorges connect the creeks with the lowlands. Part of the surface water is stored in peat bogs or percolates through the sedimentary rocks and reaches endokarstic conducts.

The Maigualida massif at the border between Amazonas and Bolívar states is the highest mountain system on granite in the Guayana Shield (1,500–2,300 m a.s.l.). Individual hills and ranges thereof, with convex summits and steep slopes, culminate often at similar elevation levels that create a general aspect of table-shaped mountains. Bare rock surfaces are frequent on the summits and lateral slopes of the granitic domes. The geologic substratum belongs to the Santa Rosalía granite of the Cuchivero Group.

3.3.3.2 Climate

Climatic data of the Guayana tepui summits are still scarce in spite of several decades of rainfall and temperature recording by the regional hydropower company CVG-EDELCA at various high-mountain summits. Air temperature and rainfall data collected during the period 1997–2009 from three stations on tepui summits in the Caroní river basin (Bolívar state), southeastern Guayana, are summarized in Table 3.2 (EDELCA 2010). Average annual rainfall increases with elevation from about 2,800 mm at 1,750 m a.s.l. to 5,300 mm at 2,600 m a.s.l. Rainfall concentrates in the period from May to October. Average annual temperature decreases with elevation from 16.5 to 11.4°C, with lowest values in the period from December to March. Temperature at tepui summits is relatively unvariable over the year with annual amplitude increasing with elevation from 1.1 to 1.9°C, while daily amplitude might be considerable (probably above 15°C).

Table 3.2 Climatic data from tepui summits in the Caroní river basin, Bolívar state; annual averages (EDELCA 2010)

		Elevation	Rainfall (mm)			Temperature (°C)		
Station	Location	m a.s.l.	Mean	Max	Min	Mean	Max	Min
Kukenán	05°12′05″N 60°49′50″W	2,598	5,262	6,747	3,853	11.4	12.5	10.6
Guaiquinimita	05°53′43″N 63°34′02″W	1,843	3,889	5,247	2,801	–	–	–
Auyantepui	05°54′44″N 62°34′49″W	1,749	2,788	3,346	2,006	16.5	17.0	15.9

There are no meteorological stations on tepuis in the Amazonas state. The following data are estimates based on the elevation of the different highland levels.

- Mesothermic (MAT 12–18°C), ombrophilous (MAR > 2,000 mm, <1 dry month with <50 mm/month) on tepui summits between 1,400/1,500 and 2,400 m a.s.l.
- Submicrothermic (MAT 8–12°C), ombrophilous (MAR > 2,000 mm, <1 dry month with <50 mm/month) on tepui summits between 2,400 and 3,000 m a.s.l.

Because of near-equatorial location, the climate of the Guayana Highlands is determined by the two main annual processes operating over the South American continent: the trade winds (TW) and the intertropical convergence zone (ITCZ). The northeastern TW, blowing constantly from the Atlantic Ocean, carry moisture-laden clouds over the hot continental surface of tropical America. In wintertime, they blow at higher elevation and rainfall is therefore less intense in northwest South America. During summertime, the presence of the ITCZ, oscillating around the equator from 25°S to 10°N, causes frequent and very intensive rainfall, both at macroclimatic (continental) and mesoclimatic (regional) scales.

The eastern tepui chain of southeastern Venezuela, from Ilú to Roraima, represents the first orographic barrier to the TW coming from the Atlantic Ocean and receives, therefore, large amounts of annual rainfall. All tepui summits above 2,000 m elevation from the west and southwest of Roraima until the Neblina massif deep in central-western Amazonia receive high amounts of annual rainfall, certainly more than 2,500–3,000 mm, in spite of the rain shadow effect.

The MAR for the overall highland area in the Amazonas state is 3,250 mm, with local maxima above 4,000 mm, according to data from CODESUR (MARNR 1979). These data are in line with those of the tepuis more to the north in Bolívar state. Annual rainfall increases from 2,200 mm in the north to 3,800 mm in the southwest and from 1,800 mm in the southeast to 3,800 mm in the northwest. Rainfall exceeds evaporation in the major part of the area. Tepui and mountain summits are constantly moist, with a little dry season (MARNR 1979). Short dry spells lasting from a few days to several weeks between December and March can cause severe water stress in the dense vegetation growing on the usually water-saturated peats. These are also the periods when occasional fires may occur, but clear evidence of such events has been documented only on a few tepui summits (Givnish et al. 1986; Huber 1995b).

MAT varies between 22 and 27°C in the lowlands. On tepui and mountain summits, the MAT ranges from 8 to 15°C depending on elevation, but can occasionally drop to lower values. Temperatures of 12–13°C were recorded using a geothermometer at 50 cm depth in some peats of the study area, providing a reasonable proxy of the mean annual soil temperature. So far, no freezing events near or below 0°C have been reported, but short freezing air conditions may likely occur during early morning hours on some higher and more exposed summits, such as Roraima (windswept plateau summit at ca 2,800 m a.s.l.), Eruoda-tepui in the Chimantá massif (ca 2,700 m a.s.l.), Marahuaka (ca 2,800 m a.s.l.), and Pico da

Neblina (ca 3,000 m a.s.l.). Gelifluction phenomena caused by freezing/thawing, frequent in the Andean páramo landscape above 3,500 m a.s.l., have not been observed in high-tepui environment.

3.3.3.3 Vegetation

Four main vegetation types grow on the uppermost elevation level of the Guayana Highlands, including forests, shrublands, meadows, and pioneer vegetation on rocks. The term "meadow" as used in this volume designates any herbaceous vegetation type. Meadows can be divided into two broad, floristically distinct classes (1) gramineous meadows or grasslands (*herbazales graminosos*, including *sabanas* or savannas), dominated by grasses and/or sedges, growing principally in the hot lowlands; and (2) nongramineous meadows (*herbazales no graminosos*), consisting mainly of herbs without gramineous morphology, growing preferably in the cooler uplands and highlands of the Guayana region. More detailed explanation of this terminology can be found in Huber (1995b).

These four formations are present on almost all higher summits, whether sandstone plateaus or igneous–metamorphic mountains, in the Venezuelan Guayana as well as on some high tepuis in Guyana (e.g., Mt. Ayanganna and Mt. Wokomong). Floristic composition is relatively homogeneous at family and genus levels, with the families Bonnetiaceae and Rapateaceae being the differential indicators. This high-elevation life zone, lying above ca 1,500 m a.s.l., has been designated as the Pantepui Province of the phytogeographic Guayana Region (Huber 1994a) (see Fig. 3.4). Pantepui includes biota of the upper montane and the alpine belts of the general classification of montane altitudinal zones (see e.g., Körner 1999; Burga et al. 2004).

Within Pantepui, further phytogeographic divisions of the tepui archipelago have been proposed based on the geographic distribution of particular genera and species assemblages (Huber 1994a; Berry et al. 1995). The peatland studies referred to in this volume are located in two Pantepui districts (1) the Jaua-Duida District that extends from the southwestern Bolívar state to the central Amazonas state, and (2) the Western Pantepui District that includes the arc of tepuis and granitic mountains along the northern and northeastern border between the two states Amazonas and Bolívar (often called shortly "N Amazonas tepuis"). Earlier peat studies by Schubert and Rull in the Chimantá massif and on Cerro Guaiquinima (Schubert and Fritz 1985; Schubert et al. 1986, 1992, 1994; Rull 1991) are located in the Eastern Pantepui District (see Fig. 8.1).

(a) Forests

Four major forest types based on rock substratum have been observed on tepui summits (1) *Bonnetia* forests on sandstone, particularly in depressions and along creeks; (2) mixed forests on soils derived from diabase intrusions in the large tepui massifs; (3) mixed forests on granite in the Sierra de Maigualida; and (4) riparian forests along permanent creeks and rivers.

- The woody genus *Bonnetia*, formerly belonging to the Theaceae but now upgraded at family level as Bonnetiaceae (Weitzman and Stevens 1997), contains 26 species that represent the most characteristic ligneous plants in almost all vegetation types of the Guayana Highlands. Usually *Bonnetia* are shrubs, but some species have arboreal growth form. One of the most common tree species forming dense high-tepui forests is *B. roraimae* with small leaves and pink flowers. These formations grow preferably in depressions of the eastern tepui summits and along creeks, but they are also found on shallow peat deposits. In a few cases, *Bonnetia* forests have been observed on mineral soils derived from the weathering of diabase intrusions, but in such places *Bonnetia* is not monodominant as it is on Roraima sandstones. Other species of *Bonnetia* are dominant in forests of the Jaua massif (*B. jauaensis*) and the Neblina massif (*B. neblinae*). High-tepui *Bonnetia* forests are usually low (6–12 m), but have a very dense canopy. The ground cover is generally discontinuous and often consists of large colonies of giant terrestrial bromeliads (e.g., *Brocchinia tatei*). Occasionally, a few other tree species occur in high-tepui *Bonnetia* forests, such as *Schefflera* div. sp. (Araliaceae), *Daphnopsis steyermarkii* (Thymelaeaceae), and *Ilex* div. sp. (Aquifoliaceae), but they play a subordinate role in the ecosystem. An ecological study of *Bonnetia* forests on the summits of the Chimantá massif has been published by Vareschi (1992).
- Low forests grow also on mineral soils derived from the diabase intrusions locally outcropping on the summits of high-tepui massifs such as Auyantepui, Chimantá, and Duida. The canopy is often open, discontinuous, and formed by trees or treelets belonging to the genera *Spathelia* (Rutaceae), and *Stenopadus* and *Gongylolepis* (Asteraceae). This forest type is at present still too poorly documented.
- A montane forest type has been reported growing on granite-derived soils on summits at around 1,800–2,000/2,100 m a.s.l. in the Sierra de Maigualida (Huber et al. 1997; Ramos 1997). This is a low (5–8 m), very dense elfin forest with many epiphytes, especially cryptogams, on stems and branches that indicate prevailing wet cloud-forest conditions at that upper montane elevation. Both the physiognomy and the floristic composition of this forest type differ notably from the *Bonnetia* forest types on the eastern sandstone tepuis (e.g., Auyantepui, Chimantá). Apparently, the genus *Bonnetia* itself is absent in most forest types so far described in the Sierra de Maigualida.
- Narrow, usually discontinuous riparian forest strips border some of the permanent water courses on sandstone plateau summits, especially in the eastern tepui massifs of Auyantepui and Chimantá. In the upper watersheds, they are replaced by bare rock surfaces or peatlands with predominantly herbaceous vegetation. In most cases, trees are *Bonnetia* species, mixed with lower trees of *Clusia* (Clusiaceae), *Stenopadus* (Asteraceae), *Schefflera* (Araliaceae), and occasionally individuals of *Geonoma appuniana*, one of the few palm species growing in Pantepui.

(b) Shrublands

Large extents of Pantepui summits are covered by shrublands, one of the most important and characteristic vegetation types of this biogeographic province. The physiognomy of the shrub types is very diverse, and the floristic composition of the shrublands is highly differentiated, reflecting geographic pattern variations within Pantepui and orographic complexity of the tepui summits. With approximately 20 shrub species, the genus *Bonnetia* is by far the most frequent and predominant member of the tepui shrublands over the entire Pantepui province. In contrast to the forest ecosystems, shrub taxa are much more diverse than tree taxa at the family, genus, and species levels. This results in a much more diversified and, at the same time, more specialized assemblage of shrub communities in the high-tepui environment.

Many shrub species of Pantepui have their foliage characteristically agglomerated towards the end of the branches ("apicirosulate"). In general, the leaves are very thick, coriaceous and concave throughout a diverse spectrum of families. Additionally, shrubs often have thick stems and branches with coriaceous, gray to dark brown bark, and many are pachycaulous. Almost all high-tepui shrubs have showy white, red, blue or yellow flowers, and many species flower synchronously either during the short dry season or during the rainy season.

Several Pantepui shrubland types have been identified. The mixed *Bonnetia* shrubland is probably the most widespread, although the predominant *Bonnetia* species may differ among tepui massifs. Other important families with characteristic high-tepui shrub taxa are Asteraceae (*Stenopadus, Gongylolepis, Quelchia*), Rubiaceae (*Pagameopsis, Aphanocarpus, Maguireothamnus, Chondrococcus, Duidania*), Rutaceae (*Raveniopsis*), Clusiaceae (*Clusia*), Ochnaceae (*Tyleria, Adenanthe, Adenarake, Poecilandra*), Melastomataceae (*Acanthella, Graffenrieda, Macairea, Mallophyton*), Tepuianthaceae (*Tepuianthus*), Malpighiaceae (*Diacidia, Blepharandra*), among others.

A unique shrubland type, first discovered by Felix Cardona in 1947 and later botanically inventoried by Julian A. Steyermark in 1953 and 1955, grows exclusively on several tepui summits within the Chimantá massif in southeastern Bolívar state. This is a paramoid tepui shrubland (Grupo Científico Chimantá 1986; Huber 1989, 1992) dominated by five caulirosulate species of the endemic genus *Chimantaea* (Asteraceae). It is physiognomically and floristically comparable to the tropical Andean páramo vegetation dominated by the similarly caulirosulate *Espeletia* s.l. It has been found only in summit areas of the Guayana Highlands. The paramoid shrubland grows preferably on deep organic soils in peatlands of the Chimantá massif at 1,900–2,300 m a.s.l. Usually, it consists of an almost monospecific shrub layer, 2–5 m tall, formed by one of the various *Chimantaea* shrub species, whereas the dense underlying herb layer is mainly made up by species of *Myriocladus*, an endemic high-tepui bamboo (Poaceae–Bambusoideae).

An uncommon shrubland type is also found in the Neblina massif at elevations of 1,700–2,000 m a.s.l. in the extreme south of Amazonas state (Brewer-Carías 1988). The dominant shrub is again a member of the Bonnetiaceae, originally described as *Neblinaria celiae* and later transferred to the genus *Bonnetia* under the specific name of *Bonnetia maguireorum* (Steyermark 1984). This peculiar shrub, up to 3 m tall, sparsely branched, thick and corky stemmed, and with showy pink flowers forms extensive colonies on tepui peat substrate. Its thick, coriaceous leaves are typically congested at the end of the branches ("apicirosulate"), giving the appearance of a giant artichoke. Givnish et al. (1986) interpreted the thick corky bark as an adaptation to recurrent fire impact. However, such interpretation is difficult to sustain because of the wet conditions prevailing on Cerro de la Neblina (literally meaning mist mountain), probably unchanged since the Last Glacial Maximum and even earlier due to its isolated, continental position near the equatorial center of the American humid tropics.

(c) Meadows and grasslands

All summits of the Guayana Highlands are covered by variable extents of herbaceous vegetation. These meadows are largely dominated by species belonging to the typical Guayana genus *Stegolepis* of the Rapateaceae family. As in the case of the woody genus *Bonnetia*, local herb species show a clear geographic differentiation according to individual tepuis or tepui massifs. In general, *Stegolepis* species are broadleaved herbs with leathery and sword-like laminas, except two species with graminoid leaf types. Broadleaved *Stegolepis* meadows can be found atop almost all tepuis. Closely related genera of the Rapateaceae can predominate locally in certain meadow types, such as *Marahuacaea* on Cerro Marahuaka, *Amphiphyllum* on Cerro Duida, and *Phelpsiella* on Cerro Parú. They grow together with many other endemic taxa, such as *Heliamphora* (Sarraceniaceae), *Xyris*, *Abolboda*, *Orectanthe* and *Achlyphila* (Xyridaceae), *Brocchinia*, *Lindmannia* (Bromeliaceae), Eriocaulaceae, Cyperaceae, Liliaceae, Santalaceae, among others. Broadleaved tepui meadows grow usually on peat deposits of variable depth, including very shallow peat mantles resting directly on sandstone or quartzite.

Grass-dominated meadows cover large areas in very few cases of high-tepui environments. So far, they have been identified locally in the Chimantá massif, where the grass species *Cortaderia roraimensis* is the main component of flooded swamps in open, flat valley bottoms, and in the Sierra de Maigualida, where the densely pubescent species *Axonopus villosus* predominates on peat bogs. Such high-tepui grasslands are infrequent and spatially restricted, in contrast to the extensive *Stipa* and *Festuca* grasslands found at higher elevations (3,000–4,800 m a.s.l.) in the Andes (puna in Perú and Bolivia, grass páramos in Ecuador, Colombia, and Venezuela). In both cases, however, the grasses exhibit the C4 photosynthetic pathway.

The following meadow types have been described on Pantepui summit areas in the Guayana Highlands (Huber 1995b, 2006).

- *Broadleaved meadows* usually consist of a dense, continuous herbaceous layer, up to 0.5 m tall, dominated by one or more members of the family Rapateaceae. In most cases, the plants belong to the genus *Stegolepis* present on almost all tepuis, or to monospecific communities of *Kunhardtia rhodantha* on the northern Amazonas tepuis and in Sierra de Maigualida, *Marahuacaea schomburgkii* on Marahuaka, *Amphiphyllum rigidum* on Duida, and *Phelpsiella ptericaulis* on Cerro Parú. In addition, the herbaceous stratum contains a variety of forbs and small herbs from the families Xyridaceae, Eriocaulaceae, Cyperaceae, Sarraceniaceae, Liliaceae, and Droseraceae, among others. A variable amount of subligneous and woody elements, mainly frutices and low shrubs, is scattered within the herbaceous cover, belonging mainly to Bonnetiaceae, Asteraceae, Rubiaceae, Malpighiaceae, Ericaceae, Melastomatacaeae, and Ochnaceae, among others.
- *Rosette meadows* are dominated by forbs and stout herbs that usually form basal leaf rosettes, a growth form typical of many Eriocaulaceae, Droseraceae, Xyridaceae, Sarraceniaceae, but especially of terrestrial Bromeliaceae. A specific type of rosette meadow grows as small, dense cushion-like communities, irregularly scattered over bare rock surfaces, especially on the eastern tepui summits (Roraima, Ilú, Yuruaní). These curious plant communities are dominated by several species of Eriocaulaceae and Xyridaceae, including the large acuminate rosettes of *O. sceptrum* together with occasional subshrubs or larger herbs. Other peculiar types of rosette meadow in high-tepui environments are formed by species of terrestrial Bromeliaceae endemic to the Guayana Highlands, as for instance the extensive *Lindmania* meadows on Chimantá or the *Brocchinia* meadows on several summits of the western and central tepuis in Amazonas state.
- *Tubiform meadows* are amongst the most impressive herbaceous plant communities of the Guayana Highlands, where a cylindrical growth habit seems to have reached high levels of specialization in at least two families. One is the genus *Heliamphora* of the Sarraceniaceae, the pitcher plant family, well known from other genera growing separately in California and Oregon (*Darlingtonia*) and along the east coast of North America (*Sarracenia*). The other is the genus *Brocchinia* of the Bromeliaceae, endemic to the Guayana region together with *Heliamphora*. Both tubiform genera are considered to be either entirely (*Heliamphora*) or at least partially (*Brocchinia*) carnivorous (Givnish et al. 1984, McPherson 2008). *Brocchinia* is widely distributed in the Guayana region including all high-tepui environments, while *Heliamphora* is circumscribed to two widely separated areas, namely the tepui summits of southeastern Bolívar state and those of central and southern Amazonas state.

All tubiform meadows grow on deep, water-saturated peat bogs of the tepui summits. Well-developed *Heliamphora* meadows are frequent on the summits of the Chimantá massif, where large colonies of the two small pitcher plant species *H. heterodoxa* and *H. minor* are interspersed within the broadleaved meadows. By contrast, the pitcher plant species *Heliamphora tatei* of the Duida and Huachamakari plateaus has a much taller, dendroid (arborescent) growth form, but does not form extensive colonies.

The most impressive tubiform meadows grow in extensive peat bogs on the plateau summits of Cerro Jaua, a huge sandstone massif located to the southeast of Sierra de Maigualida (Steyermark and Brewer-Carías 1976). They consist of millions of individuals of the terrestrial bromeliad *B. hechtioides*, a most peculiar cylindrical plant up to one meter tall, which grows in very dense colonies difficult to traverse. The characteristic bright yellow color of *Brocchinia* makes the meadows visible from far away.

(d) Pioneer saxicolous vegetation

Bare rock surfaces are extensive on almost all tepui summits and the vertical walls that usually separate uplands from highlands. Pioneer vegetation growing on rocky sites, whether sedimentary (sandstone, quartzite) or igneous–metamorphic (granite, diabase), forms the saxicolous biome (or lithobiome) of Pantepui. Further divisions, such as chasmophytic or chomophytic (rock crevices), speleophytic (caves), and other specific rock sites and habitats are not considered here, because comparative ecological studies of the lithobiome in the Guayana Highlands are still lacking.

High-tepui rock surfaces are not exposed to extreme variations in daily and annual temperature and rainfall regimes as is the case in the lowlands. The progressive colonization of bare surfaces shows successions of different plant communities (*lithoseres*) that are obviously different from the species assemblages observed in the lowlands. The initial colonizing phase is determined by the weathering and erosional activity of Cyanobacteria on the open rock (Büdel et al. 1994), but the coverage of these organisms is much less extensive than in the lowlands. The subsequent colonization phases are similar to those described for the lowlands, with terrestrial and lithophytic bromeliads playing the primary role (e.g., *Racinea, Navia*, and *Brocchinia*). On Chimantá at ca 2,000 m a.s.l., endemic species of Cyperaceae (*Cephalocarpus, Everardia*), Xyridaceae, and Eriocaulaceae achieve the final colonization stages for the establishment of perennial meadows, while Asteraceae (*Chimantaea, Stomatochaeta*) and Ericaceae (*Tepuia, Gaultheria*) are the main components of woody pioneer shrub communities in depressions and shallow incisions.

3.4 Study Areas

Three study areas were selected within the geo-ecological context described in the previous sections. They are located in the territory of the Amazonas state, the southernmost administrative entity of the country (0°45′–6°N and 63°20′–67°50′W) (Fig. 3.8). The Amazonas state covers 175,750 km^2 of which

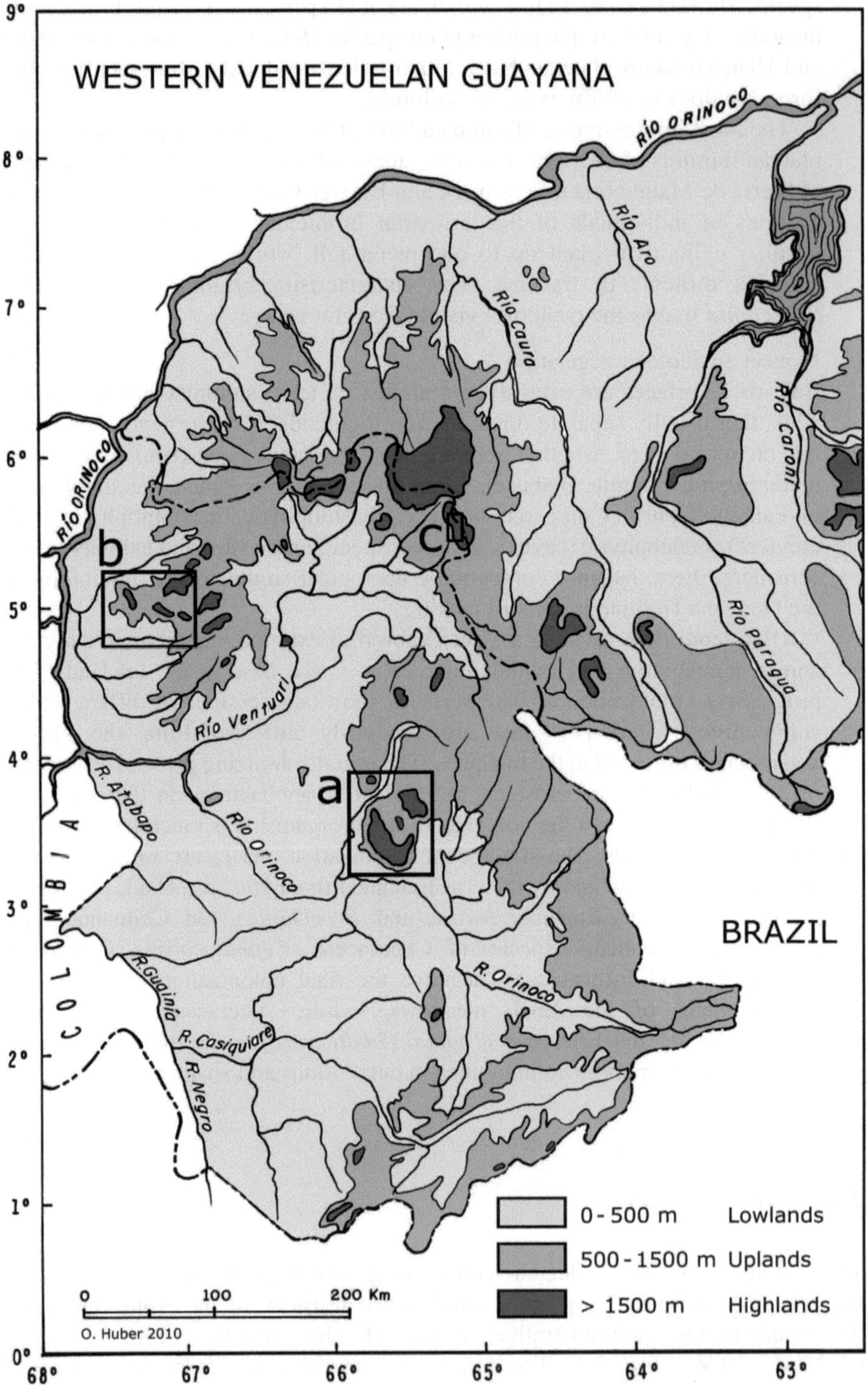

Fig. 3.8 Location of the study areas in the Amazonas state (a = Duida-Marahuaka-Huachamakari massif; b = Cuao-Sipapo massif; c = Maigualida massif)

nearly 50% are wide plains below 200 m a.s.l., surrounded in the north, east, and south by mountain and plateau landscapes reaching elevations of 2,400 m in the north (Cerro Yaví), 2,300 m in the northeast (Cerro Yudi in the Sierra de Maigualida), 2,800 m in the center-east (Cerro Marahuaka), and 3,014 m in the extreme south (Pico da Neblina). Mountains and plateaus are drained by a complex river system controlled mainly by the Orinoco, Ventuari, and Río Negro rivers.

The climate of the Amazonas state is typically equatorial, with a fairly constant day length of 11.5–12.5 h. MATs are always above 27°C in the lowlands below 100 m a.s.l. Annual rainfall is abundant, increasing from north to south (2,000–4,000 mm) and from east to west (2,500–4,000 mm). The TW blow with moderate intensity from northeast between January and March, whereas most of the state is under the impact of the ITCZ during the rest of the year, causing frequently heavy storms and intense rainfalls.

More detailed information on the Amazonas state is available in the Atlas de la Región Sur (MARNR 1979), in the Soil Atlas of the Territorio Federal Amazonas (MARNR-ORSTOM 1987), and in Huber (1995a).

3.4.1 Duida-Marahuaka Massif and Cerro Huachamakari

The Duida-Marahuaka massif is a large plateau complex of quartzites and sandstones belonging to the Roraima Group, in the central-eastern section of Amazonas state. It is formed by three distinct relief units, namely Cerro Duida, Cerro Marahuaka, and Cerro Huachamakari (Gleason et al. 1931; Steyermark and Maguire 1984; Huber 1995a) (Fig. 3.9).

3.4.1.1 Cerro Duida

Cerro Duida, also called *Yennamadi* by Ye'kwana people, is one of the largest sandstone plateaus of the Guayana region (Fig. 3.10), measuring roughly 160 × 60 km. The summit area of ca 1,000 km^2 is concave, sloping from south (2,358 m a.s.l.) to north (500–800 m a.s.l.). It is mainly drained by the Caño Negro that flows into the Cunucunuma river near the northern flank of the plateau. According to MARNR-ORSTOM (1987), the core area of the plateau consists of sandstones and quartzites of the Roraima Group, with important lutite inclusions and diabase intrusions.

Cerro Duida was the second tepui, after Mt. Roraima, to be visited by an exploration mission. In 1928–1929, G. H. H. Tate of the American Museum of Natural History, New York, and a group of scientists ascended on the plateau from the Esmeralda savanna located at the southeastern corner of the massif. The Esmeralda locality had been visited previously by Humboldt and Bonpland in

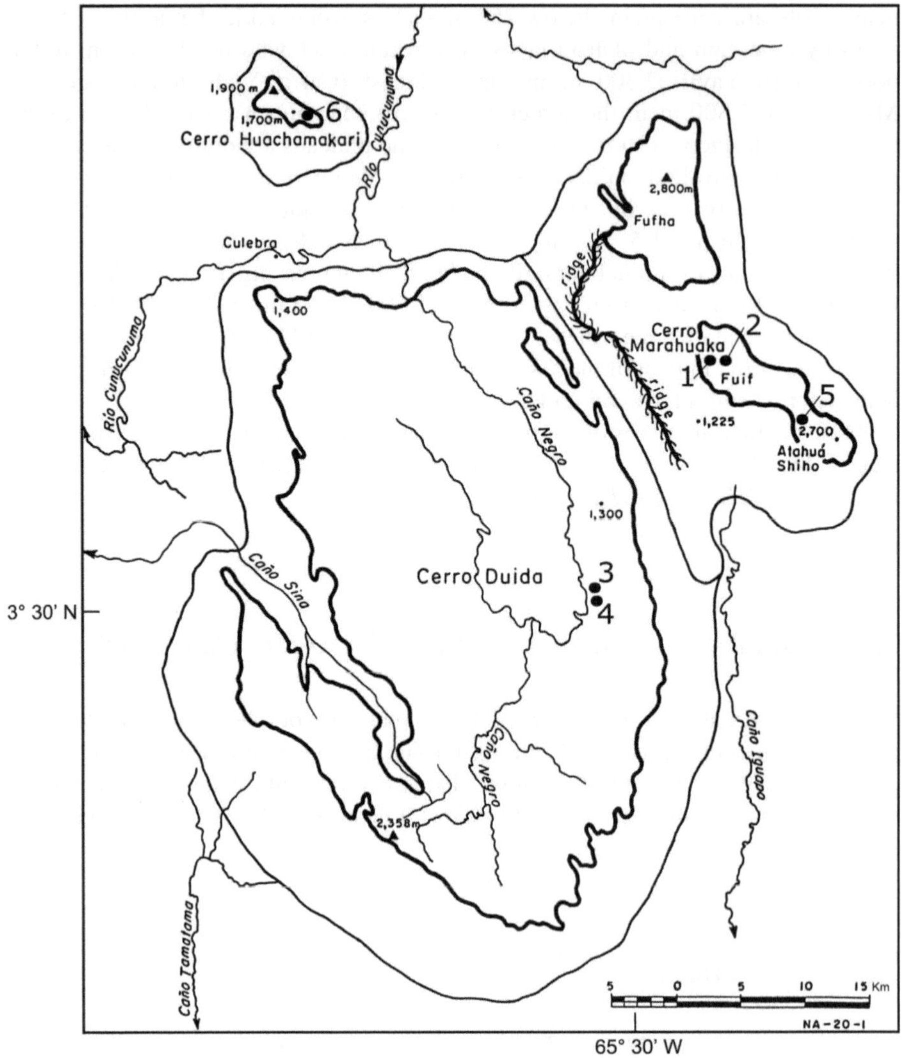

Fig. 3.9 Location of the study sites (Profiles 1–6) in the Duida-Marahuaka-Huachamakari area. *Light contour* = sloping area; *dark contour* = summit area (usually >1,500 m a.s.l.). Based on side-looking airborne radar image NA-20-1 (AEROSERVICE Corporation 1972)

1800, Robert Schomburgk in 1839, and Richard Spruce in 1853, but none of them climbed Cerro Duida.

Most of the plateau summit of Cerro Duida is covered by dense submontane and montane forests that extend from 500–600 m a.s.l. in the north to 1,900–2,000 m a.s.l. in the south (Dezzeo and Huber 1995). The slightly sloping areas at higher elevation (2,000–2,350 m a.s.l.) in the southern interior plateau are

Fig. 3.10 Marahuaka and Duida massifs. (**a**) Internal relief variation within the Marahuaka massif with tepui mesetas at different levels and tectonically controlled valleys; peatlands on tepui summits. (**b**) Southeastern escarpment of the Duida massif seen from the Esmeralda settlement (photos Zinck)

covered by very dense, tall tepui shrubland on compact peat. The uppermost sector along the southern and southwestern rim has broadleaved meadows on friable peat with a considerable amount of dispersed shrubs. Shrublands are dominated by *Bonnetia crassa* (Bonnetiaceae) and *Neotatea longifolia* (Clusiaceae) (Fig. 3.11), together with many other shrubs of Rubiaceae, Ochnaceae, Melastomataceae, and Asteraceae. Meadows are formed mainly by *Stegolepis pungens, S. linearis*, and *A. rigidum* (Rapateaceae), Xyridaceae, Eriocaulaceae, mixed with gnarled, low shrubs of Asteraceae (*Gongylolepis, Duidaea*), and *B. crassa*.

Fig. 3.11 Endemic plants of the Duida-Marahuaka massif. (**a**) *Neotatea longifolia* (Clusiaceae) is a common shrub in the open meadows and in the shrublands of the Duida massif. (**b**) *Marahuacaea schomburgkii* (Rapateaceae) is one of the most characteristic broadleaved herbs on the rocky mesetas of the Marahuaka massif (photos Huber)

3.4.1.2 Cerro Marahuaka

Cerro Marahuaka, also spelled Marahuaca, is a much smaller but not less impressive plateau consisting of two separate, tower-like high-tepui mesetas emerging directly from the northeastern border of the Duida massif. The first one extends to the north and is known as Fufha (Huha) by the Ye'kwana Amerindians. The second one, larger, located to the southeast and connected to the former by a low narrow ridge, is called Fuif (Fhuif). Adjacent to the Fuif meseta, in southwestern direction, is a very characteristic peak with a widely visible conical summit, named Atahua'shiho by the Ye'kwana and *La Churuata* by the Spanish speaking locals due to its resemblance with the conical roof of a typical indigenous hut. Fufha and Fuif are approximately 2,800 m high, whereas Atahua'shiho is slightly lower (ca 2,700 m a.s.l.), making the Marahuaka the second highest tepui massif in Amazonas state. The plateau summits of the massif cover an area of ca 120 km^2. The northern meseta is markedly inclined from northeast to southwest, while the southern one is slightly inclined from southeast to northwest.

The flanks of the Marahuaka massif are very steep, rising from 200–400 m a.s.l. at the base to 2,800 m a.s.l. at the summit (Fig. 3.10). The upper parts are inaccessible vertical walls of more than 1,500 m relative elevation. For that reason, the first exploration of the summit areas took place as recently as 1975 using helicopter. Since then, these tepuis have been visited by a dozen of scientific expeditions (Colonnello 1984; Delascio and Steyermark 1989; Michelangeli 1989), but large areas remain unexplored.

The vegetation of the Marahuaka summits presents striking differences between north and south. The northern (Fufha) plateau is almost continuously covered by dense, tall broadleaved meadows dominated by *Stegolepis terramarensis*. The southeastern (Fuif) plateau, being less inclined, has a more diversified plant cover including meadows, shrub islands, and low forests along creeks or in larger depressions. In contrast to the meadows on Fufha, the meadows on Fuif are dominated by *M. schomburgkii,* a distinct herb also member of the Rapateaceae (Fig. 3.11). Similar distribution patterns have been observed by Steyermark with two endemic members of Bromeliaceae, *Brewcaria marahuacae* common in rocky meadows of the northern summits and *Steyerbromelia discolor* in open tepui summit vegetation of Fuif. In general, Fufha meadows are much wetter, dense, and almost monospecific as compared to Fuif meadows, even when both are separated by only a few kilometers. The woody flora also seems to be more diverse in the south than in the north, with *Cyrilla, Gongylolepis,* and *Tyleria* being frequent shrub genera, and *Ledothamnus, Mycerinus*, and *Sauvagesia* common subfruticose plants.

3.4.1.3 Cerro Huachamakari

Cerro Huachamakari (also spelled Huachamacari) is an isolated tepui at 13 km from the northwestern corner of Cerro Duida, separated by the valley of the Cunucunuma river (Fig. 3.12). It is a small meseta of about 9 km^2, inclined from 1,900 m a.s.l. in the northwest to 1,500 m a.s.l. in the southeast. High vertical walls made this tepui long inaccessible, until Bassett Maguire and his colleagues of the New York Botanical Garden were able in 1950 to reach the plateau summit.

The largest part of the Huachamakari summit is covered by magnificent, dense broadleaved meadows and shrublands growing on deep peat. A patch of submontane palm forest occurs in the central depression of the meseta. The shrubby meadows are dominated by *Stegolepis grandis* and *S. membranacea*, together with *B. tatei, O. sceptrum*, Eriocaulaceae, and the curious dendroid pitcher plant *H. tatei*. Several endemic shrubs grow within the herbaceous stratum, including *Duidania*

Fig. 3.12 Huachamakari massif. (**a**) Shows the table-shaped configuration of the tepui meseta seen from the Culebra settlement. In spite of the overall flat aspect of the tepui summit, there is evidence that the sandstone layers have been locally deformed by folding (**b**) (photos Zinck)

montana, Maguireothamnus tatei (Rubiaceae), and *Stenopadus huachamacari* (Asteraceae).

3.4.2 *Cuao-Sipapo Massif*

The Cuao-Sipapo massif consists of a large, mainly granitic upland and a few small sandstone highland mesetas irregularly dispersed over its northwestern section (Figs. 3.13 and 3.14). This mountain system, located at the northwesternmost corner of the Guayana Shield, extends over 1,000 km^2 between the Orinoco river to the west and south, and the Ventuari river to the east and southeast. It is still largely unexplored, especially the wide, dissected Cuao-Marieta uplands, in spite of

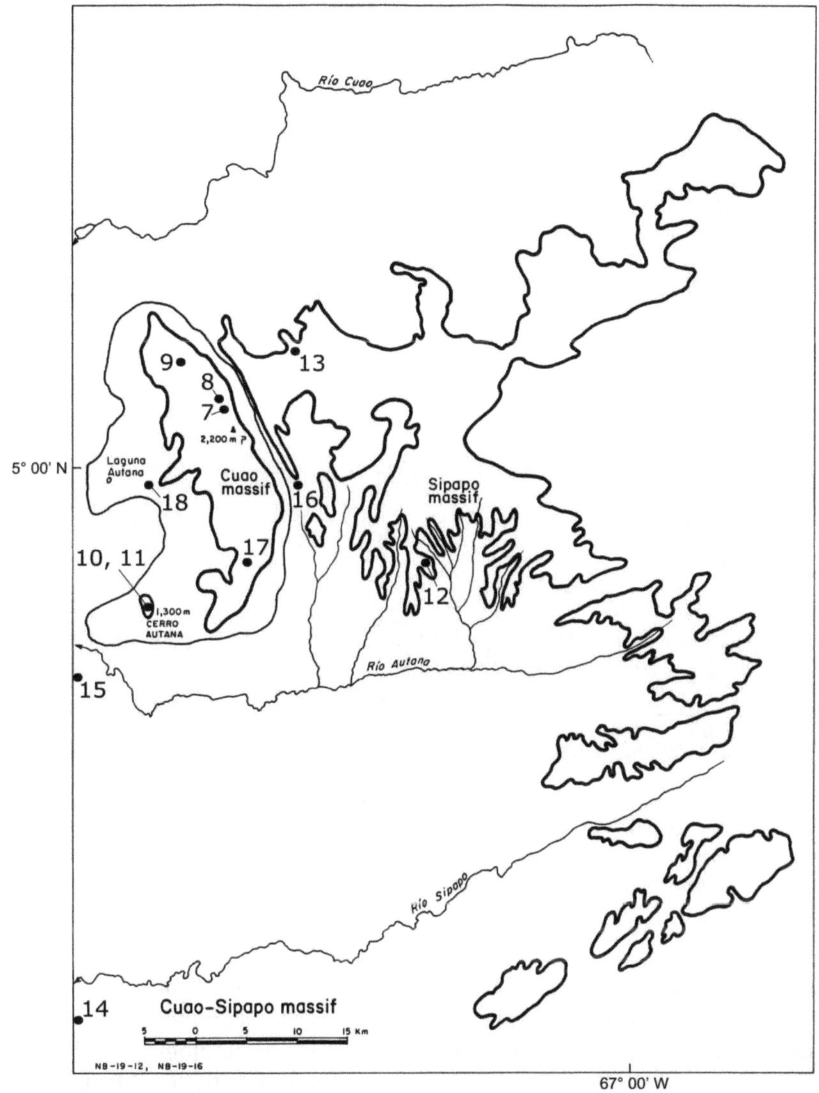

Fig. 3.13 Location of the study sites (Profiles 7–18) in the Cuao-Sipapo area. *Light contour* = sloping area; *dark contour* = summit area (usually >1,500 m a.s.l.). Based on side-looking airborne radar images NB-19-12 and NB-19-16 (AEROSERVICE Corporation 1972)

Fig. 3.14 Cuao-Sipapo massif. This area is a mosaic of two regional landscape types: (1) high-plateau landscape showing several levels of tepui mesetas on sandstone-quartzite; peatland on the foreground meseta (**a**), and (2) dissected mountain landscape including ranges and dome-shaped hills developed from igneous–metamorphic rocks of the Guayana Shield basement (**b**) (photos Zinck)

relative proximity (less than 100 km straight) to Puerto Ayacucho, the capital of Amazonas state. The first explorations of the northeasternmost meseta of Cerro Paraque in the Cuao massif were made in 1946 by the ornithologist W. Phelps Sen., followed in 1948–1949 by a botanical expedition led by B. Maguire of the New York Botanical Garden.

The summit area of the Cuao-Sipapo massif is not yet well defined and delimited. Elevations are probably not higher than 2,000 m and, in general, decrease from east to west, only interrupted by several rocky steps. The extensive Cuao-Marieta uplands, with an average elevation of 1,200–1,500 m, drain mainly to the east into the middle Ventuari river, while the high mesetas of the Cuao massif

Fig. 3.15 Endemic plants of the Cuao-Sipapo massif. (**a**) The scandent shrub *Bonnetia kathleenae* (Bonnetiaceae) is restricted to the sandstone areas in the uplands and highlands of Cerro Cuao. (**b**) *Graffenrieda versicolor* (Melastomataceae) is a curious low shrub with thickened (pachycaulous) stems and showy white flowers, growing in small vegetation islands on open rock surfaces (photos Huber)

drain into the Cuao and Sipapo rivers, tributaries of the Orinoco river along the border with Colombia.

Cerro Autana is a small, isolated, tower-like sandstone meseta with a summit area of less than 2 km^2. This beautiful and emblematic mini-tepui, reaching barely

1,300 m a.s.l., is located near the northwestern foot of Cerro Cuao. It is traversed from side to side by a large cave, relict of a former underground drainage system (Colvée 1973; Steyermark 1974).

So far, no information on the vegetation and other geo-ecological characteristics of the Cuao-Sipapo massif has been published. Botanical collections have been gathered in summit areas by B. Maguire and colleagues in the late 1940s and by O. Huber in 1993 during a helicopter-supported expedition (Fig. 3.15). Some unpublished collections were also made by CVG-TECMIN in 1990.

The largest part of the Cuao-Sipapo massif is covered by dense, almost continuous submontane and montane forests. Some open areas occur on granitic domes and hills in the uplands (up to 1,500/1,600 m a.s.l.) and on sandstone and quartzite mesetas in the highlands. These areas are discontinuously covered by tubiform tepui meadows and associated woody elements, whereas tepui shrublands seem to be less common. From the first botanical explorations, it was clear that further exploration would reveal a variety of taxonomic novelties and high level of endemism in this massif.

3.4.3 Sierra de Maigualida

Sierra de Maigualida is part of a large mountain system made up of granitic and gneissic rocks, more than 600 km long, which forms a north–south divide approximately along the 65° meridian between the Caroní and Caura watersheds to the east and the upper Orinoco and Ventuari watersheds to the southwest. Sierra de Maigualida corresponds to the northern part of this mountain range, followed by the Sierra Uasadi in the center and the Sierra Parima in the south, extending until the Orinoco headwaters. The overall elevation of the summit areas decreases from 1,800–2,300 m a.s.l. in the north to 800–1,400 m a.s.l. in the south. This core mountain area, in the center of the Guayana region, has been documented scientifically only during the last 40 years, beginning with the first radar inventory in 1969–1971 (AEROSERVICE 1972) and the first biological explorations on the northern summits in 1987 (Huber et al. 1997).

Sierra de Maigualida is the largest and highest individual mountain range on igneous–metamorphic rocks in the entire Guayana region, 200 km long and 20–40 km wide, with a summit area of about 500 km^2 (Fig. 3.16). Its rugged physiography (Fig. 3.17) contrasts with the flat sandstone mesetas that are found near the southeastern border (Cerros Guanacoco, Jaua, and Sarisariñama) and near the northwestern corner (Cerro Yaví, Cerro Yutajé, etc.).

Sierra de Maigualida was the last mountain area of the Venezuelan Guayana to be explored because of limited accessibility (practically only by helicopter). Explorations started in 1987 and continued irregularly until 1995, providing the first biophysical field information. Submontane forests (200–1,000 m a.s.l.) inhabited by Hoti Amerindians have been studied by Zent and Zent (2004) in the Iguana river valley, Amazonas state. Dense forests cover the steep massif

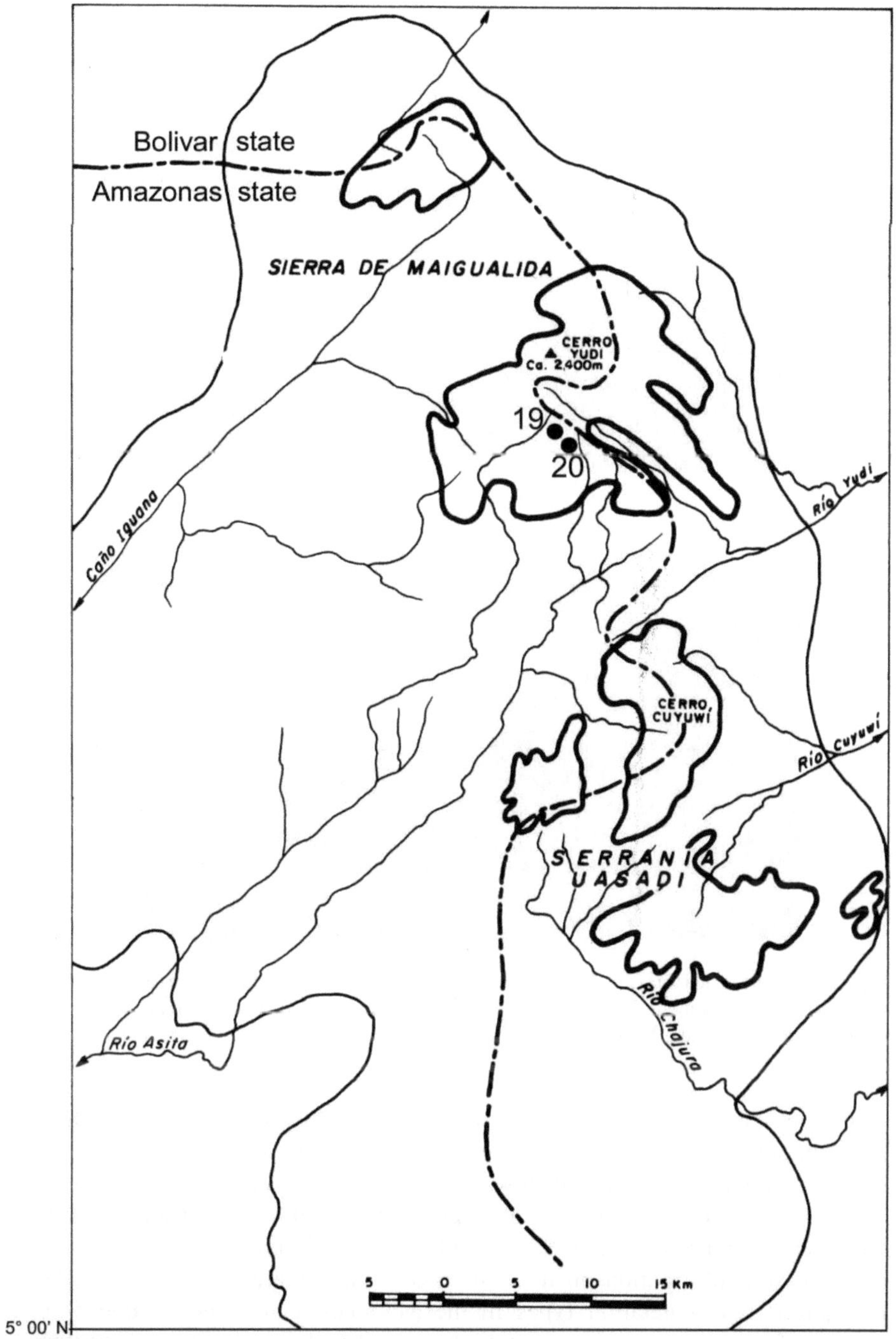

Fig. 3.16 Location of the study sites (Profiles 19 and 20) in the Sierra de Maigualida area. *Light contour* = sloping area; *dark contour* = summit area (usually >1,500 m a.s.l.). Based on side-looking airborne radar image NB-20-9 (AEROSERVICE Corporation 1972)

Fig. 3.17 Maigualida massif. The Sierra de Maigualida is a granitic massif belonging to the Guayana Shield basement. Two main relief types dominate: elongated mountain ranges (**a**) and dissected dome hillands (**b**) (photos Zinck)

slopes up to approximately 1,800–2,000 m a.s.l. At higher elevation, the summit areas integrate into an undulating plateau landscape, deeply dissected by an intricate drainage system that flows mostly into the Ventuari river watershed. The highest peak is seemingly Cerro Yudi with an elevation of about 2,300 m. Huber (1995b), Huber et al. (1997), and Ramos (1997) provide first-hand vegetation and soil information on this still poorly known area. Huber et al. (1997) recognized five vegetation types in the Maigualida highlands, namely upper montane forest, montane grassland, high-tepui shrubland, high-tepui broadleaved meadow, and pioneer vegetation.

3.4.3.1 Upper Montane Mesothermic Low Forest (Cloud Forest, Elfin Forest)

At 1,800–2,000 m elevation, the dense forest cover of the submontane and montane slopes becomes discontinuous. In the upper montane belt above ca 2,000 m elevation, only one forest type has so far been investigated. It occurs in patches dispersed over the rolling highlands between granitic outcrops, sometimes occupying small valleys, but also expanding over moderately steep slopes. This is a very low forest, usually 5–7 m tall, with a few emergent trees reaching 10 m or more. Many trees have gnarled or tortuous trunks and crowns, and start branching at 1–2 m above ground. The foliage is micro- to mesophyllous, entirely evergreen, and predominantly leathery or coriaceous. Surficial roots are frequent, probably in response to shallow soil depth that rarely exceeds 25 cm. The whole forest mass, from the base of the trunks up to the branches and crowns, is massively covered by epiphytes, in particular musci and many vascular epiphytes, including ferns, peperomias, gesneriads, and orchids, among others.

A few tree species predominate in this elfin forest type, belonging mostly to two species of *Clusia* (*C. pachyphylla* and *C. chiribiquetensis*, Clusiaceae), together with *Cyrilla racemiflora* (Cyrillaceae), *Ecclinusa ulei* (Sapotaceae), *Byrsonima concinna* (Malpighiaceae), *Ilex retusa* (Aquifoliaceae), *Phyllanthus obfalcatus* (Euphorbiaceae), *Perissocarpa umbellifera* (Ochnaceae), *Podocarpus roraimae* (Podocarpaceae), and *Dugandiodendron* cf. *ptaritepuianum* (Magnoliaceae) (Fig. 3.18). The understory consists of large colonies of the gigantic tank bromeliad *B. tatei*, together with the characteristic tepui sedge *Didymiandrum stellatum* (Cyperaceae) and large clumps of *Neurolepis* sp. (bambusoid Poaceae). One of the striking aspects is the absence of any species of *Bonnetia*, so common and ubiquitous on sandstone tepuis.

3.4.3.2 Upper Montane Shrubland

Shrublands have been found only occasionally in small islands on rocky sites and along creeks. Dominant species are *Ilex* sp. (Aquifoliaceae), *Gongylolepis jauaensis* (Asteraceae), *Podocarpus* sp. (Podocarpaceae), *Marcetia taxifolia* (Melastomataceae), *Coccochondra laevis* subsp. *maigualidae* (Rubiaceae), *Clusia* div. spp. (Clusiaceae), *Byrsonima huberi* (Malpighiaceae), and others. Small colonies of the broadleaved herb *Kunhardtia rhodantha* (Rapateaceae) with its showy purple-red inflorescences are common amongst the shrubs (Fig. 3.18).

These plant communities are usually rather dense and can be 1–3 m high. As in the case of the upper montane forest, *Bonnetia* species are absent here also.

3.4.3.3 Montane Grassland

One of the most peculiar vegetation types found over large extents in the Maigualida highlands is a montane grassland type so far reported only for this mountain range in

Fig. 3.18 Characteristic plants of the Sierra de Maigualida highlands. (**a**) The rare *Dugandiodendron* cf. *ptaritepuianum* belongs to the magnolia family and grows occasionally in the low elfin forests. (**b**) *Kunhardtia rhodantha* (Rapateaceae) is found only on the northern Amazonas tepuis and reaches its eastern limit of distribution in the shrublands and meadows of the Sierra de Maigualida (photos Huber)

the Guayana region. It mainly grows on flat, water-saturated peat bogs on small valley bottoms. The *Axonopus villosus* grass, with pubescent leaves arranged distichously from the base, forms wide and dense colonies, together with numerous leaf rosettes of *Orectanthe sceptrum* (Xyridaceae) and other rosette plants such as Eriocaulaceae, Droseraceae, and Bromeliaceae. The unexpected massive presence of this grass with C4 photosynthetic pathway at more than 2,000 m a.s.l., restricted to granitic mountains in the Guayana region, calls for further ecophysiologic investigations. Furthermore, several new species have been identified, such as for instance the small subshrubs *Macrocentrum huberi, Leandra gorzulae*, and *Tibouchina huberi* (Melastomataceae). Some of them are not yet fully described, including new species of *Anthurium* (Araceae) and *Siphocampylus* (Campanulaceae).

3.4.3.4 Upper Montane Broadleaved Meadow

Dense meadows with broadleaved herbs of Rapateaceae, isolated shrubs and shrublets, that are similar to high-tepui meadows, grow on peatlands surrounded by granitic outcrops and domes. The herbaceous layer is dominated by *Stegolepis albiflora* (Rapateaceae), together with *Panicum chnoodes* (Poaceae), *O. sceptrum, Xyris* div. spp. (Xyridaceae), and *Racinea spiculosa* var. *stenoglossa* (Bromeliaceae). The most frequent woody elements are *Spathelia* sp. nov. (Rutaceae), *Clusia* sp. (Clusiaceae), *I. retusa* (Aquifoliaceae), *C. racemiflora* (Cyrillaceae), *Gongylolepis jauaensis* (Asteraceae), and *Retiniphyllum scabrum* (Rubiaceae).

3.4.3.5 Pioneer Vegetation

Bare rock surfaces on granite outcrops are frequent in the Maigualida highlands. They are covered by black crusts of Cyanobacteria as in the case of the granitic "lajas" in the Amazonas lowlands. Small colonies of pioneer plants grow in shallow grooves and depressions on the decomposing rock, including the very frequent *Racinea spiculosa* var. *stenoglossa* (Bromeliaceae), *Comolia* sp. (Melastomatacaeae), *Rhynchospora* sp. and *Everardia* sp. (Cyperaceae), and several small terrestrial orchids.

3.5 Conclusions

3.5.1 Geologic History: The Making of a Continent

Because of its long geologic history and stability, the Guayana Shield is a core area of the South American subcontinent. A steady process of land formation has agglomerated around the primeval Amazon Craton a sequence of orogenetic phases in the Proterozoic, starting with the building of the geologic province of the Imataca Complex, followed by the province of the Greenstone Belt and then by the province of the Cuchivero granite. This igneous basement of the early Guayana Shield has been covered progressively from northeast to southwest and south by sand deposits about 3,000 m thick, which gave rise to the quartzite and sandstone formations of the Roraima Group. After conclusion of this long-lasting sedimentation period (ca 1.9–1.5 Ga), plutonic rocks intruded the sedimentary cover and formed granite batholiths such as the Parguaza batholith that extends along the western and southwestern section of the Guayana Shield. As a result of this evolution, three main lithologies are found in tepui environment (1) igneous–metamorphic basement rocks, (2) sedimentary flatbedded quartzites and sandstones of the Roraima Group, and (3) diabase sills and dikes intruding the sedimentary cover.

3.5.2 Physiography

Highlands correspond to the elevation belt from 1,400/1,500 m to 3,014 m a.s.l. They include large plateaus cut into individual mesetas (tepuis) on sandstones/quartzites, and dissected mountains and hillands on granites with fairly even summit areas. In general, the physiography of the highlands is controlled by the relatively horizontal sandstone and quartzite beds of the Roraima Group. Resulting from long-lasting erosion, the current landscape consists of a series of stepwise topographic levels ("Stufenlandschaft") from plains and peneplains in the lowlands, to large plateaus at intermediate elevation, to the high-elevation tepui mesetas. This table-shaped pattern alternates with rugged mountains and hillands of dome-shaped relief developed from igneous–metamorphic Guayana Shield basement.

The tepuis, more than 50 in southern Venezuela, and their surroundings harbor an extraordinarily rich, specialized, and diversified assemblage of biota and ecosystems not easily found in other tropical mountains. Together, they constitute the biogeographic province of Pantepui, one of the four provinces forming the biogeographic region of Guayana. Thus, tepui summits and other highland landscapes above ca 1,500 m elevation can be considered to be true benchmark ecosystems for comparative ecological studies in tropical high mountains, such as for instance the study on peat ecology developed in this volume. Peatlands are a characteristic feature throughout these areas; their preservation is relevant because they control the regime of most lowland rivers.

3.5.3 Paleoecology

The Guayana Shield has been relatively stable since at least the late Precambrian. It is assumed that the Amazon Craton and with it the Guayana Shield were part of the Gondwana supercontinent during the late Precambrian, Paleozoic, and Mesozoic. The Guayana Shield was almost invariably in near-equatorial position during these times.

With the rise of terrestrial ecosystems at the beginning of the Paleocene (ca 65 Ma ago), the land mass of the ancient Guayana Shield was most likely covered by a sequence of proto-floras that evolved subsequently under continental conditions atop the sedimentary cover of the Roraima Group. After the Gondwana breakup during the Cretaceous (ca 130 Ma ago), the early continental macroclimate was increasingly influenced by the opening of the Mid-Atlantic Ocean, which started bordering the Guayana Shield by the northeast since the beginning of the Tertiary. Subsequently, long-lasting erosion under the impact of the trade wind system and the occurrence of large tectonic events contributed to creating the present landscape of deep valleys alternating with sandstone mesetas and granitic mountain ranges. The uplift of the Andean Cordillera in western South America since mid-Tertiary has influenced significantly the entire tectonic and physiographic setting in the northeast of South America.

The Guayana Shield, contrary to the bordering lowland regions, has never been completely submerged since at least early Tertiary. It is thus a relict, long-standing land mass with a biota that evolved under undisturbed conditions during the last 60–100 Ma.

3.5.4 Climate

There are very few climatic data for the Guayana Shield in Venezuela and adjacent countries. Located approximately 1°–8°N, the Venezuelan Guayana has an equatorial climate with high temperatures near sea level (MAT always >27°C), day/night length between 11:30 and 12:30 daily hours, and high solar radiation during the year. The rainfall regime is controlled by the eastern trade winds carrying moisture-laden clouds from the warm tropical Atlantic Ocean over the Guayana inland. During summertime (April to November), the trade wind influence is overridden by the much heavier rainfalls caused by the ITCZ that oscillates north and south around the American Equator.

Annual rainfall is moderate to high in the lowlands, ranging between 1,200 mm in the northeast and 4,000 mm or more in the center and southwest. The entire region draining northwards into the Orinoco river represents therefore a large water reservoir, with 70% of the area still protected by an almost continuous, dense forest cover. The high evapotranspiration rates of this extensive forest mantle originate a set of water cycles, starting on the flat tepui summits of the highlands and propagating downwards to the mid-elevation uplands and wide lowland basins before returning to the atmosphere.

The Guayana region comprises three altitudinal life zones corresponding to the topographic lowland (0–500 m a.s.l.), upland (500–1,500 m a.s.l.), and highland (1,500–3,000 m a.s.l.) levels. Each level is characterized by a specific MAT regime: macrothermic (27°C) to submesothermic (24°C) in the lowlands, mesothermic (20–12°C) in the uplands and lower highlands, and submicrothermic to microthermic (<12°C) in the upper highlands. Temperatures below 0°C have not been documented or recorded during the past century of exploration at highland summits. Thus, the present high-tepui flora is most likely not adapted to freezing temperatures, in contrast to the páramo flora in the high Andes or similar tropical alpine floras of the world.

Rainfall on tepui summits is mostly high to very high during the largest part of the year. However, periods of a few days to several weeks with no or very little rainfall are common between December and March. These dry spells cause the upper peat layers to dry up and make them prone to fire hazard.

3.5.5 Vegetation

Four main vegetation formations have been recognized in the Guayana region: (1) forests, (2) shrublands, (3) meadows and savannas, and (4) pioneer plant communities colonizing essentially open rock surfaces. Forests form by far the largest cover estimated at about 70% of the Venezuelan Guayana surface area.

Forests are best developed between 100 and 2,200 m a.s.l., both on the flat to undulating lowland plains and on the steep tepui and mountain slopes. Several forest types have been recognized. Tall evergreen rainforests predominate in the interfluvial "tierra firme" and alternate with generally lower, flooded palm forests and swamps in the depressions and along the extensive river floodplains. Dense, evergreen submontane and montane forests of many kinds cover all lower sections of the tepui piedmonts and mountain slopes. In montane cloud forests that are immersed in clouds and mist during most of the year, the floristic richness in tree, climber, and epiphyte species is overwhelmingly high and still far from being fully explored. At the foot of the upper walls of the tepui mesetas, stunted low forest types form very dense, almost impenetrable plant communities, with stems and twigs covered by a dense mantle of slippery epiphytes including musci, filmy ferns, orchids, and tank bromeliads.

Small forest patches are also growing on the summits of some larger tepui massifs, especially in depressions, rock crevices, and narrow valleys. The trees belong almost exclusively to the tepui genus *Bonnetia* of the homonymous family Bonnetiaceae, originally classified as a member of the tea family. Throughout the entire Guayana region, almost every isolated tepui massif has its own species assemblage of *Bonnetia* trees, but these often inaccessible forest communities have not yet been classified in a consistent manner.

High-tepui summit forests are associated either with rocky areas in depressions or ultramafic diabase sills. Forests on soils developed from diabase have a characteristic floristic assemblage and a less dense arboreal cover, but with a very well developed forest floor vegetation of herbs and shrubs. Diabase forests are the only vegetation types that grow on true mineral soil at tepui summits. Most of the other vegetation types on sandstone mesetas are associated with peat soils.

Guayana shrublands are highly diversified ecosystems with a large number of endemic genera and species that are especially adapted to the harsh life conditions on the sandstone plateaus and granitic hills and mountains. Some of the most impressive adaptations include the pachycaulous growth of stems and lower branches, leathery and corky barks, agglomeration of the leaves towards the end of the branches, coriaceous to thick leathery leaves, leaf shape featuring a concave cup, agglomerated inflorescences with very showy, widely visible flowers, among others. A páramo-like growth form of monocaulous stem rosette, like in the case of the well-known genus *Espeletia* (Asteraceae) in the tropical Andes, has developed in the genus *Chimantaea*, with seven species described so far from very restricted localities in the Auyantepui-Chimantá District of the Pantepui Province.

Shrublands occur in patches in the lowlands and uplands of the Guayana region, with many endemic taxa that include tiny shrublets only a few centimeters high as well as giant shrubs several meters tall. However, the highest levels of endemicity occur in the highlands, above ca 2,000 m elevation, where many members of the Bonnetiaceae, Ericaceae, Melastomataceae, Ochnaceae, Asteraceae, and others can be found in large colonies forming an astonishing assemblage of different plant communities with endemics specific to the various tepui summits.

The herbaceous vegetation of the Guayana region can be divided into two large groups: (1) the graminiform savannas of the lowlands and uplands (0–1,600 m a.s.l.), similar to the adjacent Llanos savannas, and (2) the much more restricted meadows that are fields covered with a variety of nongramineous herbs, usually with broad sword-like leaves or rosette leaves or other growth forms, but always being herbaceous plants. The unexpectedly high diversity of herbaceous ecosystems found during recent explorations in the different life zones led to establish special classification schemes for the Guayana region alone.

As in the case of shrubs, herbaceous endemism is highest at the upper levels of the highlands where meadows and fields are widely distributed over an archipelago of ca 50 tepui summits. Like most of the shrublands, the tepui meadows as well as the few grass-dominated high-mountain ecosystems in Sierra de Maigualida grow on peat soils. Many of the study sites of the present volume are located in such environments.

Finally, the pioneer vegetation is also well differentiated in the Guayana region. Many saxicolous plant communities grow in their initial stages on bare rock surfaces from lowlands to highlands. A specialized flora has developed on granitic domes and inselbergs not higher than about 500 m a.s.l. in the lowlands along the northwestern and western border of the Guayana Shield. Similar plant communities grow on sandstone mesetas and granitic upland hilltops between 600 and 1,500 m a.s.l. The diversity of saxicolous plant communities is largest on the sheer walls and rocky summit terrains of the highlands.

The floristic composition of the saxicolous communities shows some relationship with the rock types as some species prefer igneous (e.g., granitic) terrains and others prefer sedimentary (e.g., sandstone) terrains. These preferences, however, are clearly not as strict as, for example, the well-documented differences between calcicole and calcifuge plants in the Alps and other mountain systems of the world. Unfortunately, the scientific literature concerning these interesting plant communities living on rocks in Guayana is still very scanty, especially at the upper altitudinal levels.

In general, the very ancient but stable biogeographic region of Guayana consists of a complex and wide array of highly evolved and specialized ecosystems, which have developed along an altitudinal gradient from the tropical lowlands to the montane highlands. The Guayana uplands and highlands are probably not as rich in flora and fauna species as the tropical Andes, but they are likely to be ecologically more diverse per surface area. This assumption may be influenced by the fact that, to our knowledge, no other tropical American mountain systems harbor such extensive peatlands in their mid-elevation levels as the Guayana tepui summits do. This highlights the relevance of the research on the Guayana peatlands, the results of which are presented in the following chapters of this volume.

References

The body of scientific literature on the Guayana Highlands is considerable, although large areas still lack basic information because either they have not yet been explored or gray information gathered by explorers has not been published (Huber and Foster 2003). There is an important set of publications related to the mineral resources (gold, diamonds) of the Guayana Shield. The literature on botany and related vegetation sciences is much less abundant. One of the most complete bibliographic lists on plant sciences of the Venezuelan Guayana was published in the first volume of the *Flora of the Venezuelan Guayana* (Steyermark et al. 1995–2005), Missouri Botanical Garden, St. Louis and Timber Press, Portland, OR, 1995.

AEROSERVICE Corporation (1972) Levantamiento radar de Venezuela sur. 3 vols, maps 1:250,000. MOP-CODESUR, Caracas

Barthlott W, Gröger A, Porembski S (1993) Some remarks on the vegetation of tropical inselbergs: diversity and ecological differentiation. Biogeographica (Paris) 69:105–124

Berry PE, Riina R (2005) Insights into the diversity of the Pantepui flora and the biogeographic complexity of the Guayana Shield. In: Friis I, Balslev H (eds) Plant diversity and complexity patterns: local, regional and global dimensions. Biol Skr 55:145–167

Berry PE, Huber O, Holst BK (1995) Floristic analysis and phytogeography. In: Steyermark JA, Berry PE, Holst BK (gen eds) Flora of the Venezuelan Guayana, vol 1; Berry PE, Holst BK, Yatskievych K (eds) Introduction. Missouri Botanical Garden and Timber Press, St. Louis, MO, Portland, OR, pp 161–191

Brewer-Carías C (ed) (1988) Cerro de la Neblina. Resultados de la expedición 1983–1987. Fundación para el Desarrollo de las Ciencias Físicas, Matemáticas y Naturales, Caracas

Briceño H, Schubert C (1990) Geomorphology of the Gran Sabana, Guayana Shield, southeastern Venezuela. Geomorphology 3:125–141

Briceño H, Schubert C (1992) Geomorfología. In: Huber O (ed) El Macizo del Chimantá, Escudo de Guayana, Venezuela. Un ensayo ecológico tepuyano. Oscar Todtmann Editores, Caracas, pp 61–74

Briceño H, Schubert C, Paolini J (1990) Table-mountain geology and surficial geochemistry: Chimantá Massif, Venezuelan Guayana Shield. J South Am Earth Sci 3:79–194

Büdel B, Lüttge U, Stelzer R, Huber O, Medina E (1994) Cyanobacteria of rocks and soils of the Orinoco Lowlands and the Guayana Uplands, Venezuela. Bot Acta 107:422–431

Burga GA, Klötzli F, Grabherr G (eds) (2004) Gebirge der Erde: Landschaft, Klima, Pflanzenwelt. Eugen Ulmer GmbH, Stuttgart

Cárdenas-L D (2007) Flora del Escudo Guayanés en Inírida (Guainía, Colombia). Instituto Amazónico de Investigaciones Científicas – Sinchi, Bogotá

Colonnello G (1984) Contribución al conocimiento del microclima y medio físico-biótico de la cima del tepuy Marahuaca. Mem Soc Ci Nat "La Salle" 122:9–35

Colvée P (1973) Cueva en cuarcitas en el Cerro Autana, Territorio Federal Amazonas. Bol Soc Venez Espeleol 4:5–13
Cordani UG, Fraga LM, Reis N, Brito-Neves BB (2010) On the origin and tectonic significance of the intra-plate events of Grenvillian-type age in South America: a discussion. J South Am Earth Sci 29:143–159
Dalton LV (1912) On the geology of Venezuela. Geol Mag 9:203–210
Delascio F, Steyermark JA (1989) Notas sobre la vegetación del Marahuaka. Acta Terramar 1:15–20
Dezzeo N, Huber O (1995) Tipos de bosque sobre el Cerro Duida, Guayana Venezolana. In: Churchill SP, Balslev H, Forero E, Luteyn JL (eds) Biodiversity and conservation of neotropical montane forests. The New York Botanical Garden, Bronx, NY, pp 149–158
Doyle AC (1912) The lost world. Hodder and Stoughton, New York, NY
EDELCA (2010) Sistema de información hidroclimatológico. Gerencia de Gestión Ambiental, Departamento de Gestión de Información Geográfica y Ambiental. Puerto Ordaz
Gaudette H, Hurley P, Fairbairn H (1977) U-Pb Zircon ages from the northern Guayana shield, Venezuela. 2do Congreso Latinoamericano de Geología 4:3045–3047
Ghosh SK (1977) Geología del Grupo Roraima en Territorio Federal Amazonas. In: Memoria V Congreso Geológico Venezolano, vol 1. Caracas, pp 167–193
Gibbs AK, Barron CN (1993) The geology of the Guiana Shield. Oxford Monographs on Geology and Geophysics, vol 22. Oxford University Press, New York, NY
Givnish TJ, Burkhardt EL, Happel RE, Weintraub JD (1984) Carnivory in the bromeliad Brocchinia reducta, with a cost/benefit model for the general restriction of carnivorous plants to sunny, moist, nutrient-poor habitats. Amer Naturalist 124:479–497
Givnish TJ, McDiarmid RW, Buck WR (1986) Fire adaptation in Neblinaria celiae (Theaceae), a high-elevation rosette shrub endemic to a wet equatorial tepui. Oecologia (Berlin) 70:481–485
Gleason HA, and collaborators (1931) Botanical results of the Tyler-Duida expedition. Bull Torrey Bot Club 58:277–516
Gosh SK (1985) Geology of the Roraima Group and its implications. In: Muñóz M (ed) I Simposium Amazónico (Puerto Ayacucho 1981). Bol Geol, Memoria, Publ Esp 10. Ministerio de Energía y Minas, Dirección de Geología, Caracas, pp 33–50
Gröger A (2000) Flora and vegetation of inselbergs of Venezuelan Guayana. In: Porembski S, Barthlott W (eds) Inselbergs – biotic diversity of isolated rock outcrops in tropical and temperate regions, vol 146, Ecological studies. Springer, Berlin, pp 291–314
Gröger A, Huber O (2007) Rock outcrop habitats in the Venezuelan Guayana lowlands: their main vegetation types and floristic components. Rev Bras Bot 30:599–609
Grupo Científico Chimantá (1986) Reconocimiento preliminar del Macizo del Chimantá, Estado Bolívar (Venezuela). Acta Ci Venez 37:25–42
Guerra F, Puig H, Chaume R (1998) The forest-savanna dynamics from multi-date Landsat-TM data in Sierra Parima, Venezuela. Int J Remote Sens 19(11):2061–2075
Hackley PC, Urbani F, Karlsen AW, Garrity CP (comp.) (2005) Geologic shaded relief map of Venezuela. US Geological Survey, Boulder, CO (scale 1:750,000, 2 sheets)
Huber O (1982) Significance of savanna vegetation in the Amazon Territory of Venezuela. In: Prance GT (ed) Biological diversification in the Tropics. Columbia University Press, New York, NY, pp 221–244
Huber O (1989) Shrublands of the Venezuelan Guayana. In: Holm-Nielsen LB, Nielsen IC, Balslev H (eds) Tropical forests: botanical dynamics, speciation and diversity. Academic, London, pp 271–285
Huber O (1992) La vegetación. In: Huber O (ed) El Macizo del Chimantá, Escudo de Guayana, Venezuela. Un ensayo ecológico tepuyano. Oscar Todtmann Editores, Caracas, pp 161–178
Huber O (1994a) Recent advances in the phytogeography of the Guayana region, South America. Mém Soc Biogéogr (3. Série) 4:53–63
Huber O (1994b) La vegetación. In: Dezzeo N (ed) Ecología de la Altiplanicie de la Gran Sabana (Guayana Venezolana) I. Investigaciones sobre la dinámica bosque-sabana en el

sector SE: subcuencas de los ríos Yuruaní, Arabopó y Alto Kukenán. Sci Guianae 4:79–80, 95–118

Huber O (1995a) Geographical and physical features. In: Steyermark JA, Berry PE, Holst BK (gen eds) Flora of the Venezuelan Guayana, vol 1; Berry PE, Holst BK, Yatskievych K (eds) Introduction. Missouri Botanical Garden and Timber Press, St. Louis, MO, Portland, OR, pp 1–61

Huber O (1995b) Vegetation. In: Steyermark JA, Berry PE, Holst BK (gen eds) Flora of the Venezuelan Guayana, vol 1; Berry PE, Holst BK, Yatskievych K (eds) Introduction. Missouri Botanical Garden and Timber Press, St. Louis, MO, Portland, OR, pp 97–160

Huber O (1995c) Guayana Venezolana – Mapa de vegetación. Venezuelan Guayana – Vegetation map. 1:2,000,000. CVG-EDELCA, Missouri Botanical Garden, Caracas

Huber O (2005) Diversity of vegetation types in the Guayana Region: an overview. In: Friis I, Balslev H (eds) Plant diversity and complexity patterns: local, regional and global dimensions. Biol Skr 55:169–188

Huber O (2006) Herbaceous ecosystems on the Guayana Shield, a regional overview. J Biogeogr 33:464–475

Huber O, Foster MN (eds) (2003) Conservation priorities for the Guayana Shield: 2002 consensus. Conservation International – CABS, Guiana Shield Initiative, UNDP, IUCN, Washington, DC, p 99

Huber O, Steyermark JA, Prance GT, Alès C (1984) The vegetation of the Sierra Parima, Venezuela-Brazil: some results of recent exploration. Brittonia 36:104–139

Huber O, Rosales J, Berry PE (1997) Estudios botánicos en las montañas altas de la cuenca del Río Caura (Estado Bolívar, Venezuela). In: Huber O, Rosales J (eds) Ecología de la Cuenca del Río Caura, Venezuela. II. Estudios especiales. Sci Guianae 7:441–468

Huber O, Febres G, Arnal H (eds) (2001) Ecological guide to the Gran Sabana. Canaima National Park – Venezuela. The Nature Conservancy – ChevronTexaco, Caracas

INE (2001) Censo INE 2001. Instituto Nacional de Estadística, Caracas

Körner C (1999) Alpine plant life. Functional plant ecology of high mountain ecosystems. Springer, Berlin

Leal RA (2010) Historia holocena de la vegetación y el fuego en bordes sabana/bosque y turberas de la Gran Sabana, Guayana Venezolana. PhD Dissertation. Universidad Simón Bolívar, Caracas, pp xvi + 277

MARNR (1979) Atlas de la Región Sur. Dirección General de Información e Investigación del Ambiente, CODESUR, Caracas

MARNR-ORSTOM (1987) Atlas del inventario de tierras del Territorio Federal Amazonas. MARNR, Dirección General Sectorial de Información e Investigación del Ambiente (DGSIIA), Caracas

Mendoza V (2000) Evolución geotectónica y recursos minerales del Escudo de Guayana en Venezuela (y su relación con el Escudo Sudamericano). Cátedra de Geología de Venezuela. Escuela Ciencias de la Tierra. Universidad de Oriente, Ciudad Bolívar

Michelangeli A (1989) Biósfera del Marahuaka y zonas adyacentes (Territorio Federal Amazonas-Venezuela). Introducción. Acta Terramar 1:1–3

McPherson S (2008) Lost worlds of the Guiana Highlands. Redfern Natural History Productions, Poole, Dorset, England, pp ix + 385

Nogué S, Rull V, Montoya E, Huber O, Vegas-Villarúbia T (2009) Paleoecology of the Guayana Highlands (northern South America): Holocene pollen record from the Eruoda-tepui, in the Chimantá massif. Palaeogeogr Palaeoclimatol Palaeoecol 281:165–173

Olmeta MA (1968) Determinación de edades radiométricas en rocas de Venezuela y su procedimiento por el método K/Ar. Bol Geol (Caracas) 10:113–208

Orme AR (2007) Tectonic provinces of South America. In: Veblen TT, Young KR, Orme AR (eds) The physical geography of South America. Oxford University Press, New York, NY

Pouyllau M (1989) Les paysages géomorphologiques de l'Amazonie et de la Guyane vénézuélienne. In: Pouyllau D, Pouyllau M, Reaud-Thomas G, Seurin M (eds) Venezuela environnements et changements, tome 2. Travaux et Documents de Géographie Tropicale 63. Centre d'Etudes de Géographie Tropicale, Centre National de la Recherche Scientifique, Bordeaux

Projeto RADAMBRASIL (1975) Folha NA.20 Boa Vista e parte das Folhas NA.21 Tumucumaque, NB.20 Roraima e NB.21. Levantamento de recursos naturais, vol 8. Ministerio das Minas e Energia, Departamento Nacional da Produção Mineral, Rio de Janeiro

Projeto RADAMBRASIL (1976) Folha NA.19 Pico da Neblina. Levantamento de recursos naturais, vol 11. Ministerio das Minas e Energia, Departamento Nacional da Produção Mineral, Rio de Janeiro

Ramos B (1997) Los suelos en las cumbres de las Sierras de Maigualida y Uasadi-Jidi, Guayana venezolana. In: Huber O, Rosales J (eds) Ecología de la Cuenca del Río Caura, Venezuela. II. Estudios especiales. Sci Guianæ 7:423–440

Reid AR (1972) Stratigraphy of the type area of the Roraima Group, Venezuela. Bol Geol (Caracas), Publ Esp 6:343–353

Reid AR, Bisque R (1975) Stratigraphy of the diamond-bearing Roraima-Group, Estado Bolívar, Venezuela. Colorado Sch Mine Q 70:61–82

Riina R, Huber O (2003) Ecosistemas exclusivos de la Guayana. In: Aguilera M, Azócar A, González JE (eds) Biodiversidad en Venezuela. Fundación Polar, Ministerio de Ciencia y Tecnología, Fondo Nacional de Ciencia, Tecnología e Inovación (FONACIT), Caracas, pp 828–861

Rinaldi M, Rull V, Schubert C (1990) Análisis paleoecológico de una turbera en la Gran Sabana (Santa Cruz de Mapaurí), Venezuela: resultados preliminares. Acta Ci Venez 41:66–68

Rull V (1991) Contribución a la paleoecología de Pantepui y la Gran Sabana (Guayana Venezolana): clima, biogeografía y ecología. Sci Guianae 2. Caracas

Rull V (1992) Successional patterns of the Gran Sabana (Southeastern Venezuela) vegetation during the last 5000 years, and its responses to climatic fluctuations and fire. J Biogeogr 19:329–338

Rull V (1999) A palynological record of a secondary succession after fire in the Gran Sabana, Venezuela. J Quatern Sci 14:137–152

Rull V (2003) An illustrated key for the identification of pollen from Pantepui and the Gran Sabana (eastern Venezuelan Guayana). Palynology 27:95–129

Rull V (2004) Biogeography of the "Lost World": a palaeoecological perspective. Earth Sci Rev 67:125–137

Rull V (2005) Palaeovegetational and palaeoenvironmental trends in the summit of the Guaiquinima massif (Venezuelan Guayana) during the Holocene. J Quatern Sci 20:135–145

Rull V (2007) Holocene global warming and the origin of the neotropical Gran Sabana in the Venezuelan Guayana. J Biogeogr 34:279–288

Rull V (2009) New evidence on the potential role of fire in the Gran Sabana (Venezuelan Guayana) and implications for early human occupation. Veg Hist Archaeobot 18:219–224

Rull V, Vegas-Vilarrúbia T, Nogué S, Montoya E (2008) Bureaucratic obstruction of conservation science in the Guayana Highlands. Conserv Biol 22:508–509

Rull V, Vegas-Vilarrúbia T, Nogué S, Huber O (2009) Conservation of the unique neotropical vascular flora of the Guayana Highlands in the face of global warming. Conserv Biol 23:1323–1327

Salgado-Labouriau ML, Villar L (1992) Contribución a la flora polínica de los tepuyes. In: Huber O (ed) El Macizo del Chimantá, Escudo de Guayana, Venezuela. Un ensayo ecológico tepuyano. Oscar Todtmann Editores, Caracas, pp 219–236

Schomburgk RH (1840) Journey from Fort San Joaquim, on the Rio Branco, to Roraima, and thence by the rivers Parima and Merewari to Esmeralda, on the Orinoco, in 1838–1839. J R Geogr Soc 10:191–247

Schomburgk R (1848) Versuch einer Fauna und Flora von Britisch-Guiana. In: Schomburgk RH (ed) Reisen in Britisch-Guiana in den Jahren 1840–1844, vol 3. Verlagsbuchhandlung von JJ Weber, Leipzig, pp 531–1259

Schubert C, Fritz P (1985) Radiocarbon ages of peat, Guayana Highlands (Venezuela). Naturwissenschaften 72:427–429

Schubert C, Huber O (1990) The Gran Sabana. Panorama of a region. Lagoven Booklets, Caracas

Schubert C, Briceño HO, Fritz P (1986) Paleoenvironmental aspects of the Caroní-Paragua river basin (southeastern Venezuela). Interciencia 11:278–289

Schubert C, Fritz P, Aravena R (1992) Investigaciones paleoambientales: resultados preliminares. In: Huber O (ed) El Macizo del Chimantá, Escudo de Guayana, Venezuela. Un ensayo ecológico tepuyano. Oscar Todtmann Editores, Caracas, pp 97–110

Schubert C, Fritz P, Aravena R (1994) Late Quaternary paleoenvironmental studies in the Gran Sabana (Venezuelan Guayana Shield). Quatern Int 21:81–90

Sidder GB, Mendoza SV (1995) Geology of the Venezuelan Guayana Shield and its relation to the geology of the entire Guayana Shield. In: Sidder GB, García AE, Stoeser JW (eds) Geology and mineral deposits of the Venezuelan Guayana Shield. US Geol Surv Bull 2124. US Government Printing Office, US Geological Survey, Corporación Venezolana de Guayana, Técnica Minera CA, Washington, DC, pp [iv], B1-B41 [separate chapter pagination]

Smith AP (1994) Introduction to tropical alpine vegetation. In: Rundel PW, Smith AP, Menzer FC (eds) Tropical alpine environments. Plant form and function. Cambridge University Press, Cambridge, pp 1–19

Steyermark JA (1974) The summit vegetation of Cerro Autana. Biotropica 6:7–13

Steyermark JA (1984). Bonnetia. In: Steyermark JA (ed) Flora of the Venezuelan Guayana-Part I. Ann Missouri Bot Gard 71:297–340

Steyermark JA, Brewer-Carías C (1976) La vegetación de la cima del Macizo de Jaua. Bol Soc Venez Ci Nat 22:179–405

Steyermark JA, Maguire B (1984) Informe preliminar sobre la flora de la cumbre del Cerro Marutaní. Acta Bot Venez 14:91–117

Steyermark JA, Berry PE, Yatskievych K, Holst BK (gen eds) (1995–2005) Flora of the Venezuelan Guayana, vol 1–9. Missouri Botanical Garden Press and Timber Press, St. Louis, MO, Portland, OR

Urbani F (2008) Revisión de la nomenclatura de las unidades de rocas ígneas y metamórficas del norte de Venezuela / A review of the nomenclature of igneous and metamorphic rock units of northern Venezuela. Bol Acad Ci Fis 68:27–43

Urbani F, Talukdar S, Szczerban E, Colvée P (1977) Metamorfismo de las rocas del Grupo Roraima, Edo. Bolívar y Territorio Federal Amazonas. Memoria V Congreso Geológico Venezolano, tomo 2. Ministerio de Energía y Minas, Sociedad Venezolana de Geólogos, Caracas, pp 623–638

US Geological Survey, CVG-TECMIN (1993) Geology and mineral resource assessment of the Venezuelan Guayana Shield. US Geol Surv Bull 2062. US Government Printing Office, Washington, DC

Vareschi V (1992) Observaciones sobre la dinámica vegetal en el Macizo del Chimantá. In: Huber O (ed) El Macizo del Chimantá, Escudo de Guayana, Venezuela. Un ensayo ecológico tepuyano. Oscar Todtmann Editores, Caracas, pp 179–188

Weitzman AL, Stevens PF (1997) Notes on the circumscription of Bonnetiaceae and Clusiaceae, with taxa and new combinations. BioLlania Edición Especial 6:551–564

Wynn JC, Cox DP, Gray F, Schruben PG (compiled) (1993) Geologic and tectonic map of the Venezuelan Guayana (scale 1:1,000,000). US Geol Surv Bull 2062. US Government Printing Office, Washington, DC

Zent EL, Zent S (2004) Floristic composition, structure, and diversity of four forest plots in the Sierra Maigualida, Venezuelan Guayana. Biodivers Conserv 13:2453–2484

Zinck JA, García P, van der Plicht J (2011) Tepui Peatlands: age record and environmental changes. In: Zinck JA, Huber O (eds) Peatlands of the Western Guayana Highlands, Venezuela, Chap. 7. Springer, Heidelberg

Appendix

Table 3.1 Plant taxa present in the three study areas of the western Guayana Highlands, with notes on their general distribution, habit and habitat

Taxon	Family	Distribution	Habit and habitat
Abolboda macrostachya Spruce ex Malme	Xyridaceae	Guayana region from low- to highlands	Rosette herb in savannas, meadows, and montane grasslands
Acanthella pulchra Gleason	Melastomataceae	Cuao-Sipapo low- and uplands	Low to medium tall shrub on rocks
Acanthella sprucei Hook. f.	Melastomataceae	Cuao-Sipapo low- and uplands	Low, partly deciduous shrub on rocks
Amphiphyllum rigidum Gleason	Rapateaceae	Duida up- and highlands, endemic	Tall herb in swampy meadows
Anthurium huberi Bunting ex Croat	Araceae	Maigualida highlands, endemic	Terrestrial, in meadows and montane grasslands
Axonopus villosus Swallen	Poaceae	Maigualida highlands and Duida uplands, endemic	Perennial tall grass with strongly equitant sheaths
Bonnetia crassa Gleason	Bonnetiaceae	N Amazonas tepuis, Duida and Parú highlands, endemic	Shrub to low tree, in meadows and tepui forests
Bonnetia jauaensis Maguire	Bonnetiaceae	Jaua-Sarisariñama highlands, endemic	Shrub to low tree, montane scrub and forests
Bonnetia kathleenae Lasser	Bonnetiaceae	Cuao-Sipapo low- and uplands, endemic	Scandent shrub to low tree, in tepui scrub and meadows
Bonnetia maguireorum Steyerm.	Bonnetiaceae	Neblina highlands, endemic	Caulirosulate, corky-stemmed shrub forming large colonies in boggy meadows
Bonnetia neblinae Maguire	Bonnetiaceae	Neblina up- and highlands, endemic	Low to medium sized tree in tepui forests
Bonnetia paniculata Spruce ex Benth.	Bonnetiaceae	Guayana, N Amazon and Andes regions, from low- to highlands	Shrub or small tree, in savannas, montane scrub and grasslands
Bonnetia roraimae Oliv.	Bonnetiaceae	E Guayana tepui highlands, endemic	Dense shrub to medium sized tree, in tepui forests
Brewcaria marahuacae L.B. Sm., Steyerm. & H. Rob.	Bromeliaceae	Marahuaka highlands, endemic	Acaulescent terrestrial rosette herb, on rocks
Brocchinia acuminata L.B. Sm.	Bromeliaceae	Marahuaka, Duida and Cuao-Sipapo up- and highlands	Caulescent terrestrial rosette herb, in forests
Brocchinia hechtioides Mez	Bromeliaceae	Guayana region up- and highlands	Tank-forming, caulescent, insectivorous terrestrial herb
Brocchinia melanacra L.B. Sm.	Bromeliaceae	Maigualida highlands	Caulescent terrestrial rosette herb, along rocky streams
Brocchinia tatei L.B. Sm.	Bromeliaceae	Pantepui	Caulescent terrestrial rosette herb, in forests

(continued)

Table 3.1 (continued)

Taxon	Family	Distribution	Habit and habitat
Burmannia foliosa Gleason	Burmanniaceae	Marahuaka and Duida highlands, endemic	Small rosette herb, moist rocky places
Byrsonima concinna Benth.	Malpighiaceae	Guayana region up- and highlands	Shrub or small tree, in tepui forest and scrub
Byrsonima huberi W.R. Anderson	Malpighiaceae	Maigualida highlands, endemic	Shrub in meadows and open woodland
Cephalocarpus confertus Gilly	Cyperaceae	Guayana region up- and highlands, endemic	Acaulescent perennial herb, in shrublands
Clusia chiribiquetensis Maguire	Clusiaceae	Guayana region, low- to highlands	Low tree in dwarf montane forest
Clusia pachyphylla Gleason	Clusiaceae	Pantepui, endemic	Shrub or low tree, in meadows and on rocks
Coccochondra laevis (Steyerm.) Rauschert subsp. *maigualidae* J.H. Kirkbr.	Rubiaceae	Maigualida highlands, endemic	Low, dense shrub in montane grassland and on rocks
Cortaderia roraimensis (N.E. Br.) Pilger	Poaceae	Guayana region highlands, endemic	Tall perennial tussocks, forming large dense clumps in meadows
Cyrilla racemiflora L.	Cyrillaceae	Widespread in the Neotropics	Shrub in shrublands and on rocks
Daphnopsis steyermarkii Nevling	Thymelaeaceae	Pantepui, endemic	Low shrub, on rocky outcrops and in meadows
Decagonocarpus cornutus R.S. Cowan	Rutaceae	Cuao-Sipapo up- and highlands	Low to medium tall shrub on rocks
Didymiandrum stellatum (Boeck.) Gilly	Cyperaceae	Guayana region up- and highlands	Perennial, cane-like, cespitose herb, common
Dugandiodendron ptaritepuianum (Steyerm.) Lozano	Magnoliaceae	Eastern Guayana and Maigualida highlands, endemic	Shrub or low tree, in montane and riverine forests
Duidania montana Standl.	Rubiaceae	Duida, Marahuaka and Huachamakari up- and highlands, endemic	Shrub or small tree, in tepui scrub or low forest
Ecclinusa ulei (Krause) Gilly ex Cronquist	Sapotaceae	Eastern Guayana and Maigualida highlands, endemic	Shrub or low tree, in montane and riverine forests
Everardia montana Ridl. ex Thurn	Cyperaceae	Guayana region up- and highlands	Small perennial rosette herb, common
Geonoma appuniana Spruce	Arecaceae	Pantepui	Low pinnate palm in forests and thickets
Gongylolepis jauaensis (Aristeg., Maguire & Steyerm.) V.M.Badillo	Asteraceae	Maigualida, Marahuaka, and N Amazonas tepui highlands, endemic	Tall, caulirosulate shrub in high tepui scrub
Graffenrieda fantastica R.E.Schult. & L. B. Sm.	Melastomataceae	Cuao-Sipapo low- and uplands	Low, pachycaulous shrub on rocks

Graffenrieda versicolor Gleason	Melastomataceae	N Amazonas tepuis, up- and highlands, endemic	Low, sometimes pachycaul shrub on rocks
Heliamphora heterodoxa Steyerm.	Sarraceniaceae	Eastern Guayana up- and highlands, endemic	Low carnivorous pitcher plant, in swampy meadows
Heliamphora minor Gleason	Sarraceniaceae	Eastern Guayana highlands, endemic	Low carnivorous pitcher plant, in swampy meadows
Heliamphora tatei Gleason var. *tatei*	Sarraceniaceae	Duida, Marahuaka and Huachamakari up- and highlands, endemic	Up to 2 m tall carnivorous pitcher plant, in swampy meadows and tepui scrub
Heteropterys oblongifolia Gleason	Malpighiaceae	Cuao-Sipapo lowlands, endemic	Shrublet in lowland meadows
Heteropterys steyermarkii W.R. Anderson	Malpighiaceae	Cuao-Sipapo up- and highlands, endemic	Shrub or small tree, on rocks and in marshy meadows
Humiria balsamifera Aubl.	Humiriaceae	Guayana and N Amazonas region	Dense shrub in shrublands and open areas
Hypericum marahuacanum N. Robson subsp. *marahuacanum*	Clusiaceae	Marahuaka highlands, endemic	Wiry shrub, in shaded streamsites
Ilex divaricata Mart. ex Reiss.	Aquifoliaceae	Guayana region low- and uplands	Shrub in savannas and on rock outcrops
Ilex retusa Kl.	Aquifoliaceae	Guayana low-, up- and highlands	Low shrub in meadows and scrub, widespread
Isidrogalvia schomburgkiana (Oliv.) Cruden	Liliaceae	Guayana region up- and highlands	Robust, fan-shaped herb in moist meadows
Kunhardtia rhodantha Maguire	Rapateaceae	N Amazonas tepuis and Maigualida, up- and highlands, endemic	Large terrestrial herb, frequent in broadleaved meadows
Leandra gorzulae Wurdack	Melastomataceae	Maigualida highlands, endemic	Low shrub in moist meadows
Ledothamnus guyanensis Meisn.	Ericaceae	Guayana region up- and highlands, endemic	Slender, wiry, erect shrub in meadows and rocky areas
Macrocentrum huberi Wurdack	Melastomataceae	Maigualida highlands, endemic	Subshrub in moist meadows and grassland
Maguireothamnus tatei (Standl.) Steyerm.	Rubiaceae	Duida, Marahuaka, Huachamakari, and Cuao-Sipapo up- and highlands, endemic	Low shrub in meadows and tepui scrub
Marahuacaea schomburgkii Maguire	Rapateaceae	Marahuaka highlands, endemic	Perennial herb with distichous leaf bases, in exposed, rocky areas
Marcetia taxifolia (A. St.-Hil.) DC	Melastomataceae	S American up- and highlands	Dense shrub in open areas
Mycerinus sclerophyllus A.C. Sm.	Ericaceae	Duida and Marahuaka highlands, endemic	Low shrub, in rocky tepui meadows

(continued)

Table 3.1 (continued)

Taxon	Family	Distribution	Habit and habitat
Neotatea longifolia (Gleason) Maguire	Clusiaceae	Duida and Marahuaka highlands, endemic	Shrub or small tree in moist meadows and shrublands
Nietneria paniculata Steyerm.	Liliaceae	Guayana region up- and highlands	Slender, fan-shaped herb in moist meadows
Orectanthe sceptrum (Oliv.) Maguire	Xyridaceae	Guayana region up- and highlands	Rosette herb in meadows and montane grasslands
Panicum chnoodes Trin.	Poaceae	Guayana region up- and highlands	Tall perennial cespitose grass, in meadows
Perissocarpa umbellifera Steyerm. & Maguire	Ochnaceae	Duida, Cuao-Sipapo, and Maigualida up- and highlands, endemic	Tree in submontane and montane forests
Phelpsiella ptericaulis Maguire	Rapateaceae	Parú, up- and highlands, endemic	Small herb in tepui meadows
Phyllanthus obfalcatus Lasser & Maguire	Euphorbiaceae	N Amazonas tepuis and Maigualida highlands, endemic	Open shrub in montane scrub and grassland
Podocarpus roraimae Pilger	Podocarpaceae	Pantepui, endemic	Small tree, in tepui scrub on rock outcrops
Pradosia schomburgkiana (A.DC.) Cronquist	Sapotaceae	Guayana region low- and uplands	Dense shrub or tree, in scrub or low open forest
Racinea spiculosa (Griseb.) M.A. Spencer & L.B. Sm. var. *stenoglossa* (L.B. Sm.) M. A. Spencer & L.B. Sm.	Bromeliaceae	Widespread in Central and S America	Terrestrial, lithophytic
Retiniphyllum scabrum Benth.	Rubiaceae	Guayana region up- and highlands	Common shrub in tepui scrub, meadows and low forest
Sauvagesia imthurniana (Oliv.) Dwyer	Ochnaceae	Pantepui, endemic	Subshrub in tepui meadows
Spathelia sp. nov.	Rutaceae	Maigualida highlands, endemic	Low, open shrub in montane grassland and meadows
Stegolepis albiflora Steyerm.	Rapateaceae	Maigualida and Jaua highlands, endemic	Broadleaved herb in meadows
Stegolepis choripetala Maguire	Rapateaceae	Maigualida and Cuao-Sipapo highlands, endemic	Tall herb in open swampy areas and scrub
Stegolepis grandis Maguire	Rapateaceae	Duida and Marahuaka up- and highlands, endemic	Giant herb up to 3 m tall, in tepui meadows and scrub
Stegolepis linearis Gleason	Rapateaceae	Duida up- and highlands, endemic	Low herb with narrow leaf blades, in open areas

Stegolepis membranacea Maguire	Rapateaceae	Marahuaka, Duida, and Huachamakari highlands, endemic	Tall herb in scrub and on rock outcrops near streams
Stegolepis pungens Gleason	Rapateaceae	Duida up- and highlands, endemic	Large, robust herb in swampy meadows and thickets
Stegolepis terramarensis Steyerm.	Rapateaceae	Marahuaka highlands, endemic	Medium sized herb forming dense colonies in meadows
Stenopadus campestris Maguire & Wurdack	Asteraceae	Cuao-Sipapo uplands	Open shrub on rock outcrops
Stenopadus huachamacari Maguire	Asteraceae	Huachamakari highlands, endemic	Low shrub in tepui scrub
Steyerbromelia discolor L.B. Sm. & H. Rob.	Bromeliaceae	Marahuaka highlands, endemic	Terrestrial rosette herb, on rocks
Tepuianthus savannensis Maguire & Steyerm.	Tepuianthaceae	Cuao-Sipapo uplands and upper Orinoco lowlands, endemic	Dense shrub in savannas and shrublands
Tibouchina huberi Wurdack	Melastomataceae	Maigualida highlands, endemic	Small shrub, in shrublands
Trachypogon spicatus (L.f.) Kuntze	Poaceae	Neotropics, widespread in low- and uplands	Perennial cespitose grass in savannas
Utricularia humboldtii R.H. Schomb.	Lentibulariaceae	Guayana region up- and highlands	Slender herb usually in water of bromeliad tanks

Taxonomy follows Steyermark et al. (eds) (1995–2005) Flora of the Venezuelan Guayana (see text for reference to geographic terms)

Chapter 4
Tepui Peatlands: Setting and Features

J.A. Zinck and P. García

4.1 Introduction

Peat deposits have been recognized in the Venezuelan Guayana over a large altitudinal range from lowland plains below 200 m a.s.l., to the vast plateau of the Gran Sabana at 700–1,350 m a.s.l., to the table-shaped mountains known as tepuis at more than 2,000 m a.s.l. In the highlands, peat types and patterns are strongly related to the peculiar geomorphic landscape that has developed at the tepui-meseta summits. The weathering of the bedrocks has conduced to a variety of karst-like relief forms from small pans to large depressions in which organic materials have accumulated. While the peatlands in the eastern Guayana Highlands have already been documented, those occurring in the western areas are still largely unknown. To fill this knowledge gap, several exploration missions were carried out in the early 1990s covering parts of three table-mountain massifs in the western Amazonas state of Venezuela: the Marahuaka-Huachamakari massif (including the northern edge of the adjoining Duida massif), the Cuao-Sipapo massif, and the western border of the Maigualida massif.

This chapter focuses on the physical setting of the peat deposits, characterizing the rock substratum and geomorphic landscape that control peatland features in terms of spatial distribution, configuration, and hydrology, and determine peat types and patterns.

4.2 Rock Substratum

The explored areas are table highlands surrounded by low-lying old peneplains and more recent alluvial plains. Figure 4.1 shows a contact area between lowlands (valley, peneplain, and piedmont landscapes) and highlands (hilland, plateau, and mountain landscapes). Elevation ranges mainly from 1,200 to 2,800 m a.s.l., but some table-shaped areas lie as low as 650–700 m a.s.l. The distinctive feature of the

J.A. Zinck and O. Huber (eds.), *Peatlands of the Western Guayana Highlands, Venezuela*, Ecological Studies 217, DOI 10.1007/978-3-642-20138-7_4,

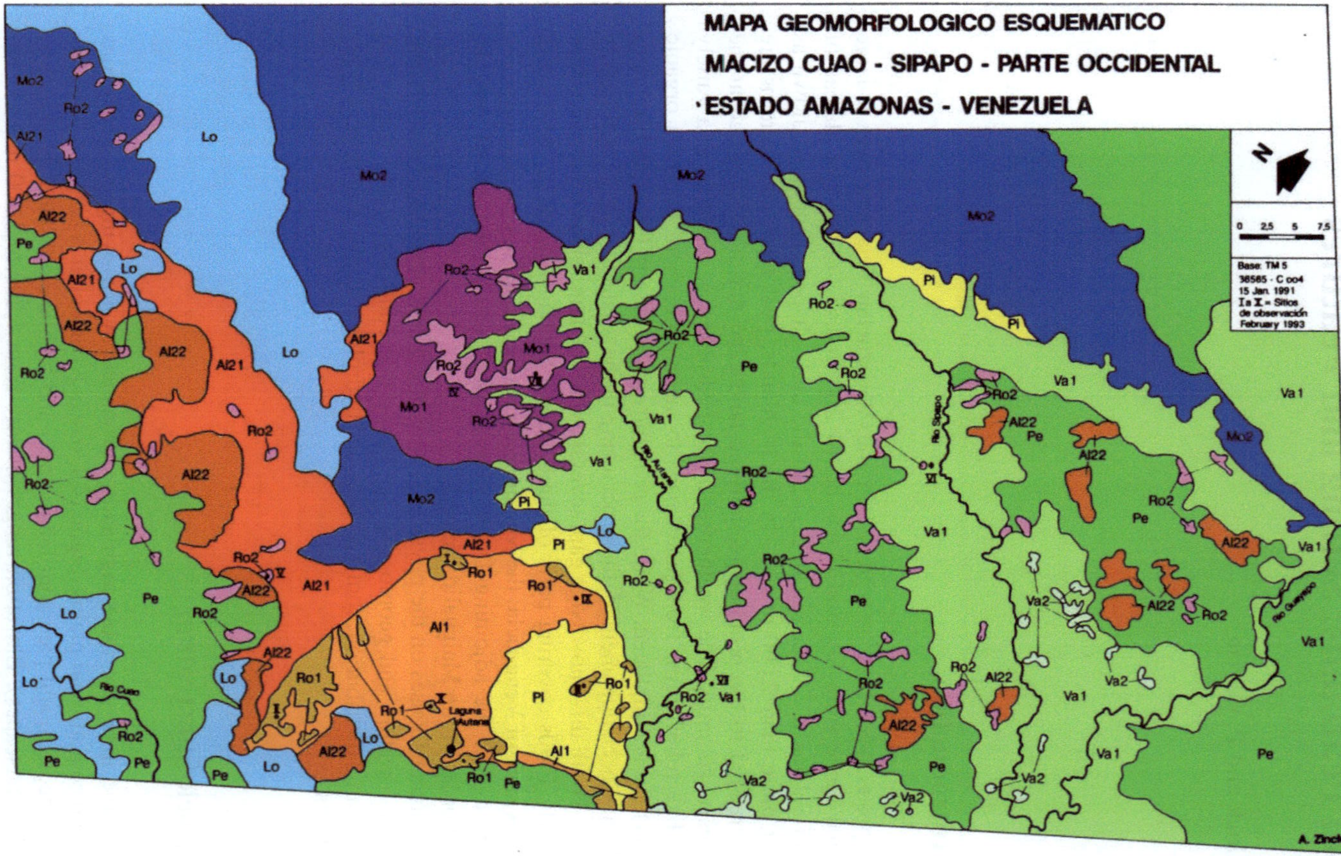

Fig. 4.1 Geomorphic map of the western Cuao-Sipapo massif (scale in km). Va1: valley; Va2: whitesand patches in valley landscape; Pe: peneplain; Lo: hilland; Al1: sandstone plateau; Al21: granite plateau (high elevation); Al22: granite plateau (low elevation); Pi: piedmont; Mo1: mountain dome; Mo2: mountain range; Ro1: bare sandstone terrain; Ro2: bare granite terrain (J.A. Zinck)

highlands is the presence of large plateaus, fragmented into mesetas called tepuis (meaning table-mountain in indigenous Pemón language). Tepuis have flat to slightly undulating, sometimes irregular, summit topography and are fringed by vertical cliffs of bare rock that can stretch over 1,000 m relative elevation. The mesetas have developed from sandstone and quartzite formations belonging to the Roraima Group. In places where this sedimentary cover has been removed by long-lasting erosion, the igneous–metamorphic basement outcrops in the form of granitic and gneissic domes and mountain ranges (Fig. 4.1). Our observation points and peat sampling sites are mainly located on sandstone and quartzite mesetas belonging to the sedimentary cover of the Roraima Group, but some are also on domes and hills developed from basement rocks of the Guayana Shield.

4.2.1 Cover Rocks: The Roraima Group

The Roraima Group is a comprehensive geologic entity of Precambrian age (1.65 ± 0.05 Ga, according to Gibbs and Barron 1993) that stretches over large areas in the northeast of South America, including the west of Guyana, south of Venezuela, and north of Brazil. According to Ghosh (1985), the east–west lateral extent is about 1,500 km, with progressively younger sediments westward. This led to the conclusion that the Roraima material has filled in a large sedimentary basin with erosion debris dragged from mountain ranges located to the east (Ghosh 1977). However, as the present distribution of the sedimentary cover is geographically fragmented, it has also been postulated that Roraima sediments may have deposited in discrete, smaller sedimentary basins separated by basement highs (Ghosh 1985).

The Roraima Group includes quartzitic conglomerates, quartzitic sandstones, quartzites, lutites, and jaspers, with intrusive dykes and sills of diabase (Ministerio de Minas e Hidrocarburos 1970). On the basis of lithostratigraphy, the Group has been subdivided into four formations in the eastern Guayana Highlands, including from bottom to top: the Uairén Formation (>850 m conglomerates, sandstones, siltstones, and shales), the Kukenán Formation (50–400 m shales and siltstones), the Uaimapué Formation (ca 650 m sandstones, arkoses, and tuff), and the Mataui Formation (600–900 m quartz sandstones and quartzites with diabase sills and dykes) (Reid 1972; Yánez 1972, 1985; Briceño et al. 1990; Gibbs and Barron 1993). The resistant Matauí rock beds form the summits and vertical walls of the tepui mesetas.

The lithological variations of the Group are reflected in the physiographic expression of the different Roraima components. In the Gran Sabana plateau, located east of our study area, Yánez (1977) distinguished three members that are underlying the three main structural surfaces recognized at regional level. The lower member, mainly conglomerates and arkosic sandstones, appears as cuesta-type monoclinal reliefs at 600–800 m elevation. The central member is composed of arkosic sandstones, lutites, and jaspers, and forms a plateau level at 1,100–1,500 m a.s.l. (i.e., the Gran Sabana level proper) that surrounds the footslopes of the tepui

Fig. 4.2 Sandstone–quartzite landscape: lithological variations (Cuao-Sipapo massif). (**a**) Tepui summit escarpment showing massif quartzite layers, incised by cascading creeks along major fractures. (**b**) Small internal meseta escarpments showing finely stratified sandstone layers (photos Zinck)

mesetas. The upper member with the most resistant rocks of the Group, especially sandstones, quartzites, and conglomerates, corresponds to the tepui summits and vertical walls that reach up to 2,500–2,800 m a.s.l. (Fig. 4.2). Although the sedimentary strata are mainly horizontal, there is evidence of tilting, short- to large-scale folding, and faulting (Fig. 4.3).

Large parts of the original rock mass of the Roraima Group have been eroded since its deposition in the Precambrian. Thus, the present relief is a residual one in which Briceño and Schubert (1990) recognized six planation surfaces. From the youngest to the oldest, they include (1) the alluvial plain of the Orinoco river (0–50 m) of Holocene age; (2) the Llanos surface (90–150 m) coinciding with the top of the Mesa Formation of Plio-Pleistocene age; (3) the Caroní-Aro surface (200–450 m) that truncates the basement rocks of the Guayana Shield (Oligo-Miocene?); (4) the Imataca surface (600–700 m) that cuts across the lower formations of the Roraima Group and the underlying basement rocks (Early Tertiary?); (5) the Wonkén surface (900–1,200 m) that constitutes the large regional plateau

Fig. 4.3 Tectonic deformation of the sedimentary rocks (Cuao-Sipapo massif). Monoclinal tilting resulting in cuesta relief (**a**); narrowly compressed synclinal fold (**b**); river bed controlled by a fracture line (**c**) (photos Zinck)

of the Gran Sabana above which rise the tepui mesetas (Mesozoic?); and (6) the Auyantepui surface (2,000–2,900 m) which correlates with the summits of the higher tepuis (Mesozoic?). Only the last four ones are actually erosion surfaces, while the two youngest ones are depositional surfaces.

Tectonically, many tepuis show a basin-ward convergent dip, resulting in slightly spoon-shaped surfaces (Van de Putte 1972; Ghosh 1985). Structurally, they are perched synclines resulting from a regional process of relief inversion between original structural highs (anticlines) that have been eroded, and structural lows (synclines) that have been preserved (Van de Putte 1972; Gansser 1973) (Fig. 4.4). The concave summit topography of the perched synclines causes the surface drainage pattern to be centripetal. A dense framework of fractures and joint planes allows water to cross the rock layers to underground pipes and galleries, and to reappear at different heights along the external walls of the tepuis forming cascades and waterfalls. Because rain and runoff waters tend to concentrate in the central parts of the individual mesetas, peat deposits are more frequent in these areas, while the meseta rims are generally bare rock surfaces. However, the specific location of peatlands depends more on micro-relief than on meso-relief; it depends in particular on the distribution of (pseudo-)karstic depressions.

Fig. 4.4 Regional-scale tectonic deformation of sedimentary rocks (Huachamakari massif). The sandstone–quartzite layers of the Roraima formations lie mostly horizontal, originating the table-shaped configuration of the tepui mesetas. However, folding can be observed at regional scale as in (**a**) showing a anticline–syncline sequence, and in (**b**) showing the concave, spoon-like summit topography of a perched syncline resulting from relief inversion. The centripetal topography of the synclinal areas favors water concentration and the accumulation of organic materials giving rise to peatland formation (photos Zinck)

The tepuis of the western Guayana Highlands, belonging actually to the Amazon region of Venezuela, have been less documented than those of the eastern Guayana Highlands. Published data concerning specifically the three visited areas are scarce. Gray information, such as the geologic and geomorphic reports and maps prepared by CVG-TECMIN (1991), is difficult to access. In the Cuao-Sipapo massif, the Roraima Group includes a monotonous sequence of fine- to medium-grained quartz arenites and feldspathic arenites, which unconformably overlies the Parguaza granite (Ghosh 1985). The Cuao-Sipapo sequence includes three members, similar to the subdivisions recognized in the Gran Sabana area. The lower 500 m section consists of quartz arenites; the middle 200 m section includes mainly arkosic arenites; and the upper 600–700 m section consists of coarse-grained quartz arenites that form the tepui cliffs (Ghosh 1985). The Marahuaka-Huachamakari tepuis have developed from sandstones of the Roraima Group, while the visited part of the Maigualida massif consists mainly of granitic rocks. No published information is available about these two areas.

4.2.2 Basement Rocks: The Parguaza Granite and Other Igneous–Metamorphic Rocks

The Roraima Group is underlain by granitic and gneissic basement rocks of the Guayana Shield. In places where the sedimentary cover formations of Roraima have been removed by erosion, the basement rocks are outcropping, especially along the axis of excavated anticlines, assisted by horst-graben fault compartments.

The most common of the basement rocks in the study area belong to the Parguaza granite, a variety of medium-grained, biotite-rich granite dated 1.5 Ga (Gaudette and Olszewski 1985). The intrusive batholith of Parguaza granite belongs to the Cuchivero/Amazonas structural province (Talukdar et al. 1973; Mendoza 1977), and corresponds to the most recent uplift event of the Guayana Shield. The Cuchivero/Amazonas province is the second most recent (1.4–1.6 Ga) of the four geologic provinces that have been recognized in the Venezuelan Guayana Shield (González de Juana et al. 1980). The Parguaza granite has a porphyritic, rapakivi-like texture, characterized by the presence of large ovoid phenocrysts of potassium feldspars (30–50%) covered by skins of sodium plagioclases.

Parguaza rocks occur in the western Cuao-Sipapo massif, as windows opened into the Roraima sandstone cover by differential erosion. Parguaza-like rocks belonging to the Santa Rosalía granite occur also in the Maigualida massif as larger basement exhumation areas.

4.3 Geomorphic Landscape

Peat deposits are mainly found on high-elevation sandstone–quartzite mesetas (eight sites visited) and, to a lesser extent, on granitic domes (four sites visited). For that reason, emphasis is given hereafter and in other parts of this chapter as well as in other chapters of this volume (Garcia et al. 2011; Zinck et al. 2011) to the setting and landscape characteristics of tepui peats.

4.3.1 Mesetas and Domes

Large tepui mesetas, reaching up to 2,800 m a.s.l., are usually fragmented into smaller units that are separated by canyons and crevices controlled by fractures and joint planes (Figs. 4.5 and 4.6). Although table-shaped in general, tepui summits have frequently rugged meso- and micro-topography resulting from differential rock weathering and erosion. Dissolution and/or mechanical removal of

Fig. 4.5 Sandstone–quartzite landscape (Cuao-Sipapo massif). (**a**) Typical tepui meseta relief showing a table-shaped structural summit area covered by shrub vegetation, a border escarpment with vertical cliffs, and a sloping debris talus with forest cover. (**b**) Plateau edge dissected along a dense fracture field that results from rock decompression in tepui borders; large karstic depression with peatland in the center of the picture (photos Zinck)

sandstone–quartzite cement and matrix cause rocks to lose coherence and finally crumble down into sand grains that are washed away by runoff. This has resulted in a ruiniform, castellated-tower landscape on top of the tepuis, similar to the karstic relief forms found on calcareous rocks. Weathering depressions are the favorite places where organic material accumulates to form peat.

The landscape of the igneous–metamorphic basement is a dissected mountain relief that comprises elongated massif ranges and groups of dome-shaped hills with highest elevations around 2,000–2,400 m a.s.l. (Fig. 4.7). Crest lines of the ranges and convex summits of the hills culminate often at fairly similar elevation levels giving the impression from distance of a dissected high-plateau landscape, a sort of "Gipfelflur." Igneous–metamorphic rocks extent over large areas, much larger than those of the siliceous tepui mesetas, upon exhumation of the shield basement from dismantled sedimentary Roraima cover. Valleys are deeply entrenched between ranges and hills. This landscape is mainly forest-covered, but individual hills have

Fig. 4.6 Sandstone–quartzite landscape (Cuao-Sipapo massif). (**a**) Shows relief variation within the plateau landscape with a high meseta in the background, a tectonic depression in front of the meseta escarpment to the right, and tectonically controlled benches in the center. (**b**) Shows a contact area between the Cuao-Sipapo highlands and the surrounding lowland peneplains including several outlier buttes (Cerro Autana to the left) that are remnants of an originally larger plateau, now dismantled by long-lasting dissection and erosion (photos Zinck)

often bare summits and backslopes with pockets of shrubs and meadows where peat has formed.

4.3.2 Karst and Pseudokarst Morphology

Karst-like landscape features and relief forms, similar in appearance to real karst morphology, have been reported to occur on noncalcareous rocks (i.e., sandstone, quartzite, granite, volcanic lava, among others) and documented in the Guayana Shield and elsewhere. As such forms were assumed for some time to not strictly result from the dissolution of the rock components except the inter-granular cement, they were conventionally termed pseudokarstic in contrast to the karst forms resulting from dissolution of carbonate bedrocks (Otvos 1976; Wray 1997a, b; Halliday 2007; Urban 2008; among others). However, there is evidence that

Fig. 4.7 Granite landscape. (**a**) Dissected mountain ranges (Maigualida massif). (**b**) Rolling hilland with hill summits at relatively similar elevations giving the impression of a sort of "Gipfelflur" at the horizon (Cuao-Sipapo massif) (photos Zinck)

dissolution affects not only the amorphous silica cement but also the skeletal quartz grains of the matrix (Chalcraft and Pye 1984; Wray 1997a, b; Doerr 1999). Although the term pseudokarst designating the formation of karst-like landscapes in noncalcareous rocks has been criticized (Otvos 1976; Jennings 1983; Twidale 1984; Urbani 1990; Doerr 1999; Doerr and Wray 2004; among others), it continues being widely used nowadays (Halliday 2007).

On the basis of the strong resemblance of the surface and underground tepui features caused by silica dissolution with those caused by carbonate dissolution, we have opted to call the former karstic instead of pseudokarstic although the rock weathering processes and dissolution rates are quite different. This is supported by several detailed studies on the weathering of quartzite and sandstone through inter-granular and granular quartz dissolution in the eastern Guayana Highlands (Chalcraft and Pye 1984; Doerr 1999; Piccini and Mecchia 2009; Lundberg et al. 2010; among others) and follows the suggestions forwarded by Doerr and Wray (2004). However, we call pseudokarstic the karst-like surface features developed on igneous–metamorphic rocks, as these features are more related to the hydrolysis of primary minerals than dissolutional weathering and are not connected to any underground drainage system.

4.3.2.1 Karst in Siliceous Rocks

A variety of small- to large-scale features, both on the terrain surface and beneath, has been described on sandstones and quartzites of the Roraima Group (Colvée 1973; Szczerban 1973; Szczerban and Urbani 1974; Szczerban et al. 1977; Pouyllau and Seurin 1985; Galán 1988; Briceño and Schubert 1990; Briceño et al. 1990; Clapperton 1993; Yanes and Briceño 1993; Yanes et al. 2006).

At micro-level, rock surfaces exposed to the action of rainwater, before it infiltrates underground, are incised by shallow flutes, runnels, clints, and grikes, resulting in a lapies-like (karren) micro-topography (Fig. 4.8). In other places, the

Fig. 4.8 Surface karst microforms in sandstone–quartzite (Cuao-Sipapo massif). Rock weathering results in the creation of lapies (**a**), anthropomorphic features (**b**), and runnel fields (**c**) (photos Zinck)

bare rock is spattered with small rounded-to-oval alveoli, pans, ponds, and kettles a few centimeters deep and a few centimeters to a few meters large. Medium-size features common on tepui summits include rounded to elongated, frequently ramified depressions, similar to karstic sinkholes (dolines), and poljes. They are in general less than two meters deep, several tens to hundreds of meters large, and connected with stream losses and resurgences. At macro-level, tower, tor, and pinnacle fields, simulating anthropo-, zoo-, and phytomorphic (mushroom-like) figures, occur in places were more bedrock has been removed than remains, especially along tepui rims (Fig. 4.9). Impressive vertical shafts (sima), with several hundreds of meters in diameter and depth, are connected with complex underground endokarst drainage systems, including galleries, caves, and caverns (Brewer-Carías and Steyermark 1976) (Figs. 4.10 and 4.11). Peat accumulates mainly in ponds, dolines, and poljes.

The origin and formation mechanisms of the karst-like features and forms do not make yet consensus among researchers (Galán 1988; Briceño et al. 1990). In sandstones, the cement uniting the sand grains is of variable nature (i.e., arkose,

Fig. 4.9 Surface karst forms in sandstone–quartzite (Cuao-Sipapo massif). Rock weathering results in the creation of mushroom-like tower fields (**a**) and zoomorphic features (**b**) (photos Zinck)

Fig. 4.10 Paleo-endokarst features in sandstone– quartzite (Cerro Autana, Cuao-Sipapo massif). (**a**) Shows several gallery mouths aligned at the same elevation on the upper part of the tepui wall, evidencing an endokarstic paleo-drainage system. (**b**) Is a close-up of some of the gallery mouths (right-hand segment) (photos Zinck)

iron oxides, clay, or mixtures thereof) and can be readily removed by physical–mechanical weathering processes such as moisture and/or heat variations (Fig. 4.12). In contrast, silica dissolution is needed to remove the siliceous cement of quartzite and quartz sandstone and to start dislocating the rock structure. Weathering of the siliceous cement through inter-granular dissolution of quartz, with liberation of sand grains, operates along weakness planes such as joints and bedding planes, and results in the formation of gallery systems and caves.

Briceño et al. (1990) discuss the issue of silica dissolution in the acid tepui environment, with rainwater pH of 4.3–5.1 and peat water pH of 3.5–3.9. They consider that acidity prevents the complexing of silicon by organic compounds and that silica weathering in sandstones is mainly controlled by inorganic solution. However, it is also recognized that the solubility of silica is largely pH-independent, although increasing rapidly above pH 9, while the presence of organic molecules greatly enhances the dissolution of silica (Wilding et al. 1977). Additionally, the tepui summits are a lixiviating environment because of the abundance of rainfall and the sustained water percolation through the rock substratum; surface water is

Fig. 4.11 Karstic features in sandstone–quartzite formations (Cuao-Sipapo massif). (**a**) Dissolution shaft cave (sima) connected to underground drainage system. (**b**) Enlarged karstic depression occupied by the Autana Lake (formerly called Leopold Lake) that collects blackwaters from the surrounding peatlands (photos Zinck)

continuously renewed so that silica saturation is possibly seldom reached. Thus, the tepui environment is not intrinsically unfavorable to silica dissolution. The occurrence of silica speleothems and crusts in caves (Urbani 1976; Galán 1988; Briceño et al. 1990; Lundberg et al. 2010), the corrosion of crystalline silica cement and quartz grains strongly interlocked by overgrowths as evidenced on thin sections (Chalcraft and Pye 1984; Doerr 1999), and the fairly high concentration of silica in highland cave streams (Piccini and Mecchia 2009) and lowland rivers (Mora-Polanco et al. 2007) are indicators of silica dissolution in the tepui environment.

Silica dissolution and precipitation consume time. Uranium–thorium dating of silica biospeleothems found in a cave of the Chimantá tepui, southeast Venezuela, shows that silica deposition from sandstone dissolution proceeds very slowly. Dates determined on a 12-cm-thick speleothem with concentric stratigraphic layers varied from 390 $\pm$ 33 ka (MIS 11) at 21 mm from the edge to 298 $\pm$ 6 ka (MIS 9) at 4 mm from the edge, with the fastest growth rate being 0.37 $\pm$ 0.23 mm ka^{-1} in MIS 9

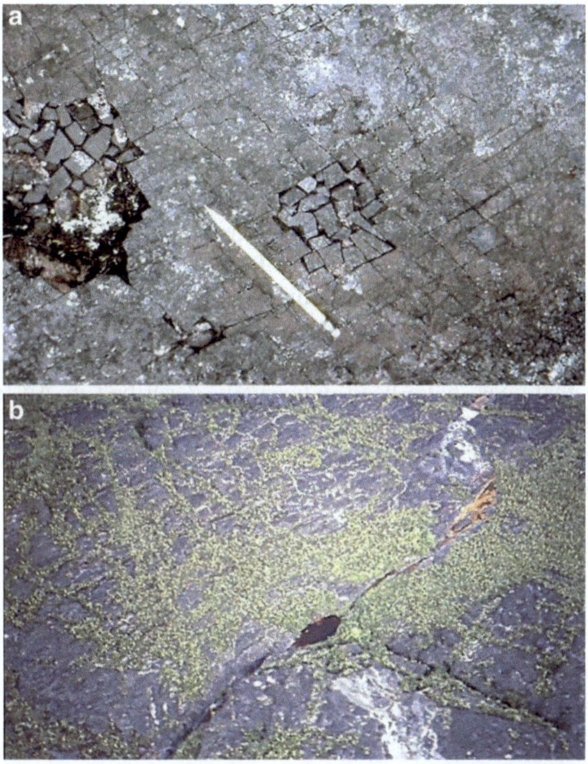

Fig. 4.12 Initial weathering and colonization stages on sandstone (Cuao-Sipapo massif). (**a**) Sandstone dislocation along an orthogonal joint frame; initialization of biotic weathering by lichen colonies. (**b**) Colonization of the widened joints by pioneer plants (photos Zinck)

(Lundberg et al. 2010). Similarly, Doerr (1999) estimates that it would take 112 ka to remove the equivalent of 1 m rock from a quartzite tepui surface, assuming an average annual rainfall of 5,000 mm and 5 mg of silica dissolved per liter. In addition to time, silica dissolution needs a set of favorable geologic, climatic, hydrologic, and topographic conditions that are met in the Guayana Highlands. The region is tectonically stable since the end of the Mesozoic, thus over a time span of about 70 Ma (Briceño and Schubert 1990). Bedding is horizontal to slightly dipping over large areas. Sandstones are fairly porous, quartzites are not, but both have plenty of joints, fissures, and fractures that allow water to penetrate and percolate. Rainfall is abundant presumably since the beginning of the Holocene and probably was also abundant during the Quaternary interglacial periods. Upon tectonic uplift, the area has been intensively dissected so that an important hydraulic gradient exists between the high-elevation tepui summits and the surrounding lowlands. Altogether, this context favors and provides continuity to deep water percolation and the exportation of dissolved and suspended substances.

4.3.2.2 Pseudokarst in Igneous–Metamorphic Rocks

Karst-like microforms and mesoforms are known to occur also in igneous–metamorphic rocks in the Guayana Highlands (Urbani and Szczerban 1975; Blancaneaux and Pouyllau 1977; Pouyllau 1989) and elsewhere (Porto-Domingues 1952; Twidale and Corbin 1963; Feininger 1969; Tricart 1969; Khobzi 1972; Twidale 1982; Watson and Pye 1985; Bremer and Sander 2000; Twidale and Vidal Romani 2005; Vidal Romani and Vaqueiro Rodriguez 2007). They are, however, much less spectacular and much less diversified than the karst features on quartzite and sandstone. Granite domes show pockets and depressions in the summit-shoulder areas, flutes and grooves incising the steep backslopes, and runnels and crevices on the footslopes (Fig. 4.13). Depressions on crystalline rocks, termed gnammas, have been extensively described in Australia (Twidale and Corbin 1963).

Granite and gneiss with contrasting grain sizes and colors are favorable to differential rock weathering and disintegration of minerals. At the origin, small

Fig. 4.13 Pseudokarstic features in igneous– metamorphic rocks (Cuao-Sipapo massif). (**a**) Cavernous weathering pits and pockets on the flank of a granitic dome where peat accumulates and vegetation develops. (**b**) Runnels grooving a bare granite slope (photos Zinck)

irregularities on the backbone surface of the granitic domes allow water to accumulate and stay there before drying up after a few hours to a couple of days according to the length of the rainfall event. These incipient pools, exposed to repeated cycles of moistening-drying, are places where specialized poikilohydric, desiccation-tolerant plants develop (Kluge and Brulfert 2000). The alternation of rock moistening and drying is probably the starting process of crystalline rock disintegration, especially of the micas that are very sensitive to moisture variations. The concave rock surface around these primary depressions is usually free of cryptogamic crust (lichens, cyanobacteria) due to rainwater runoff. This allows thermoclastism, based on the differential heating between dark- and light-colored minerals, to play an additional role in mechanical rock weathering, especially in the upper highland zone where the diurnal thermic amplitude is appreciable. The loosening of the mineral assemblage causes the rock structure to dislocate. This favors the alteration of the mafic minerals via hydrolysis, leaving on the terrain a felsic residuum of quartz grains and especially feldspar phenocrysts from the rapakivi granite. Part of this coarse-sandy and fine-gravelly clastic material is washed away by rainwater or flushed out by pond overflow during heavy storms. The depressions resulting from differential crystalline rock weathering are usually small and have regular configuration, reflecting the coherent, massive granitic substratum, while depressions on sandstone and quartzite are rather elongated and frequently ramified responding to the underlying joint framework. They are colonized by pioneer vegetation, followed by higher plants, with accumulation of organic material in the water-filled ponds (Gröger 2000; Gröger and Huber 2007) (Fig. 4.14).

Fig. 4.14 Initial weathering and colonization stages on granite (Cuao-Sipapo massif). Rounded dissolution alveoli in massif rock (**a**) and elongated alveoli along fracture lines (**b**); colonization of alveoli and runnels between alveoli by pioneer plants (**c**) (photos Zinck)

Arenization takes place not only on the terrain surface (Galán 1988) but also under the peat mantle along joints acting as virtual fissures. Small strata (2–5 cm) of coarse sand and fine gravel are frequently found along the interface between the base of the peat deposit and the underlying rock surface. Underneath the peat cover, irregular karren micro-topography develops, with enlargement of the joints favoring water percolation. Underground, some chemical and physical weathering probably takes place along discontinuities, joints, and fractures, as porosity in igneous rocks in usually low, but no connected subsurface drainage system was observed.

4.3.3 Origin of the Micro-relief

Peat forms in low-lying, closed or nearly closed landforms where water stagnates and organic material accumulates faster as it decomposes. Pre-existing micro-topographic irregularities, including concave-shaped areas, are thus indispensable to the initiation and development of peat bogs. There is a strong relationship between surface (pseudo-)karst relief and peat formation.

Micro-topography under the peat mantles is very irregular, with sharp highs and narrow troughs, as shown by transects across selected peatlands (see Sect. 4.5.1). Similar rugged micro-relief was also observed on freshly exposed rock surfaces after sliding of the peat cover. In contrast, on long-time exposed rock surfaces, the micro-relief is in general less deeply incised and less intensively carved. Sub-aerial rock weathering causes less irregular micro-topography than sub-peat rock weathering, and attenuates the sharpness of the micro-topography initially formed under peat. Pouyllau and Seurin (1985) describe exposed lapies areas with incisions not more than 10–20 cm deep in the Gran Sabana. Even in ruiniform summit areas, the terrain surface between towers and tors is less irregular than the crypto-topography we have found in peat bogs. It seems thus that the micro-relief at bog bottom partly forms and changes beneath the peat through differential rock weathering. In sandstones and quartzites, weathering proceeds along joints and fractures, deepening the incisions and enhancing the original micro-relief; this results into an intricate crypto-lapies (crypto-karren) with rectangular or trellis pattern. In granites, the micro-relief underneath the peat is smoother, slightly undulating, where rock weathering is less controlled by joints and more by rock mineralogy. Thus active micro-morphogenesis takes place under the peat mantle, possibly because of acid organic matter and acid peat water. At the peat–rock interface, pH values as low as 3.2 were measured. This supports the idea that peatlands act as strong agents of weathering in and underneath them (Bennett et al. 1991; Le Roux and Shotyk 2006).

Few data on the solutional weathering rate of quartzite in tepui environment are available so far. Studying the Aonda cave system on Auyantepui in the eastern Guayana Highlands, Piccini and Mecchia (2009) determined SiO_2 concentrations of 0.19 mg l^{-1} at the sink point and 0.27 mg l^{-1} close to the resurgence of the cave system. Thus part of the exit solute load is provided by rock weathering at the terrain surface. Exposed rock outcrops are usually epigenized by hard silica crust which prevents further weathering. As a consequence, active rock disintegration takes place mainly in poorly drained (pseudo-)karstic depressions. Piccini and Mecchia (2009) report SiO_2 concentrations as high as 0.19 mg l^{-1} in peat water and as low as 0.02 mg l^{-1} in solution pan and pond waters. The presence of organic molecules increases quartz solubility, and quartz–sand grains collected from depressions on Mt. Roraima summit show surface etch pits (Chalcraft and Pye 1984). This suggests that the tepui places most favorable to dissolutional rock weathering are the peat bogs, specifically the interface between the bottom of the organic deposit and the underlying quartzite or quartzose sandstone. This also explains why under-peat micro-topography is much more irregular than that of the terrain surface.

Organic matter under permanent water table seems to accelerate rock weathering and arenization, resulting in the occurrence of sand strata under the peat cover. The weathering debris is partly washed down along joints and fractures, contributing to the formation of whitesand areas in the lowlands. Rock disintegration under peat, with differential intensity along joints and other fissures, accentuates the micro-relief. In contrast to these observations, Briceño et al. (1990), working on several tepuis in the eastern Guayana Highlands, did not notice any difference in micro-topography between rocky riverbeds and adjacent riverbanks covered by peat, and concluded that organic matter did not play a role in rock weathering.

4.4 Peatland Types

The summits of the 50 major tepui massifs in the Guayana Highlands, from Mt. Neblina in the southwest to Mt. Roraima in the northeast, are covered in about 70% by a variety of vegetation types including forests, scrublands, meadows, and grasslands, with the remaining surface area being bare rocky terrains (25%) and water bodies (5%) (O. Huber personal communication). Bare summits are particularly frequent on small mesetas and on the easternmost tepuis. Some large tepuis such as the Duida massif are forested in 80% of their extent. In general, peatlands cover about 30% of tepui summits, making up ca 1,500 km^2 for the whole Guayana Highlands (Huber 1995).

Large parts of the tepui summits are bare sandstone or quartzite outcrops colonized by algae, fungi, and lichens. A variety of small tree and shrub formations occurs in slope crevices, block heaps, relief niches, and rock fissures where shallow skeletal soils have formed. Herbaceous formations, including meadows and grasslands, colonize in general lower positions in the landscape where water concentrates and organic material accumulates, leading to peat formation (Huber 1992; see also Huber and Garcia 2011).

Topography plays a fundamental role by creating conditions favorable to water concentration, vegetation development, and accumulation of organic materials. There is a variety of landscape positions where peat occurs: peat in depressions, peat on slopes, peat in small valleys, peat in narrow tectonic crevices and fissures, and peat in narrow floodplains (Figs. 4.15 and 4.16).

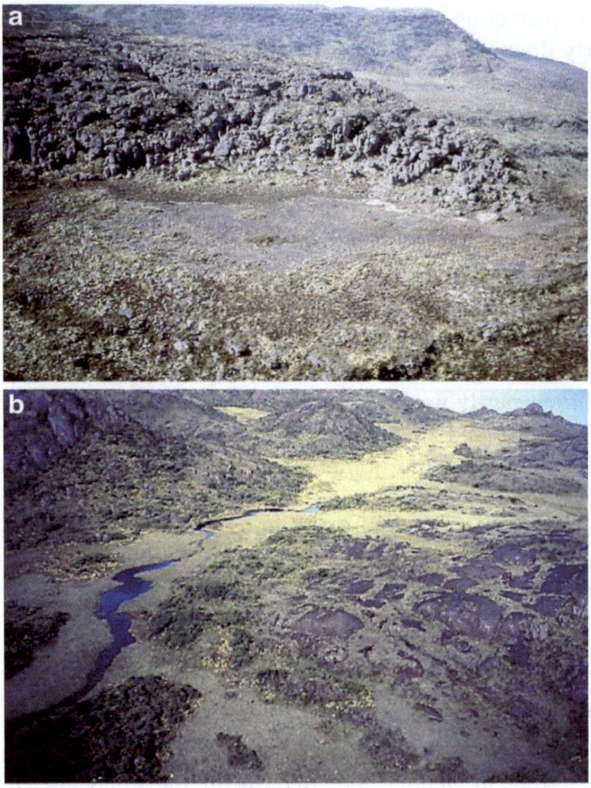

Fig. 4.15 Peatland types. (**a**) Peat in a doline-like karstic depression on sandstone (Cuao-Sipapo massif). (**b**) Peat in a polje-like karstic depression on granite with blackwater outlet (Maigualida massif) (photos Zinck)

4.4.1 Depression Peats (Bog Peats)

Typically, peat occurs in depressions of (pseudo-)karstic origin, surrounded by rims of rock outcrop. The process of organic matter accumulation starts in small pans and ponds colonized by pioneer vegetation (algae, lichens, and cyanobacteria). This initial geogenic accumulation of organic materials in temporarily waterlogged conditions corresponds to peat inception via paludification (Andriesse 1988). As the depressions enlarge and deepen through rock weathering along joints and fractures, higher plants including herbs, forbs, shrubs, and small trees, develop. Peat deposits rest usually directly on the rock substratum. Quaking bogs with floating vegetation carpets were not found. Thus, peat formation at tepui summits responds to the peat-forming-upwards mechanism (i.e., paludification), also called kettle-hole-mire mechanism (Gaudig et al. 2006), that better explains rapid peat accumulation than the commonly assumed peat-forming-downwards mechanism (i.e., terrestrialization).

Fig. 4.16 Peatland types. (**a**) Valley peatland on granite (Maigualida massif). (**b**) Slope peatland on sandstone with tubiform meadow cover (*Brocchinia, Stegolepis*) (Cuao-Sipapo massif) (photos Zinck)

The surface of the peat mantle is flat, sometimes with a few small mounds, but it is not always horizontal. Slopes of 2–5% were observed in bogs that merge at the depression rims with surrounding glacis peats. In general, depressions are elongated when controlled by fractures, or they have more ramified configuration or trellis pattern when sitting on a more or less orthogonal grid of joints. Walking on the peat mantle is usually uneasy as the water level is at the surface or close to it. The general crypto-relief under peat deposits is concave, but the micro-topography is very irregular. Depth to the bedrock was measured at a number of sites along detailed cross-sections that revealed the alternation of crests and troughs as a reflection of the underlying framework of joints and micro-fractures. Maximum depth recorded under the surface of the peat mantle was around 2 m. On average, depressions are about 100 m long and 40–50 m wide. Maximum size recorded was 325 m $\times$ 100 m. In general, the length–width ratio is 3:1.

Sandstone–quartzite disintegration takes place under the peat mantle, favored by permanent water table and possibly the presence of aggressive organic acids. Rock weathering produces in general a coarse-sand residue, from a few millimeters to a few centimeters thick, lying on top of the substratum. A large part of this weathering product is probably washed out through joints and fractures and exported to the lowlands. Conspicuous iron reduction features were not found at any of the places visited, meaning that water saturation does not generate permanent anaerobic conditions at the bottom of the depressions. Peat hydrology is controlled by rock joints and fractures that drain water to underground pipes and to small creeks and cascades along the walls of the tepuis, contributing to feed blackwater rivers in the lowlands. Thus, the water table in peat bogs seems to be regularly renewed, especially during the rainy season. This would mean that two of the remarkable features of the surrounding lowland plains, i.e., whitesand and blackwater, originate on and come at least partly from the top of the tepuis. In this sense, tepui peat mantles play an important role in lowland hydrology, collecting and storing rainwater in small retardation basins (i.e., the peat sites) and releasing it at slow rate to the drainage system.

Bog peat occurs not only on sandstone mesetas of the Roraima Group but also on granite and gneiss domes of the Parguaza granite basement. On granite, the configuration and crypto-topography of the depressions are less irregular than on sandstone: depressions are rounded or oval bowls with undulating concave bottom. Weathering is more controlled by rock mineralogy than by rock structure.

4.4.2 Slope Peats (Glacis Peats, Blanket Peats)

Less frequent than depression peats are peat mantles on slopes that are dammed downward by outcropping rock stubs forming barriers. Peat starts accumulating in the gutter where two opposite slopes meet. Peat thickening in such places conduces to upslope plant colonization and stepwise aggradation of organic materials. Downslope water circulation within the peat mantle and seepage along the contact plane between the peat base and the rock substratum concentrate water in the gutter that leads to the initiation of small creeks. During exceptionally rainy periods, high water levels cause the creek channel to shift laterally and, by doing so, to cut a scarp into the lower edge of the adjoining slope peat. This process results in the formation of a hanging peat mantle that is unstable and frequently slides downslope following shear failure at the peat–mineral interface. Differential sliding speed contributes to breaking down the peat cover into fragments, showing peat scars and patches of freshly exposed rock surface (see Fig. 7.5). Thus in slope positions, peat formation may not be constant as it is interrupted by events of erosion (edge cutting, differential sliding), followed by periods of reconstitution.

4.4.3 Narrow-Valley Peats

Peat also accumulates in narrow tectonic grabens and blind valleys, and along tectonic crevices and fissures that have been enlarged through sandstone weathering.

Such narrow, elongated canyons and valleys are generally drained by small creeks that run on bare rock surface without fixed channel. On the banks of the creeks, peat profiles frequently include strata of mineral material, mainly sand but sometimes also finer grained mineral sediments.

4.5 Peatland Patterns

4.5.1 Spatial Variability

Peat cover on tepui summits shows essentially a discontinuous mosaic pattern. Peat occurs in discrete landscape units surrounded and fragmented by rock outcrops. To document the variability of peat thickness in individual units and that of the micro-topography of the rock surface buried under the peat mantle, the depth to the rock substratum was measured along several transects using a marked wooden stick. Longitudinal transects were located along the main axis of the bogs and were crossed by one or more transversal transects (Figs. 4.17–4.21).

In general, bog and valley peats have elliptic configuration, while slope peats are rather rectangular. Rounded peatland units are usually doline-related (Fig. 4.22). Length varies from a few meters to several hundreds of meters. From the five peatlands that have been measured, average length is 150–300 m and average width is 50–100 m. The extent of the bogs varies from a few square meters in the case of small alveoli to several hectares in the case of polje-like depressions. The surface area of the larger bogs is about 0.5–3 ha.

In the case of bog peats, surface topography is concave, with 1–2% slope from both margins toward the axis of the depression. The longitudinal slope of the peat surface varies from 1 to 5%. Usually, a small creek runs on bare rock approximately in the center of the depression (Fig. 4.23). Isolated rock outcrops are frequent across the peat surface. In the case of slope (glacis) peats, the inclination of the peat surface decreases from 8–16% in the headwater alveoli, to 3–5% in the central part, and 0.5–1% in the lower part of the cross-section. Bogs and the distal parts of the glacis are water-saturated during the largest part of the year, with patches of stagnating water at the peat surface.

Peat thickness varies from a few centimeters to several meters. In our exploration area, maximum depth measured was about 2 m. At the five sites where full transects were recorded from margin to margin, maximum peat depth was 40, 95, 120, 170, and 200 cm, respectively.

Size and depth of the peat units formed on sandstone and granite, respectively, are not clearly different. However, the topography of the rock surface buried underneath the peat mantle is significantly different. The sandstone profile under peat is usually very irregular, with indented micro-topography of sharp highs (peaks) alternating with narrow troughs (crevices) that reflect preferential rock weathering along a dense grid of joints and fractures. In contrast, the granite profile

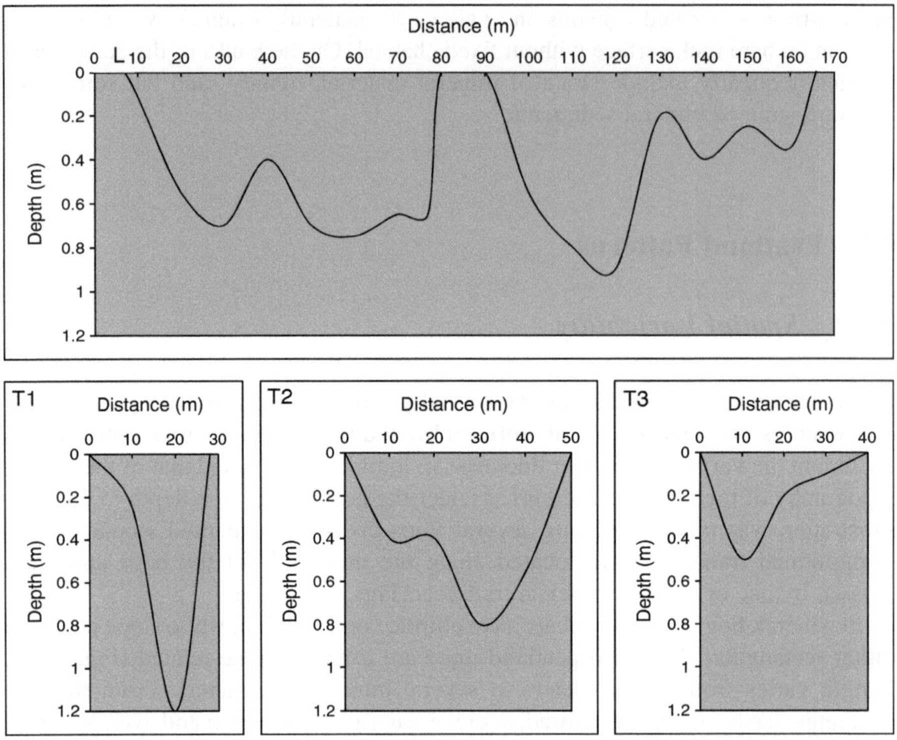

Fig. 4.17 Peatland transect at site Cuao 3 in the northwestern sector of the Cuao-Sipapo massif (N 05°04′, W 67°24′; 1,600 m a.s.l.). Transect T1 crosses the longitudinal section L at about 30 m from the origin, transect T2 at about 60 m, and transect T3 at about 140 m. The landscape is a highland plateau (600–1,800 m a.s.l.), developed on horizontally lying to slightly folded or tilted sandstone and quartzite formations of the Roraima Group (Precambrian), with areas of hilly dome-like relief corresponding to igneous–metamorphic basement rocks of the Guayana Shield. General topography is gently undulating (3–5% slope), with rugged ruiniform rock outcrops and deep crevices. The site corresponds to a narrow, elongated karstic depression (165 × 50 m) at the top of a sandstone tepui meseta, water-saturated during the largest part of the year, with patches of stagnating water at the peat surface and free water table at 40 cm depth. The peat area is covered by tubiform high-tepui meadow (*Brocchinia* sp. dominant) with shrubs. Peat surface topography is concave with 1–2% slope from both margins toward a small creek that runs on bare rock approximately in the center of the depression (rock outcrop at 80–90 m from the origin). Isolated rock outcrops are frequent throughout the peat surface. Crypto-topography at the bottom of the bog is very irregular. Sharp highs alternate with narrow troughs that reflect preferential rock weathering along the dense joint framework of the sandstone substratum. Loose sand accumulation (5–10 cm thick), resulting from the weathering of the sandstone, occurs at the bottom of some troughs. Thickness of the peat mantle varies at short distance with a maximum of 120 cm. Peat thickness was measured at 10 m intervals, with 5 m intervals in the proximity of rock outcrops. Profile 9 is located at about 110 m from the origin of the longitudinal cross-section: a 115-cm-deep Lithic Terric Haplofibrist (USDA 2006) or Rheic Fibric Histosol (Hyperdystric) (WRB 2006). Radiocarbon age is: ^{14}C 165 BP at 40–60 cm depth

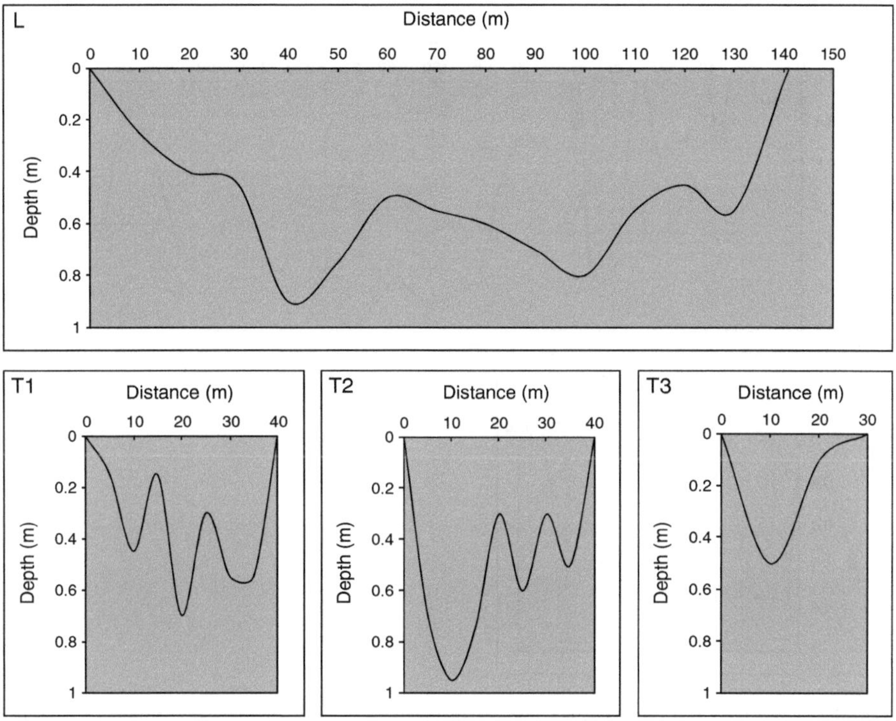

Fig. 4.18 Peatland transect at site Cuao 7 in the central area of the Cuao-Sipapo massif, northwest of the Autana river (N 04°55′, W 67°11′; 1,370 m a.s.l.). Transect T1 crosses the longitudinal section L at about 45 m from the origin, transect T2 at about 65 m, and transect T3 at about 120 m. The landscape is a deeply incised hilland with rocky domes developed from granodiorite containing feldspar phenocrysts (rapakivi texture). The site corresponds to a narrow, elongated pseudokarstic depression (140 × 40 m), water-saturated during the largest part of the year, with patches of stagnating water at the peat surface, located at the summit of a bare convexe dome, steeply dissected (>100% slope). The peat area is covered by shrubby broadleaved meadow. The longitudinal topographic profile buried under the peat mantle is slightly undulating, but the transversal profiles show sharp irregularities at short distance. The longitudinal slope of the peat surface is 3–5%, while the transversal slope is about 1%. Peat thickness was measured at 10 m intervals on the longitudinal cross-section and 5 m intervals on the transversal ones, and reaches a maximum of 95 cm. Profile 12 is located at the intersection of T2 with the longitudinal cross-section, at 65 m from the origin. It is a 60-cm-deep Lithic Haplosaprist (USDA 2006) or Endoleptic Ombric Hemic Histosol (Hyperdystric) (WRB 2006). Radiocarbon age is: ^{14}C 850 BP at 15–40 cm depth and ^{14}C 3,110 BP at 40–60 cm depth

under peat is smoother, slightly undulating, where rock weathering is controlled more by the rock mineralogy and rapakivi texture than by structural joints (Fig. 4.24). In general, the depressions have abrupt boundaries, with very steep to vertical side (border) slopes, frequently 60–100 cm high. This highlights that rock weathering under the peat mantle is more intensive than on the exposed terrain surface.

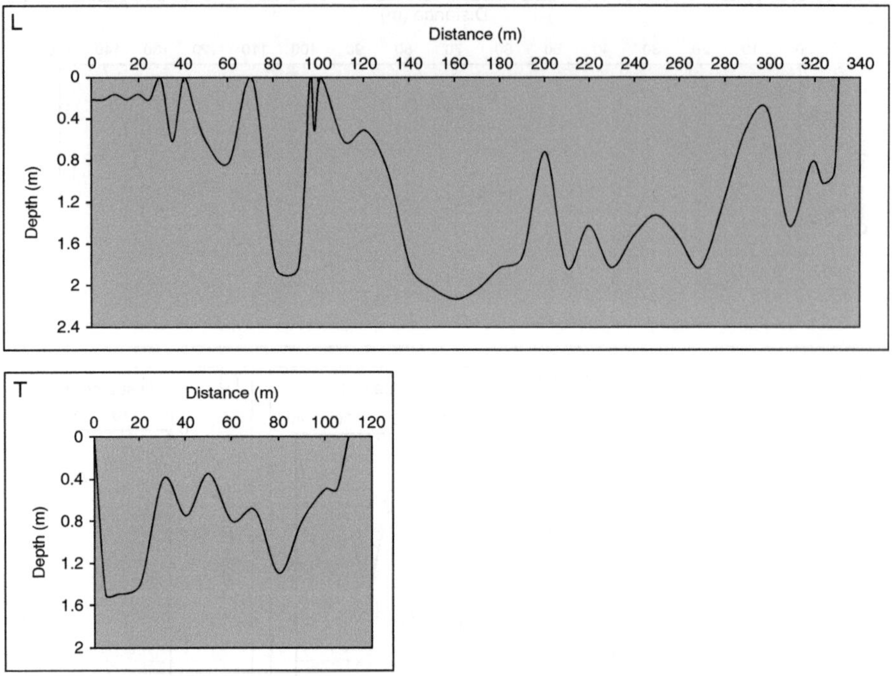

Fig. 4.19 Peatland transect at site Cuao 12 in the center-south of the Cuao-Sipapo massif (N 04°54′, W 67°21′; 1,500 m a.s.l.). The landscape is a highland plateau (700–2,000 m a.s.l.) developed on horizontally lying to slightly folded or tilted sandstone and quartzite formations of the Roraima Group (Precambrian). General topography is gently undulating, with rugged ruiniform rock outcrops and deep crevices. Longitudinal (L) and transversal (T) peat sections cross an elongated (330 × 110 m), totally closed karstic depression at the top of a sandstone tepui meseta, water-saturated during the largest part of the year, with patches of stagnating water at the peat surface and free groundwater level from 30 cm downwards. The peat area is covered by tubiform tepui meadow with shrubs. Peat surface topography is concave with 1–2% slope from both margins toward a small creek running approximately in the center of the depression. Isolated rock outcrops are frequent throughout the peat surface. The depression is dominated by rocky slopes (3–5% slope along the rims of the depression and 18–20% slope above), with many rock fragments on the terrain surface and local peat patches in small alveoli between rock outcrops. Crypto-topography at the bottom of the bog is very irregular. Sharp highs alternate with narrow troughs that reflect preferential rock weathering along the dense joint framework of the sandstone substratum. Thickness of the peat mantle varies at short distance with a maximum of 2 m. Peat thickness was measured at 10 m intervals, with 5 m intervals in the proximity of rock outcrops. Profile 17 is located approximately in the middle of the longitudinal cross-section: a 150-cm-deep Typic Haplosaprist (USDA 2006) or Ombric Hemic Histosol (Hyperdystric, Thaptosapric) (WRB 2006). Radiocarbon age is: ^{14}C 2,990 BP at 40–80 cm depth and ^{14}C 7,030 BP at 80–100 cm depth

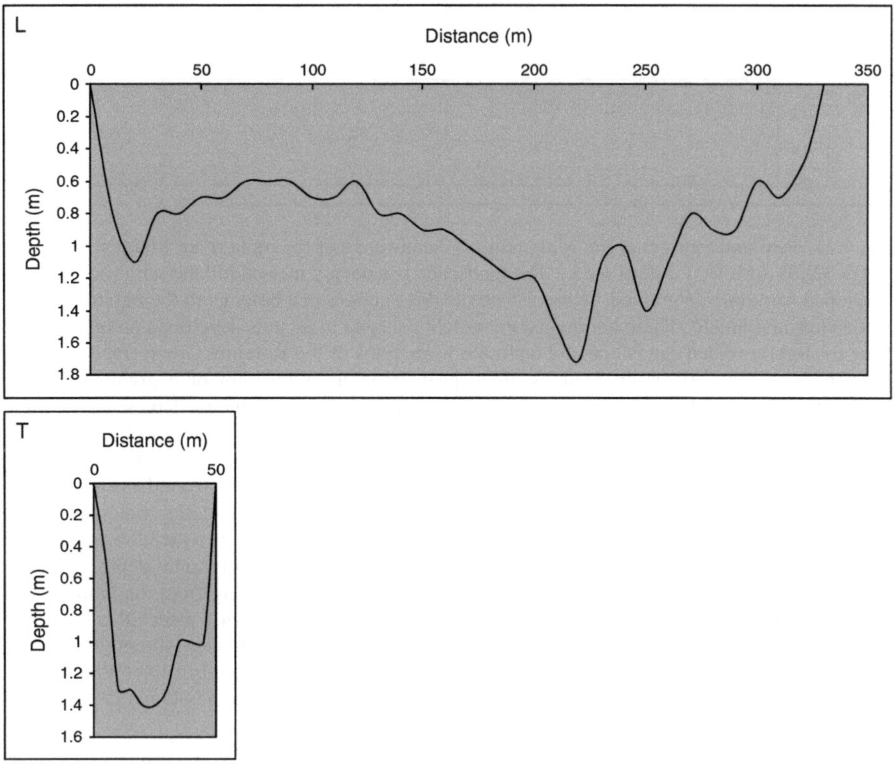

Fig. 4.20 Peatland transect at site Maigualida 1, located in Cerro Yudi, Maigualida massif (N 05°34′, W 65°13′; 2,150 m a.s.l.). Transect T crosses the longitudinal section L at about 220 m from the origin. The landscape is a deeply incised hilland with rocky domes developed from coarse-grained, feldspar-rich rapakivi granite that belongs to the basement rocks of the Guayana Shield. There are inclusions of table-shaped highlands developed on horizontally lying to slightly folded sandstone and quartzite formations of the Roraima Group (Precambrian). The site corresponds to a narrow, elongated pseudokarstic depression (325 × 60 m), water-saturated during the largest part of the year, with patches of stagnating water at the peat surface, receiving smaller lateral vales and being flanked by granite outcrops with 20–30% slope. At both extremities, the bog narrows and includes rock outcrops and boulders. The peat area is covered by high-tepui grassland with short grasses, forbs and isolated shrubs; pioneer vegetation colonizes joints and cracks in granite. The longitudinal topographic profile (L) buried under the peat mantle is asymmetrical, with a smooth convex bottom terrain in the first half from the origin and several crevices in the second half. The buried transversal profile (T) shows the concave trough-like bottom of the vale. The border slopes of the depression are sharp and steep. The longitudinal slope of the peat surface is 3–5%, while the transversal slope is 0.5–1%. Peat thickness was measured at 10 m intervals on the longitudinal cross-section and 5 m intervals on the transversal one. Peat reaches a maximum depth of 170 cm. Profile 19 is located close to the intersection between the longitudinal and transversal cross-sections, at about 220 m from the origin. It is a 130-cm-deep Typic Haplosaprist (USDA 2006) or Ombric Sapric Histosol (Hyperdystric) (WRB 2006). Radiocarbon age is: ^{14}C 5,190 BP at 60–80 cm depth and ^{14}C 6,580 BP at 110–130 cm depth

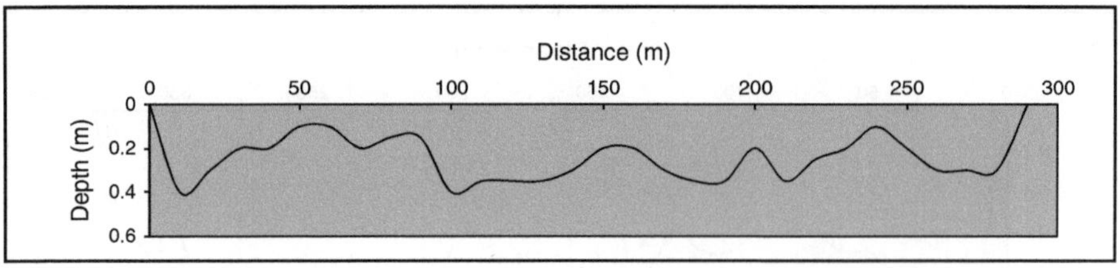

Fig. 4.21 Peatland transect at site Maigualida 2, headwaters of the Asita river, Maigualida massif (N 05°32′, W 65°09′; 1,700 m a.s.l.). The landscape is a deeply incised hilland with rocky domes developed from coarse-grained, feldspar-rich rapakivi granite that belongs to the basement rocks of the Guayana Shield. There are inclusions of table-shaped highlands developed on horizontally lying to slightly folded sandstone and quartzite formations of the Roraima Group (Precambrian). The site is a rocky glacis covered by slope peat, hanging at the foot of a granite hill. The topographic profile buried under the shallow peat mantle is an undulating lapies-like (karren) micro-relief, with steep border slopes. The slope of the peat surface is 8–16% in the headwater alveolus, 5–8% in the upper part of the cross-section, 3–5% in the central part, and 0.5–1% in the lower part. The peat is moist in the proximal part of the glacis (until 120 m from the origin), while it is water-saturated during the largest part of the year, with patches of stagnating water at the peat surface, in the distal part. The peat area is covered by high-tepui broadleaved meadow mixed with shrubs. Peat thickness was measured at 10 m intervals and reaches a maximum depth of 40 cm. A 5–10 cm thick layer of coarse sand and fine gravel (1–5 mm) resulting from the granite weathering lies between the bottom of the peat cover and the rock substratum. Profile 20 is located in the distal part of the peat glacis at about 210 m from the origin. It is a 40-cm-deep Lithic Haplosaprist (USDA 2006) or Epileptic Rheic Sapric Histosol (Hyperdystric) (WRB 2006). Radiocarbon age is: ^{14}C 2,530 BP at 20–35 cm depth and ^{14}C 3,610 BP at 35–45 cm depth

4.5.2 Stratification of the Peat Mantles

Peat mantles show elementary stratification in layers according to the degree of decomposition of the organic material (see Figs. 6.1–6.4). Strongly decomposed sapric material is less frequent than other types of organic material. It was identified in seven of the 15 peat sites and occurs generally in the lower layers. The most common material is slightly to moderately decomposed fibric and hemic peat. Hemic material is the most frequent among all visited sites, and this could be explained in different ways. (1) Environmental factors, such as long-standing water saturation, nearly anaerobic conditions and/or strong acidity, could make it difficult to reach the stage of sapric material by preventing or limiting the activity of micro-organisms. (2) Sapric material is composed of fine-grained organic substances (i.e., humic and fulvic acids) that go easily in suspension in percolating water, constantly renewed by the addition of rainwater. Thus, sapric material could have been exported within the blackwaters that percolate through joints and fractures to underground water circulation systems or via the surface flow of the creeks that drain the peat bogs. In both cases, the rainwater collected at the tepui summits ends up feeding the lowland blackwater rivers. In this way, peat mantles

Fig. 4.22 Peatland patterns. Tepui summit showing several rounded peat-filled depressions clearly aligned in response to structural control in the sandstone substratum and in connection with an endokarstic drainage system. Several gallery mouths can be observed beneath the dolines at variable elevations on the tepui wall, often evidenced by vegetation clumps growing along the wall (photo Zinck)

would have been depleted preferentially in sapric components, especially at the bottom of the bogs. These strongly decomposed components can be assumed to be among the oldest ones, and their depletion might cause the peat ages to be younger than they would be in a closed system.

In most tropical peats, surface layers are more decomposed than the underlying ones (Driessen and Rochimah 1976). In contrast to this general trend, the upper peat layers in tepui environment are less decomposed than the deeper ones. Peat inception takes place in karstic pans and ponds through paludification. Initially, these small depressions are only seasonally flooded, mainly by surface runoff (topogenous-rheophilous peat), allowing the organic debris to decompose relatively fast to reach the stage of advanced hemic and, in some cases, sapric material. As the peat cover grows vertically and horizontally, more rain- and runoff water accumulates and stays longer in the enlarged depressions, delaying the decomposition of the organic materials (ombrogenous peat). Simultaneously, vegetation

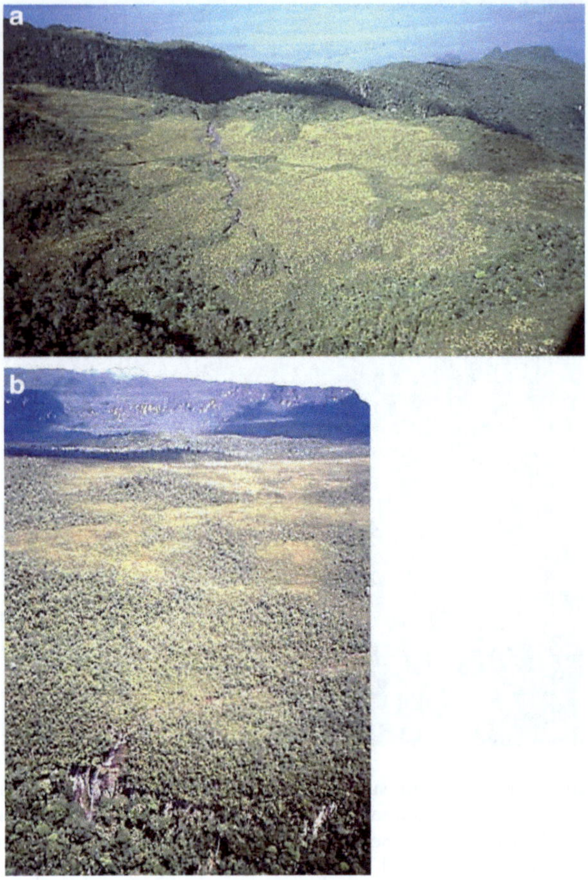

Fig. 4.23 Peatland patterns. (**a**) Large homogeneous peatland with tubiform meadow vegetation (yellow dots are *Brocchinia* plants) that has developed on a slightly concave tepui summit drained by a creek. (**b**) Mosaic of peat depressions with shrubby meadow cover and slightly higher areas, often with rock outcrops, covered by woodland (photos Zinck)

becomes shrubbier on aging peats, with a lot of coarse fibers that decompose more slowly than the initial forb–herb cover.

Only in two cases (Profile 2 and Profile 17, see Appendix) does the layer sequence go from fibric (Oi) to hemic (Oe) to sapric (Oa) with increasing depth, as it can be expected on the basis of the relationship between depth, age, and degree of decomposition of the organic material. In general, the profiles are composed of a variety of fibric and hemic layers. The most common layer sequences are Oi–Oe and Oe–Oa. Although Oi layers overlay usually Oe layers, some sequences include deeper fibric layers that are less decomposed than the overlying ones. In the majority of profiles, peat rests directly on the substratum, whether sandstone–quartzite or granite. In some cases, however, thin strata of coarse sand and fine gravel, resulting from the weathering of the bedrock, underlie the peat cover.

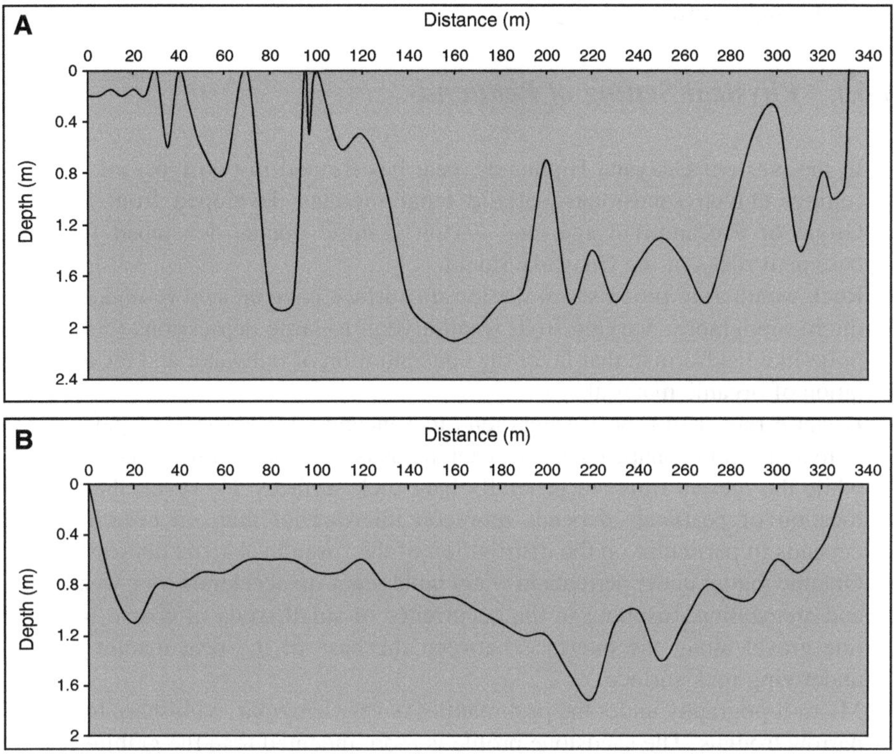

Fig. 4.24 Peatland transects at sites Cuao 12 and Maigualida 1. Comparison of under-peat crypto-topography in polje-like karstic depressions. Transect A (Cuao 12) is on sandstone, while transect B (Maigualida 1) is on granodiorite. In general, the depressions have abrupt boundaries, with very steep to vertical side (border) slopes, frequently 60–100 cm high. The sandstone profile under the peat mantle is very irregular, with indented micro-topography of sharp highs (peaks) alternating with narrow troughs (crevices) that reflect preferential rock weathering along a dense grid of joints and fractures. In contrast, the granodiorite profile under peat is smoother, slightly undulating, where rock weathering is controlled more by rock mineralogy than by structural joints

In a few sites, medium-to-fine grained mineral material, reworked and transported by local creeks, was found underneath the organic material or burying it. The thickness of the profiles above the substratum, including organic and mineral materials, varies from 45 to 150 cm. There are no significant differences between the three study areas in terms of profile depth and layer sequence.

The density and structure of the peat is strongly influenced by the nature of the vegetation cover providing the organic material. Under forb and herb cover, peat is denser and more decomposed than under shrub vegetation (e.g., *Bonnetia* forest).

4.6 Conclusions

4.6.1 Physical Setting of Peatlands

- In the western Guayana Highlands, peat has formed in two types of physical setting: (1) on sandstone–quartzite tepui mesetas developed from Roraima Group of Precambrian age, and (2) on granitic domes developed from the basement rocks of the Guayana Shield.
- Rock weathering processes at the terrain surface have created (pseudo-)karstic micro-topography, varying from small alveoli to large depressions (doline and polje-like landforms), that favor the concentration of rainwater and the accumulation of organic material.
- Because rain- and runoff waters tend to concentrate in the central parts of the individual spoon-shaped mesetas, peat deposits are more frequent in these areas, while the meseta rims are generally bare rock surfaces. However, the specific location of peatlands depends more on micro-relief than on meso-relief; it depends in particular on the distribution of the (pseudo-)karstic depressions.
- Organic matter under permanent water table seems to accelerate rock weathering and arenization, resulting in the occurrence of small strata of coarse sand and fine gravel along the interface between the base of the peat mantle and the underlying rock surface.
- Micro-topography under the peat mantles is very irregular, with sharp highs and narrow troughs. The sandstone profile is more indented than the granite profile.

4.6.2 Peatland Types and Patterns

- Peat cover on tepui summits shows essentially a discontinuous mosaic pattern. Peat occurs in discrete landscape units surrounded and fragmented by rock outcrops.
- There is a variety of landscape positions where peat occurs: in depressions, on slopes, in small valleys, in narrow tectonic crevices and fissures, and on narrow floodplains. The peat mantle is thicker in bogs (up to 2 m deep) than on slopes. Slope peats are unstable and exposed to fragmentation through differential sliding.
- The extent of the bogs varies from a few square meters in the case of small alveoli to several hectares in the case of polje-like depressions. The surface area of the larger bogs is about 0.5–3 ha.
- The upper peat layers in tepui environment are less decomposed than the deeper ones, in contrast to what happens in most tropical peats.
- The most common peat material is slightly to moderately decomposed fibric and hemic organic material.

- Strongly decomposed sapric material is less frequent. Sapric material goes easily in suspension/solution and could have been exported within the blackwaters that percolate through rock joints and fractures to underground water circulation systems and end up feeding the lowland blackwater rivers.
- The most common layer sequences are Oi–Oe and Oe–Oa. Although Oi layers overlay usually Oe layers, some sequences include deeper fibric layers that are less decomposed than the overlying ones. In the majority of profiles, peat rests directly on the substratum, whether sandstone–quartzite or granite.

References

Andriesse JP (1988) Nature and management of tropical peat soils. FAO Soils Bulletin 59. Food and Agriculture Organization of the United Nations, Rome

Bennett PC, Siegel DI, Glaser PH (1991) Fate of silica minerals in a peat bog. Geology 19:328–331

Blancaneaux Ph, Pouyllau M (1977) Formes d'altération pseudokarstiques en relation avec la géomorphologie des granites précambriens du type Rapakivi dans le territoire fédéral de l'Amazone, Vénézuela. Cahiers ORSTOM, série Pédologie XV(2):131–142

Bremer H, Sander H (2000) Inselbergs: geomorphology and geoecology. In: Porembski S, Barthlott W (eds) Inselbergs: biotic diversity of isolated rock outcrops in tropical and temperate regions, vol 146, Ecological Studies. Springer, Berlin, pp 7–35

Brewer-Carías Ch, Steyermark J (1976) Simas de Sarisariñama y su vegetación. Boletín Sociedad Venezolana de Ciencias Naturales 22(132–133):179–405

Briceño H, Schubert C (1990) Geomorphology of the Gran Sabana, Guayana Shield, southeastern Venezuela. Geomorphology 3:125–141

Briceño H, Schubert C, Paolini J (1990) Table-mountain geology and surficial geochemistry: Chimantá Massif, Venezuelan Guayana Shield. J S Am Earth Sci 3(4):179–194

Chalcraft D, Pye K (1984) Humid tropical weathering of quartzite in southeastern Venezuela. Z Geomorphol 28:321–332

Clapperton C (1993) Quaternary geology and geomorphology of South America. Elsevier, Amsterdam

Colvée P (1973) Cueva en cuarcitas en el Cerro Autana, Territorio Federal Amazonas. Boletín Sociedad Venezolana de Espeleología 4(1):5–13

CVG-TECMIN C.A. (1991) Informes de avance. Clima, geología, geomorfología, suelos, vegetación. Proyecto Inventario de los Recursos Naturales de la Región Guayana, Corporación Venezolana de Guayana, Ciudad Bolívar

Doerr SH (1999) Karst-like landforms and hydrology in quartzites of the Venezuelan Guyana shield: pseudokarst or "real" karst? Z Geomorphol 43:1–17

Doerr SH, Wray R (2004) Pseudokarst. In: Goudie AS (ed) Encyclopedia of geomorphology, vol 2. Routledge, London, pp 814–816

Driessen PM, Rochimah L (1976) The physical properties of lowland peats from Kalimantan and their significance for land suitability appraisal. In: Research on peat and podzolic soils in Indonesia and their potential for agriculture. Soil Research Institute, Bogor

Feininger T (1969) Pseudo-karst on quartz diorite, Colombia. Z Geomorphol NF, Bd 13, Hft 3:287–296

Galán C (1988) Cavernas y formas de superficie en rocas silíceas precámbricas del Grupo Roraima, Guayana, Venezuela. Boletín Sociedad Venezolana de Espeleología 23:1–12

Gansser A (1973) The Roraima problem (South America). Verhandlungen der Naturforschende Gesellschaft vol 84, Basel, pp 81–97

Garcia P, Schargel R, Zinck JA (2011) Properties and classification of the tepui peats. In: Zinck JA, Huber O (eds) Peatlands of the Western Guayana Highlands, Venezuela, Chap. 6. Springer, Heidelberg, doi: 10.1007/978-3-642-20138-7_6

Gaudette HE, Olszewski WJ (1985) Determination of radiometric ages, Amazonas Territory, Venezuela. In: Muñoz MI (ed) Memoria 1er Simposium Amazónico. Boletín de Geología, Publicación Especial 10. Ministerio de Energía y Minas, Caracas, pp 733–746

Gaudig G, Couwenberg J, Joosten H (2006) Peat accumulation in kettle holes: bottom up or top down? Mires and Peat 1: art. 6. http://www.mires-and-peat.net/

Ghosh SK (1977) Geología del Grupo Roraima en Territorio Federal Amazonas, Venezuela. Memoria V Congreso Geológico Venezolano Tomo I. Ministerio de Energía y Minas y Sociedad Venezolana de Geólogos, Caracas, pp 167–193

Ghosh SK (1985) Geology of the Roraima Group and its implications. In: Muñoz MI (ed) Memoria 1er Simposium Amazónico. Boletín de Geología, Publicación Especial 10. Ministerio de Energía y Minas, Caracas, pp 33–50

Gibbs AK, Barron CN (1993) The geology of the Guiana Shield. Oxford University Press, New York, NY

González de Juana C, Iturralde de Arozena JM, Picard X (1980) Geología de Venezuela y de sus cuencas petrolíferas. Ediciones FONINVES, Caracas

Gröger A (2000) Flora and vegetation of inselbergs of Venezuelan Guayana. In: Porembski S, Barthlott W (eds) Inselbergs: biotic diversity of isolated rock outcrops in tropical and temperate regions, vol 146, Ecological Studies. Springer, Berlin, pp 291–314

Gröger A, Huber O (2007) Rock outcrop habitats in the Venezuelan Guayana lowlands: their main vegetation types and floristic components. Rev Bras Bot 30(4):599–609

Halliday WR (2007) Pseudokarst in the 21st century. J Cave Karst Stud 69(1):103–113

Huber O (1992) El macizo del Chimantá: la vegetation. In: Huber O (ed) El macizo del Chimantá, Escudo de Guayana, Venezuela. Un ensayo ecológico tepuyano. Oscar Todtmann Editores, Caracas, pp 161–178

Huber O (1995) Geographical and physical features. In: Steyermark JA, Berry PE, Holst BK (gen eds) Flora of the Venezuelan Guayana, vol 1; Berry PE, Holst BK, Yatskievych K (eds) Introduction. Missouri Botanical Garden and Timber, St. Louis, MO and Portland, OR, pp 1–61

Huber O, Garcia P (2011) The Venezuelan Guayana region and the study areas: geo-ecological characteristics. In: Zinck JA, Huber O (eds) Peatlands of the Western Guayana Highlands, Venezuela, Chap. 3. Springer, Heidelberg, doi: 10.1007/978-3-642-20138-7_3

Jennings JN (1983) Sandstone pseudokarst or karst? In: Young RW, Nanson GC (eds) Aspects of Australian sandstone landscapes. Australian and New Zealand Geomorphology Group, Special Publication 1, Wollongong, pp 21–30

Khobzi J (1972) Erosion chimique et mécanique dans la genèse de dépressions “pseudo-karstiques” souvent endoréiques. Rev Géomorphol Dynam, XXIe année 2:57–70

Kluge M, Brulfert J (2000) Ecophysiology of vascular plants on inselbergs. In: Porembski S, Barthlott W (eds) Inselbergs: biotic diversity of isolated rock outcrops in tropical and temperate regions, vol 146, Ecological Studies. Springer, Berlin, pp 143–174

Le Roux G, Shotyk W (2006) Weathering of inorganic matter in bogs. In: Martini IP, Martinez-Cortizas A, Chesworth W (eds) Peatlands: evolution and records of environmental and climate changes, vol 9, Developments in earth surface processes. Elsevier, Amsterdam, pp 197–216

Lundberg J, Brewer-Carías C, McFarlane DA (2010) Preliminary results from U-Th dating of glacial-interglacial deposition cycles in a silica speleothem from Venezuela. Quatern Res 74:113–120

Mendoza V (1977) Evolución tectónica del Escudo de Guayana. Boletín de Geología de Venezuela, Publicación Especial 7(3):2237–2270

Ministerio de Minas e Hidrocarburos (1970) Léxico Estratigráfico de Venezuela (Segunda Edición). Boletín de Geología, Publicación Especial 4. Editorial Sucre, Caracas

Mora-Polanco A, Sánchez-Calderón L, Lasso-Alcalá C, McQuhae-Romero C (2007) Parámetros físicoquímicos de algunos cuerpos de agua adyacentes a la confluencia de los ríos Orinoco y Ventuari, Estado Amazonas, Venezuela. Boletín del Centro de Investigaciones Biológicas 41(1):44–59. Universidad del Zulia, Maracaibo

Otvos EG (1976) Pseudokarst and pseudokarst terrains; problems of terminology. GSA Bull 87(7):1021–1027

Piccini L, Mecchia M (2009) Solution weathering rate and origin of karst landforms and caves in the quartzite of Auyan-tepui (Gran Sabana, Venezuela). Geomorphology 106:15–25

Porto-Domingues A (1952) Provável origen das depressões observadas no Sertão do Nordeste. Rev Bras Geogr XIV:305–315

Pouyllau M (1989) Les paysages géomorphologiques de l'Amazonie et de la Guyane vénézuelienne. In: Venezuela: environnements et changements. Tome 2. Travaux et Documents de Géographie Tropicale 63. CEGET-CNRS, Bordeaux, pp 39–75

Pouyllau M, Seurin M (1985) Pseudo-karst dans des roches gréso-quartzitiques de la formation Roraima. Karstologia 5:45–52

Reid AR (1972) Stratigraphy of the type area of the Roraima Group, Venezuela. Memoria de la Novena Conferencia Geológica Inter-Guayanas. Boletín de Geología, Publicación Especial 6. Ministerio de Minas e Hidrocarburos, Caracas, pp 343–353

Szczerban E (1973) Cavernas y simas en areniscas precámbricas del Territorio Federal Amazonas. Memoria Segundo Congreso Latinoamericano de Geología, Tomo II. Boletin de Geología, Publicación Especial 7. Editorial Sucre, Caracas (1976), pp 1055–1072

Szczerban E, Urbani F (1974) Carsos de Venezuela. Parte 4: Formas cársicas en areniscas precámbricas del Territorio Federal Amazonas y Estado Bolívar. Boletín Sociedad Venezolana de Espeleología 5(1):27–54

Szczerban E, Urbani F, Colvée P (1977) Cuevas y simas en cuarcitas y metalimolitas del Grupo Roraima, Meseta del Guaiquinima, Estado Bolívar. Boletín Sociedad Venezolana de Espeleología 8(16):127–154

Talukdar S, Szczerban E, Granati A (1973) Precambrian geological evolution of Territorial Federal Amazonas, Venezuela. A workable structural and petrological Framework. Memoria Segundo Congreso Latinoamericano de Geología, Tomo II. Boletin de Geología, Publicación Especial 7. Editorial Sucre, Caracas (1976), pp 633–634

Tricart J (1969) Le modelé des régions sèches. SEDES, Paris

Twidale CR (1982) Granite landforms. Elsevier, Amsterdam

Twidale CR (1984) So-called pseudokarst in granite. Boletín Sociedad Venezolana de Espeleología 12(21):3–12

Twidale CR, Corbin EM (1963) Gnammas. Rev Géomorphol Dynam 14:1–20

Twidale CR, Vidal Romani JR (2005) Landforms and geology of granite terrains. Balkema, Leiden

Urban J (2008) Commission for Pseudokarst of the International Union for Speleology (UIS). Newsletter. http://web.utanet.at/speleoaustria/pseudokarst.htm

Urbani F (1976) Opalo, calcedonia y calcita en la cueva del Cerro Autana, Territorio Federal Amazonas, Venezuela. Boletín Sociedad Venezolana de Espeleología 7(14):129–146

Urbani F (1990) Algunos comentarios sobre terminología kárstica aplicada a rocas silíceas. Boletín Sociedad Venezolana de Espeleología 24:5–6

Urbani F, Szczerban E (1975) Formas pseudocarsicas en granito rapakivi precámbrico, Territorio Federal Amazonas. Boletín Sociedad Venezolana de Espeleología 6(12):57–70

Van de Putte HW (1972) Contribution to the stratigraphy and structure of the Roraima Formation, State of Bolivar, Venezuela. Memoria de la Novena Conferencia Geológica Inter-Guayanas. Boletín de Geología, Publicación Especial 6. Ministerio de Minas e Hidrocarburos, Caracas, pp 372–394

Vidal Romani JR, Vaqueiro Rodriguez M (2007) Types of granite cavities and associated speleothems: genesis and evolution. Nat Conserv 63:41–46

Watson A, Pye K (1985) Pseudokarstic micro-relief and other weathering features on the Mswati Granite (Swaziland). Z Geomorphol, N.F. 29(3):285–300

Wilding LP, Smeck NE, Drees LR (1977) Silica in soils: quartz, cristobalite, tridymite, and opal. In: Dixon JB, Weed SB (eds) Minerals in soil environments. Soil Science Society of America, Madison, WI, pp 471–552

Wray RAL (1997a) A global review of solutional weathering forms on quartz sandstones. Earth Sci Rev 42(3):137–160

Wray RAL (1997b) Quartzite dissolution: karst or pseudokarst? Cave Karst Sci 24(2):81–86

Yanes CE, Briceño HO (1993) Chemical weathering and the formation of pseudo-karst topography in the Roraima Group, Gran Sabana, Venezuela. Chem Geol 107:341–343

Yanes C, Alvarez H, Jaffe R (2006) Geochemistry of a tropical lake (Lake Leopoldo) on pseudo-karst topography within the Roraima Group, Guayana Shield, Venezuela. Appl Geochem 21(6):870–886

Yánez GA (1972) Geología del área de Santa Elena de Uairén. Memoria de la Novena Conferencia Geológica Inter-Guayanas. Boletín de Geología, Publicación Especial 6. Ministerio de Minas e Hidrocarburos, Caracas, pp 660–667

Yánez GA (1977) Provincia geológica de Roraima, geología estructural y geomorfología de su parte sur-oriental (Venezuela). Memoria V Congreso Geológico Venezolano Tomo II. Ministerio de Energía y Minas y Sociedad Venezolana de Geólogos, Caracas, pp 845–854

Yánez GA (1985) Geología y geomorfología del Grupo Roraima en el sureste de Venezuela. Memorias VI Congreso Geológico Venezolano 2:1243–1306

Zinck JA, Garcia P, van der Plicht J (2011) Tepui Peatlands: age record and environmental changes. In: Zinck JA, Huber O (eds) Peatlands of the Western Guayana Highlands, Venezuela, Chap. 7. Springer, Heidelberg, doi: 10.1007/978-3-642-20138-7_6

Chapter 5
Laboratory Methods for Characterization of Peat Materials

R. Schargel, P. García, and D. Jiménez

5.1 Introduction

Organic soils have characteristics that distinguish them markedly from mineral soils. The analytical procedures used for their characterization and classification are relatively recent. Not many organic soils in the world have been analyzed by the methods of the USDA Soil Taxonomy (Soil Survey Staff 1999, 2006), and those that have been are mostly in the United States of America. Although Venezuelan Histosols have been mapped quite extensively in the Orinoco river delta, the analytical procedures that have been applied on air-dry <2 mm soil material are similar to those that are used for mineral soils. The decomposition stage of the organic material has been based on the field appraisal. The Guayana Highland peats and four benchmark soils used to train the laboratory personnel are the first organic soils characterized in Venezuela by most of the methods proposed by the USDA Soil Taxonomy. These methods were originally developed on soils occurring in geomorphic environments very different from that of the Guayana Highlands. Consequently, adjustments of the analytical procedures were necessary.

This chapter first highlights the specificity of organic soils in general, as their characterization and taxonomic classification require the use of criteria and laboratory methods different from those applied to mineral soils. Then, the testing of existing procedures on selected organic soils from a variety of settings in Venezuela is described and discussed. Finally, the application of the calibrated procedures to the Guayana Highland peats is addressed.

5.2 Criteria Used for the Recognition of Organic Soils in the Soil Taxonomy

The classification of organic soils lagged during the early stages of development of the Soil Taxonomy. Histosols were only defined at the highest categorical level (i.e., the order level) in the first published version of this classification system (Soil

J.A. Zinck and O. Huber (eds.), *Peatlands of the Western Guayana Highlands, Venezuela*, Ecological Studies 217, DOI 10.1007/978-3-642-20138-7_5,

Survey Staff 1960). Criteria for the definition of suborders, great groups, subgroups, and families were proposed but not fully applied. Instead, the degree of decomposition of the organic materials traditionally used in the definition of soil series and types (i.e., muck, peat, and mucky peat) was selected as the main criterion for the definition of suborders. A supplement dealing with the classification of Histosols including criteria applicable from the suborder to the family level was presented by the Soil Survey Staff in 1968. A publication by Farnham and Finney in 1965 was used as a base for developing the classification of these soils (McKinzie 1974).

Basic kinds of organic soil material were defined based on the amount of fibers. These fibers are fragments of plant tissue retained on a 100-mesh sieve (0.15 mm openings). Wood fragments larger than 2 cm in cross section, which cannot be crushed and shredded with the fingers, are not included with the fibers. They are considered as coarse fragments comparable to gravel in mineral soils. The percentage of fibers that do not break down with rubbing is a more realistic estimate of the degree of decomposition of the plant remains than the amount of fibers present in undisturbed condition. Bulk density values and subsidence upon drainage are closely related to rubbed fiber content. The latter can be evaluated in the field by rubbing a small piece of the organic material ten times between thumb and forefinger with firm pressure. After rubbing, the material is molded and broken for examination with a hand lens (10-power or more) to estimate the rubbed fiber content. In the laboratory, the rubbed material is washed onto a 100-mesh sieve. The Munsell color designation of a sodium pyrophosphate extract on white filter paper is used as a complementary measure of decomposition. The volume occupied by unrubbed and rubbed fiber and the color of the sodium pyrophosphate extract on white filter paper allow distinguishing several groups of organic soil materials, according to the degree of decomposition. Fibric soil material is the least decomposed, with a high fiber content, very low bulk density (commonly less than 0.1 Mg m^{-3}), and very high water content at saturation (about 850% or more). Sapric soil material is the most decomposed, with the lowest fiber content, highest bulk density (commonly 0.2 Mg m^{-3} or more), and lowest water content at saturation (less than 450%). Hemic soil material is of intermediate decomposition, bulk density, and water content. Hemilluvic material consisting of illuvial humus, which accumulates in the lower part of some drained acid organic soils, and limnic materials (coprogenous earth, diatomaceous earth, and marl) were used for definition at the great group and subgroup levels.

Changes in Histosol recognition and classification occurred in later publications of the Soil Taxonomy as more knowledge was gathered on these soils. However, the principal criteria for the definition of the taxa at the order level did not change much. Unrubbed fiber was excluded from the definition of the fibric, hemic, and sapric materials in later keys, although in the 1975 and 1999 editions of the Soil Taxonomy unrubbed fiber is included as complementary information to characterize these materials, together with bulk density and water content. Table 5.1 shows the suborders and great groups in the Histosol order in versions of the Soil Taxonomy published in 1968, 1975, and 1999. Only minor changes below the great group level were introduced in later keys.

Table 5.1 Changes of great groups within the four suborders of the Histosols between 1968 and 1999 (Soil Survey Staff 1968, 1975, 1999)

1968	1975	1999
Folists	*Folists*	*Folists*
No great groups	Cryofolists	Cryofolists
	Tropofolists	Torrifolists
	Borofolists	Ustifolists
		Udifolists
Fibrists	*Fibrists*	*Fibrists*
Cryofibrists	Cryofibrists	Cryofibrists
Sphagnofibrists	Sphagnofibrists	Sphagnofibrists
Borofibrists	Borofibrists	Haplofibrists
Tropofibrists	Tropofibrists	
Medifibrists	Medifibrists	
Luvifibrists	Luvifibrists	
Hemists	*Hemists*	*Hemists*
Cryohemists	Sulfohemists	Sulfohemists
Borohemists	Sulfihemists	Sulfihemists
Tropohemists	Luvihemists	Luvihemists
Medihemists	Cryohemists	Cryohemists
Luvihemists	Borohemists	Haplohemists
	Tropohemists	
	Medihemists	
Saprists	*Saprists*	*Saprists*
Cryosaprists	Cryosaprists	Sulfosaprists
Borosaprists	Borosaprists	Sulfisaprists
Troposapristss	Troposapristss	Cryosaprists
Medisaprists	Medisaprists	Haplosaprists

The Soil Taxonomy of 1975 included great groups for the Folist suborder and added the sulfo- and sulfi-great groups to the Hemists. Sulfohemists have a sulfuric horizon within 50 cm of the soil surface, and Sulfihemists have sulfidic materials within 1 m soil depth. Greater changes were introduced in the Soil Taxonomy of 1999. The boro-, medi-, and tropo-great groups, defined on the basis of the soil temperature regime, were grouped into the haplo-great groups. This eliminates redundancy because the soil temperature regime is also used at the family level. Only the cryic temperature regime is preserved as a criterion for grouping organic soils at the great group level. Organic soils with permafrost are included into the Histel suborder of the Gelisols. The great groups of the Folists are defined by soil moisture regime and cryic temperature regime.

5.3 Laboratory Methods Specific for Organic Soils

The soil survey laboratory methods published in 1972 by the Soil Conservation Service provided only a few statements on the sampling of organic soils. The following tests on moist organic soil materials were included in Appendix III of the Soil Taxonomy (Soil Survey Staff 1975): pyrophosphate color test, unrubbed and rubbed fiber percentage, and pH in 0.01 M $CaCl_2$. Lynn et al. (1974) added the determination of the mineral content after heating a dry sample to 400°C. These methods were described in the publication on soil analysis of the Soil Conservation Service in 1984. Sample collection and preparation of organic soils for laboratory determinations on field-moist and air-dry samples were included in later publications of the Soil Conservation Service (1992) and by Burt (2004). The methods currently used are briefly described hereafter. More details can be found in Burt (2004) available online.

5.3.1 Sample Preparation

The field-moist soil sample is weighed and placed on a tray, and live roots are removed. The moist, air-dry, and oven-dry weights are obtained for roots and subsamples of the moist material. The soil material is thoroughly mixed, and a subsample is stored for the determination of fiber volume, pyrophosphate color, pH in 0.01 M $CaCl_2$ solution, and moist 1,500 kPa water content. The remaining sample is air dried and passed through a 2-mm sieve. This material is used for determining the water content at 1,500, 200, and 10 kPa. Weight measurements are made on 20–75, 5–20, and 2–5 mm fractions. A subsample of air-dry fine-ground material (<80 mesh, 180 μm openings) is used for chemical analysis.

5.3.2 Mineral Content

The mineral content consists of the plant residual ash and the particles that remain after the organic matter has been removed on ignition. The loss on ignition is a better estimator of organic matter content in Histosols than the Walkley–Black method (acid dichromate digestion) traditionally used for organic carbon determination in soils. Data by the latter method are generally considered invalid when organic carbon is higher than 8%. The sample is heated to 110°C overnight, cooled, and weighed. It is then placed in a cool muffle furnace and heated to 400°C for 16 h. The ratio of the weights at 400°C/110°C gives the mineral content percentage.

5.3.3 Pyrophosphate Color

Decomposed organic materials are soluble in sodium pyrophosphate, originating a solution color that correlates with the degree of decomposition. Dark colors are characteristic of sapric materials and light colors are of fibric materials. This determination is usually performed in the field office, using moist samples. A syringe (6 mL) cut in half is used to measure the volume of the organic material. To obtain comparable results in this and the following methods, a standardized packing of the material is necessary. Paper towels are used to remove excess water until the sample is firm and saturated. The material is then packed into the half-syringe with the plunger at the 5 mL mark to obtain 2.5 cm^3. This volume of material is placed into a vial with sodium pyrophosphate solution (1 g dissolved in 4 mL of water), mixed, and left to stand overnight. After mixing, a strip of chromatographic paper is inserted vertically into the sample to 1 cm depth and removed when the paper strip has wetted to 2 cm height above the slurry surface. Excess solution is removed by contact with blotting paper, and the color of the strip is compared with Munsell color charts.

5.3.4 Fiber Volume

The water-dispersed fiber volume characterizes the physical decomposition state of organic materials. The determination is performed on a 2.5 cm^3 sample placed upon a 100-mesh sieve (0.15 mm openings) to retain the fibers. The unrubbed fiber content is determined in three steps of increasingly vigorous treatments designed to remove the sapric material. All three steps may not be necessary. Following each step, the percentage of sapric material remaining is estimated using a microscope or a hand lens. If the sapric material is $>10\%$, the next step is required.

The following successive treatments are applied:

1. The sample is washed under a stream of tap water, adjusted to deliver 200–300 mL in 5 s until the water passing through the sieve appears clean.
2. The residue of the previous step retained on the sieve is placed into a 500 mL plastic container half full with tap water and stirred vigorously with an eggbeater for 1 min. The material is transferred to the 100-mesh sieve, and procedural step 1 is repeated.
3. The residue of the previous step is placed into the container of an electric mixer, filled about two-thirds with water, and mixed for 1 min. The material is transferred to the 100-mesh sieve, and procedural step 1 is repeated.

Once the sapric material is $<10\%$, the residue is washed to one side of the sieve and blotted from underneath with absorbent tissue to withdraw water. The residue volume is measured by packing into a half-syringe. To determine the rubbed fiber content, the material remaining from the previous treatments is rubbed between thumb and fingers under a stream of tap water, adjusted to deliver 150–200 mL in 5 s, until water passing through the sieve is clean. The sample is blotted to remove excess water, and the volume is measured with the half-syringe.

5.3.5 *Soil Reaction (pH)*

The pH determined in a 0.01 M $CaCl_2$ solution is used in the Soil Taxonomy to distinguish the reaction classes at family level in Histosols. The determination is performed on a 2.5 cm^3 sample placed into a 30-mL plastic container with 4 mL of 0.015 M $CaCl_2$ solution. The final concentration is approximately 0.01 M with most packed moist organic materials. The determination is made with pH paper or meter after mixing and equilibrating for at least 1 h.

5.4 Testing Laboratory Methods on Selected Venezuelan Histosols

5.4.1 *The Test Soils*

The laboratory methods used to characterize organic soils had not yet been applied in Venezuela when the soil samples from western Guayana Highlands were received in the laboratory of the Ministry of Environment in Guanare city. To test the methods and train the laboratory personnel, four easily accessible organic soils from western Venezuela were selected (Jiménez 1995). Thirteen organic matter rich layers were sampled with auger and described according to the norms of the Soil Survey Division Staff (1993). Site elevation varied from 50 m to over 3,000 m a.s.l. One soil was drained and cultivated; two were under natural herbaceous vegetation; and one under forest vegetation. In three profiles, the thickness of the organic layers was less than 1.5 m and in one more than 3.5 m. Organic materials rested on mineral substrata of variable origin (alluvial, eolian, and glacial). The moist color was black (10YR 2/1) in most layers. Roots were common in the surface layers but few in the deeper layers. Additional characteristics are summarized in Table 5.2.

5.4.2 *Laboratory Determinations*

The disturbed samples were analyzed following the methods indicated above with some ad hoc adaptations. The pyrophosphate color was not determined because sodium pyrophosphate was not available in the laboratory. Field moisture and bulk density were obtained by oven drying 100 mL of field-moist soil. Moisture retention was determined at 10 kPa and 1,500 kPa on field-moist and air-dry samples (<2 mm). Rubbed and unrubbed fiber treatments were applied. Organic matter was determined by loss on ignition and by Walkley–Black. The pH in water was measured on field-moist samples and on air-dry fine-ground soil material mixed

Table 5.2 Some characteristics of the test soils

Soil	Geographic region	Elevation (m a.s.l.)	Depth of organic materials (cm)	Taxonomic classification
LP1	Llanos	50	120	Terric Haplohemist loamy, siliceous, dysic, isohyperthermic[a] Hemic Rheic Histosol (Hyperdystric)[b]
LP2	Llanos	55	140	Typic Haplosaprist dysic, isohyperthermic[a] Sapric Histosol (Hyperdystric)[b]
VL	Andes	1,300	360+	Typic Haplosaprist euic, isothermic[a] Sapric Rheic Histosol (Eutric? Drainic)[b]
M	Andes	3,680	125	Terric Haplosaprist loamy, mixed, euic, isomesic[a] Sapric Histosol (Dystric)[b]

[a]USDA: Keys to soil taxonomy (Soil Survey Staff 2006)
[b]WRB: World reference base for soil resources (IUSS Working Group 2006)

with sufficient water to reproduce the original moisture content. The determination of exchangeable cations and that of available phosphorus extracted with 0.5 M sodium bicarbonate solution at pH 8.5 (Olsen) and 0.025 N HCl and 0.03 N NH_4F solution (Bray1) were performed in triplicates on air-dry fine-ground material and on field-moist material with similar dry matter contents. Organic matter was also analyzed in triplicates. Exchangeable calcium, magnesium, potassium, and sodium were extracted with neutral 1 N ammonium acetate and measured with an atomic absorption spectrophotometer for Ca and Mg, and by flame photometry for K and Na. Extracts obtained from soil samples leached with $BaCl_2$-TEA at pH 8.2 and 1 N KCl solutions were titrated to determine extractable (i.e., total) acidity and exchangeable (i.e., salt-replaceable) acidity, respectively (Soil Science Society of America 2008). Phosphorus was measured by colorimetric procedures.

5.4.3 Assessing the Tested Methods

The methods used to characterize organic soil materials did not require much equipment and could be easily performed in a field office. The main objection is the rather laborious procedures to determine unrubbed and rubbed fiber, considering that the unrubbed fiber is even not used in the taxonomic keys. Step 3 of the unrubbed and rubbed fiber includes quite vigorous treatments that could cause the fragmentation of weak fibers. During these analytical steps, part of the volume decrease could result from fiber fragmentation in addition to the removal of sapric material. Furthermore, sand grains were found in the rubbed fiber of some soils. Because of the much higher particle density of the mineral matter, the volume occupied by mineral grains is small compared with that of the organic material. However, rubbing organic material together with sand grains might contribute to the breakdown of slightly altered fibers. The 100-mesh sieve retains part of the fine sand and coarser sand fractions. Although these fractions are not common in most organic soils, they can be significant in soils exposed to flooding, colluvial wash, or sand storm.

A more efficient procedure to assess the fiber content is desirable. Boelter (1969) determined unrubbed fiber content on field-moist samples with a wet sieving process. The sample was soaked in a Calgon solution (dispersion agent) before performing the wet sieving through a nest of sieves. Working with a large number of peat samples, Boelter found that bulk density and unrubbed fiber content had a curvilinear relation in all the fractions tested (>0.1, >0.25, >0.5, >1, and >2 mm), with coefficients of multiple determination (R^2) ranging from 0.85 to 0.87. Decreasing fiber volume was related to increases in bulk density. He also determined a curvilinear relationship of water content at saturation and at suctions of 0.5, 10, and 1,500 kPa, to unrubbed fiber content (>0.1 mm) and bulk density with R^2 ranging from 0.66 to 0.88. Regression analysis of the logarithm of hydraulic conductivity on fiber content and bulk density indicated a linear relationship ($R^2 = 0.54$). These results show that the unrubbed fiber determination is useful for the characterization of organic soil materials. In our experiment, the relation between bulk density and unrubbed fiber volume was not that clear. Layers with similar bulk density values had very different unrubbed fiber contents. Some layers with high bulk densities had higher unrubbed fiber volumes than soils with low bulk densities. The bulk density of the layers studied is strongly influenced by variable quantities of mineral soil material incorporated into the organic soil during its formation.

Malterer et al. (1992) reviewed selected national methods that measure degree of decomposition and fiber content, and evaluated their capacity and precision to distinguish between classes of peat. The Von Post field test (1924) was used in the identification and sampling of ten peat classes with increasing degrees of decomposition (H_1–H_{10}). Replicate determinations (10) by different methods, including the Von Post procedure, were performed on the samples. Statistical analysis revealed that only eight Von Post classes were significantly different, grouping classes H_2 and H_3, as well as H_6 and H_7. The USSR centrifugal method identified all these eight Von Post classes. The unrubbed and rubbed fiber contents by the USDA procedures identified five of the eight Von Post classes in the less decomposed material (H_1, H_{2+3}, H_4, H_5, H_{6+7}) with overlap in the remaining classes. The USDA pyrophosphate color method uniquely identified only three of the less decomposed Von Post classes. This research shows that the Von Post scale of humification should not be discarded, and that there is room for improvement of the laboratory procedures proposed by the Soil Taxonomy to measure the decomposition of the organic material.

For testing purposes, we propose a one-step unrubbed fiber procedure, followed by rubbed fiber determination. In the first step, a volume of the organic soil material (the half-syringe measuring device could be used) is placed in a 500 mL plastic container half-full with water and shaken gently in a reciprocal shaker for a time to be determined. The use of a dispersant should be tested. After shaking, coarse mineral grains that might be present should be separated from the organic material by decantation. The organic soil material is then washed through the 100-mesh sieve under a stream of tap water, adjusted to deliver 200–300 mL in 5 s until the water passing through the sieve appears clean. The residue volume is measured by packing into a half-syringe, and the rubbed fiber determination is applied.

Physical and chemical determinations should be performed on field-moist soil samples when possible. In addition to changes in moisture retention on air drying, wetting the air-dry material is time consuming. The pH determination in water is also more expedient on the field-moist soil. Extractable acidity shows large differences between determinations performed on air-dry and field-moist samples. Significant differences in the determination of exchangeable bases and acidity on air-dry and field-moist samples can be expected to occur in some organic soils. Air-dry material should be used for total elemental analysis only. Fine grinding the organic soil material would not be necessary. To standardize these procedures for chemical determinations, excess moisture should be removed and the sample should be packed into a sufficiently large volume. A similarly packed sample should be taken for dry matter determination. The results would be expressed on the basis of oven-dry material.

5.5 Laboratory Determinations Performed on the Guayana Histosols

5.5.1 Soil Sampling

Soil samples were taken from pits at randomly selected sites, chosen by helicopter overflight. Soil pits were dug until the bedrock or, in deeper soils, to 150 cm depth. Water was removed from the pit with a bucket to allow soil profile deepening and layer identification. Because of rapidly rising water table, soil layer identification, description, and sampling had to be done in expedite manner. Soil samples were taken by layer following morphological characteristics such as color, mottles, root content, root stage (living or dead) and distribution pattern, structure, and changes in texture. Samples were put in 2 kg plastic bags and transported by helicopter to the EDELCA headquarter in Puerto Ordaz, and then by car to the laboratory of the Ministry of Environment and Natural Resources in Guanare city (around 1,200 km from Puerto Ordaz) where chemical and physical analyses were carried out.

Field sampling faced several limitations that could affect the laboratory data:

- All peat sites were at remote locations that were reached by helicopter in 30–60 min flights. Opening of soil pits, description, and sampling had to be done very fast, which causes inaccuracies in layer identification and sampling. In some cases, site selection was strongly constrained by the landing conditions (i.e., lack of flat surfaces, dense vegetation cover, stony or rocky terrains, and swampy conditions).
- High water table and water saturation of the peats were the most limiting factors for description and sampling. Water flow through the profiles during the opening

and sampling may have caused some contamination in deeper layers by mobilization of chemical and physical constituents from upper layers.

- Collapse of pit walls due to high water saturation and weak consistency of the organic materials was very common, affecting the delineation and sampling of the soil layers.
- The looseness of the peat material made it difficult to collect truly undisturbed samples, especially in folic and fibric layers, affecting the accuracy of physical determinations such as for instance bulk density.
- Soil samples remained stored for a long time in the EDELCA headquarter at air temperature and, later, under refrigeration in the laboratory before determinations were undertaken. This may have contributed to alter some physical and chemical properties.

Our experience in describing and sampling peats with high water table, especially in remote areas, leads to the following recommendations. Using a special probe with a minimum diameter of 8 cm would provide samples less mixed and altered than those obtained from pits. The upper root mat is better sampled separately with a sharp spade. Processing and description of the cores, extraction of live roots, physical tests, and pH determination should be done in a field office. The equipment of the field office should include an analytical balance ±0.01 g, an oven for heating up to 110°C, a pH-meter, sieves, glassware, sodium pyrophosphate, chromatographic paper, and 0.01 M calcium chloride solution. The samples sent to the laboratory should be kept under refrigeration until chemical determinations, loss on ignition, and water retention are performed. If mineral grains are retained on the 100-mesh sieve (0.15 mm openings) after step 1 treatment of unrubbed fiber, decantation should be applied before proceeding to step 2.

5.5.2 Analytical Methods Applied

On the basis of the experience gained from testing a variety of methods on four selected organic soils as reported earlier, we proceeded to characterize the Guayana Histosols. The disturbed organic soil samples collected in the field in plastic bags were placed in a refrigerator to reduce decomposition. Sample preparation and determination of mineral content, fiber volume, and pH in 0.01 M $CaCl_2$ solution followed the outline given in Sect. 5.3. For some samples, field material was insufficient to carry out all the determinations with replicates.

The percent of moist roots was reported on field-moist soil basis. After root removal, the remaining moist soil material was taken as 100% for further laboratory determinations. Dry matter was determined on field-moist soil. For some samples, a volume of the field-moist soil was oven dried to obtain a surrogate measure of bulk density (Bd2). The weight and bulk density (Bd1) of a 2.5 cm^3 moist soil material packed into a half-syringe were determined weighing the empty and loaded half-syringe during fiber determination. The values of bulk density Bd1 range from

0.48 to 0.76 Mg m^{-3}. These variations are related to the mineral content of the soil material and the initial moisture content, which influences the amount of air trapped in the packed material.

Organic matter content was obtained from weight loss on ignition and by the Walkley–Black procedure that determines organic carbon using acid dichromate digestion. Organic matter percent is calculated by multiplying by 1.724 the organic carbon percent.

The air-dry fine-ground organic soil material was used for chemical analysis. The extract obtained using a 0.5 N $BaCl_2$-triethanolamine pH 8.2 solution was titrated with standardized 0.2 N HCl to determine the extractable or total acidity (Acd1). To determine the 1 N KCl exchangeable acidy or salt-replaceable acidity (Acd2), the extract obtained with the former solution was titrated with a standard NaOH solution to a phenolphthalein end point (pink color). Exchangeable aluminum was obtained by adding 10 mL of 1 N KF to the previously titrated solution; this solution was then titrated with standard H_2SO_4 until the pink color disappeared. Exchangeable bases were extracted with neutral 1 N ammonium acetate and measured using atomic absorption spectrometry (Ca, Mg) and flame photometry (K, Na). Available phosphorus extracted with 0.5 M sodium bicarbonate solution at pH 8.5 (Olsen) was measured by colorimetric procedure. The sum of the exchangeable bases with the $BaCl_2$-TEA extractable acidity gives the cation exchange capacity by sum of cations (CEC1). The effective cation exchange capacity (CEC2) was calculated adding the 1 N KCl exchangeable acidity to the sum of bases (Fermín 1974a, b; Soil Conservation Service 1972, 1992).

The mineral soil layers were air dried, crushed to break up clods, and passed through a 2-mm sieve. All analyses of the mineral soil material were performed on the <2 mm material. The particle size distribution of the mineral soil was determined with the pipette procedure and dry sieving of the sand fractions. The organic carbon by Walkley–Black, the pH in 0.01 M $CaCl_2$ solution and in soil–water ratio 1:2, and the chemical determinations were determined as indicated earlier. Total nitrogen was determined on some mineral soils by the micro-Kjeldahl digestion procedure.

5.6 Conclusions

The Guayana Highland peats and four benchmark soils used to train the laboratory personnel are the first organic soils characterized in Venezuela by most of the methods proposed by the USDA Soil Taxonomy. These methods were originally developed on soils occurring in environments very different from that of the Guayana Highlands. Consequently, adjustments of the analytical procedures were necessary.

The vigorous treatments used to determine unrubbed and rubbed fiber and the presence of sand grains found in the rubbed fiber of some soils might have contributed to the fragmentation of weak fibers. The relation between bulk density and unrubbed fiber volume was not clear. Organic layers with similar bulk density values had very different unrubbed fiber contents because of variable quantities of mineral soil material incorporated into the organic soil during its formation.

Physical and chemical determinations should be performed on field-moist soil samples whenever possible. The pH determination in water is more expedient on the field-moist soil. Extractable acidity shows large differences between determinations performed on air-dry and field-moist samples. Significant differences in the determination of exchangeable bases and acidity on air-dry and field-moist samples can be expected to occur in some organic soils. Air-dry material should be used for total elemental analysis only.

Field sampling faced several limitations that could affect the laboratory data. It was difficult to collect truly undisturbed samples, especially in folic and fibric layers, because of high water table, risk of contamination in deeper layers by mobilization of chemical and physical constituents from upper layers, and the collapse of pit walls due to high water saturation and weak consistency of the organic materials. Using a special probe would provide less mixed and altered samples than those collected in open pits.

References

Boelter DH (1969) Physical properties of peats as related to degree of decomposition. Soil Sci Soc Am Proc 33:606–609

Burt R (ed) (2004) Soil survey laboratory methods manual. Soil Survey Investigations Report 42, version 4.0. Natural Resources Conservation Service, U.S. Department of Agriculture, Washington, DC

Fermín B (1974a) Compendio de métodos para análisis de calicatas. División de Edafología, Dirección General de Recursos Hidráulicos, Ministerio de Obras Públicas, Guanare

Fermín B (1974b) Compendio de métodos para análisis de rutina. División de Edafología, Dirección General de Recursos Hidráulicos, Ministerio de Obras Públicas, Guanare

IUSS Working Group (2006) World reference base for soil resources, 2nd edn. World Soil Resources Report 103. FAO, Rome

Jiménez D (1995) Aplicación de la taxonomía de suelos en la caracterización de cuatro suelos orgánicos venezolanos. MSc Thesis, Universidad Nacional Experimental de los Llanos Occidentales, Guanare

Lynn WC, McKinzie WE, Grossman RB (1974) Field laboratory tests for characterization of Histosols. In: Histosols: their characteristics, classification and use. SSSA Special Publication 6, Soil Science Society of America, Madison, WI, pp 11–31

Malterer TJ, Verry ES, Erjavec J (1992) Fiber content and degree of decomposition in peats: review of national methods. Soil Sci Soc Am J 56:1200–1211

McKinzie WE (1974) Criteria used in soil taxonomy to classify organic soils. In: Histosols: their characteristics, classification and use. SSSA Special Publication 6, Soil Science Society of America, Madison, WI, pp 1–10

Soil Conservation Service (1972) Soil survey laboratory methods and procedures for collecting soil samples. Soil Survey Investigations Report 1. U.S. Department of Agriculture, Washington, DC

Soil Conservation Service (1984) Procedures for collecting soil samples and methods of analysis for soil survey. Soil Survey Investigations Report. U.S. Department of Agriculture, Washington, DC

Soil Conservation Service (1992) Soil survey laboratory methods manual. Soil Survey Investigations Report 42, version 2.0. U.S. Department of Agriculture, Washington, DC

Soil Science Society of America (2008) Glossary of soil science terms. Soil Science Society of America, Madison, WI

Soil Survey Division Staff (1993) Soil survey manual. Handbook 18. U.S. Department of Agriculture, Washington, DC

Soil Survey Staff (1960) Soil classification, a comprehensive system (7th Approximation). Soil Conservation Service, U.S. Department of Agriculture, Washington, DC

Soil Survey Staff (1968) Supplement to soil classification system (7th Approximation). Soil Conservation Service, U.S. Department of Agriculture, Washington, DC

Soil Survey Staff (1975) Soil taxonomy, a basic system of soil classification for making and interpreting soil surveys. Agriculture Handbook 436. Soil Conservation Service, U.S. Department of Agriculture, Washington, DC

Soil Survey Staff (1999) Soil taxonomy, a basic system of soil classification for making and interpreting soil surveys. Agriculture Handbook 436, 2nd edn. Natural Resources Conservation Service, U.S. Department of Agriculture, Washington, DC

Soil Survey Staff (2006) Keys to soil taxonomy, 10th edn. Natural Resources Conservation Service, U.S. Department of Agriculture, Washington, DC

Von Post L (1924) Das genetische System der organogenen Bildungen Schwedens. Mémoires sur la nomenclature et la classification des sols. International Committee of Soil Science, Helsinki, pp 287–304

Chapter 6
Properties and Classification of the Tepui Peats

P. García, R. Schargel, and J.A. Zinck

6.1 Introduction

Tropical lowland peats develop in environments characterized by high rainfall, evapotranspiration, and temperature. Their distribution, formation, and characteristics are fairly well known from studies carried out in Southeast Asia, especially in coastal environments, and have been documented in a few comprehensive papers and documents including, among others, those by Farnham and Finney (1965), Driessen and Rochimah (1976), EKONO (1981), Lucas (1982), Bord na Mona (1984), Andriesse (1988), Mohamed et al. (2002), and Satrio et al. (2009). In contrast, knowledge about the origin and features of tropical highland peats is still lagging behind.

Tropical highland peats form and evolve in climatic environments close to temperate, with surplus rainfall, but lower temperature and evapotranspiration than in the lowlands. In the Guayana Highlands, peats have formed along an elevation gradient spanning roughly from 600 m to 2,900 m a.s.l. They have been studied in the eastern highlands from a paleogeographic point of view using radiocarbon dating (Schubert and Fritz 1985; Rull 1991; Schubert et al. 1994; Nogué et al. 2009) and pollen analysis (Rull 1991, 2004, 2005; Nogué et al. 2009). However, little is known about their morphological, physical, and chemical characteristics, and their taxonomic classification. So far, only a few isolated highland peat profiles have been systematically described in the field and characterized in laboratory (e.g., CVG 1991; Ramos 1997). This study contributes to filling this information gap.

Peats in the Guayana Highlands show a fragmented distribution pattern. They are located in depressions and on gentle slopes receiving regular water supply that ensures that the sites are permanently waterlogged. Peatlands are mosaics comprising many small peat areas that have important contribution to local and regional hydrology. The accumulation of organic materials is favored by high rainfall, water stagnation or restricted flow, type of vegetation cover and development, very low water pH, and relatively low temperature. Under these environmental conditions,

J.A. Zinck and O. Huber (eds.), *Peatlands of the Western Guayana Highlands, Venezuela*, Ecological Studies 217, DOI 10.1007/978-3-642-20138-7_6,

the production, accumulation, and preservation of biomass are greater than its chemical breakdown. Organic soils are often associated with organomineral soils, both resting on lithic substratum at variable depth and surrounded by large rocky terrains.

Highland peats are formed by organic debris coming from a variety of plants (grasses, forbs, shrubs, and small trees) and include roots, branch and stem fragments, leaves, flowers, and seeds, with variable addition of mineral materials. They usually have dark brown to black color, distinctive odor common of decaying vegetation, and spongy and very friable consistence. Root content decreases with depth. Stages of organic matter decomposition vary from the surface to the bottom. Commonly, the top layers are made up of younger, fibrous, and very highly compressible materials with little breakdown. Subsurface layers are slightly to moderately decomposed, and basal layers are usually the most decomposed.

This study reports on the properties of peatsoils that have developed on sandstone–quartzite mesetas (tepuis) and igneous–metamorphic massifs in the western Guayana Highlands. Visited sites are distributed over three study areas, all located in the western portion of the Amazonas state: (1) the Marahuaka-Huachamakari massif, (2) the Cuao-Sipapo massif, and (3) the western border of the Maigualida massif. Twelve organic soils and three organomineral soils were described in the field following the FAO guidelines for profile description (FAO 1990), with ad hoc adaptations for the description of organic soils. Site features and profile characteristics are reported in the Appendix of this volume. Samples were analyzed in laboratory for physical and chemical characterization according to the methods described and discussed in Schargel et al. (2011). Full determination results are presented in the Appendix. Soils were classified using the Soil Taxonomy (Soil Survey Staff 2006) and the WRB systems (FAO 2006). Univariate statistical analysis was performed for individual layers or full profiles (average estimates) depending on sample size. Statistical parameters were calculated for sample sizes of five or larger. Some soil properties with incomplete datasets, referring for instance to organic matter content, organic matter ratio, and base+aluminum saturation, were estimated for whole profiles. Simple correlation analysis was carried out on physical and chemical properties to establish affinity relationships.

This chapter describes and discusses the morphological, physical, and chemical properties of the peatsoils in the western Guayana Highlands together with their spatial variations. Taxonomic classification of the soils is presented and issues related with the classification are discussed.

6.2 Morphological Features

Most Guayana Highland peats show gradual layer variations that result from increasing organic matter decomposition with depth. Other variations in profile layering and layer features are controlled by the position on the landscape, the drainage conditions depending on the nature of and depth to bedrock, and differences in age of the peat materials. Most profiles consist of several organic layers differentiated by the degree of organic matter decomposition or humification. Peatsoils were described and classified in the field according to visual features such as color, texture, degree of wetness, root and fiber content, thickness, and layer boundaries. Root description was given special attention in terms of color, size, amount, state (live and dead), distribution pattern (vertical and horizontal), and degree of decomposition, using visual and touch appraisal. Morphological features were described according to the Soil Survey Manual (Soil Survey Staff 1993). Master horizons and layers (i.e., Oo, Oi, Oe, Oa, and A) were identified following criteria from the Soil Taxonomy (Soil Survey Staff 2006) and Soil Survey Manual.

6.2.1 Layer Types and Sequences

In the field, organic layers were identified using the following criteria:

- Folic Oo layers are made up of fresh organic residues, particularly leaves with different degrees of decomposition, live and dead roots, and fragments of branches and stems, forming a mat of variable thickness and spatial continuity.
- Fibric Oi layers are distinguished by their brownish color, poorly decomposed plant residues, live and dead roots with white and light yellow colors, and readily identifiable plant and root tissues.
- Hemic Oe layers are formed of partly decomposed plant residues that are at a stage of decomposition intermediate between those of Oi and Oa layers; with very dark colors; lower amount of roots, most of them being dead and having whitish colors; and slightly sticky organic materials, sometimes with admixtures of mineral sediments. Although water-saturated, these layers show very friable consistence and massive structure.
- Sapric Oa layers are formed of well-decomposed plant residues whose origin can no longer be identified; root content is very low in most soil profiles and the few identifiable roots are dead; most layers are very dark, friable, massive, and slightly adhesive.

Usually, organic layers have compact structure, but are only slightly plastic, providing foothold for the kind of mainly low vegetation stand they support. More plastic layers always contain larger amounts of mineral material (organomineral horizons and soils).

Rubbed and unrubbed fiber contents were taken into account for final identification of the organic layers. For many samples, rubbed and unrubbed fiber contents as determined in laboratory were relatively lower than field contents. For that reason, the field data were often preferred to identify the layers and group them for further analysis. The difference between field and laboratory data may have resulted from fiber decomposition during the relatively long time elapsed between sampling and determination. Also quartz-sand particles present in many samples possibly contributed to fiber fragmentation during the treatments. When determining fiber content in organic materials high in mineral matter, rubbed fibers were mostly quartz sand. On the basis of field observations, the following criteria were used to separate layers:

- Oi layer: more than 66% unrubbed fibers by volume
- Oe layer: 33–66% unrubbed fibers by volume
- Oa layer: less than 33% unrubbed fibers by volume

Variations in layer arrangement are largely controlled by profile position on the landscape. Peatsoils on (pseudo-)karstic landforms show commonly an Oo–Oi–Oe layer sequence. Some soil profiles exhibit Oi–Oe or, more locally, Oi–Oe–Oa sequences. Organic layers rest on sandy and fine gravelly material derived from rock weathering (Cr horizons) or directly on the bedrock (R). At slightly better drained sites such as colluvial glacis and stepped landforms, umbric A horizons sometimes overlie or underlie Oe layers, or rest on C horizons (A–Oe, Oe–A, and A–C sequences). Soils on narrow floodplains have Oi layers underlain by mineral materials with variable textures, including sand, silt loam, and silty clay loam (Oi–C). The thickness of the soil profiles above the lithic substratum, including organic and mineral materials, varies from 40 to 150 cm. Figures 6.1–6.4 illustrate the layer and horizon sequences of the soil profiles described at the different study sites, with distinction between sequences of Oo–Oi–Oe layers (Fig. 6.1), Oo–(Oi) Oe–Oa layers (Fig. 6.2), and Oo–Oi–Oe–Oa layers (Fig. 6.3). Figure 6.4 shows the horizonation in organomineral profiles.

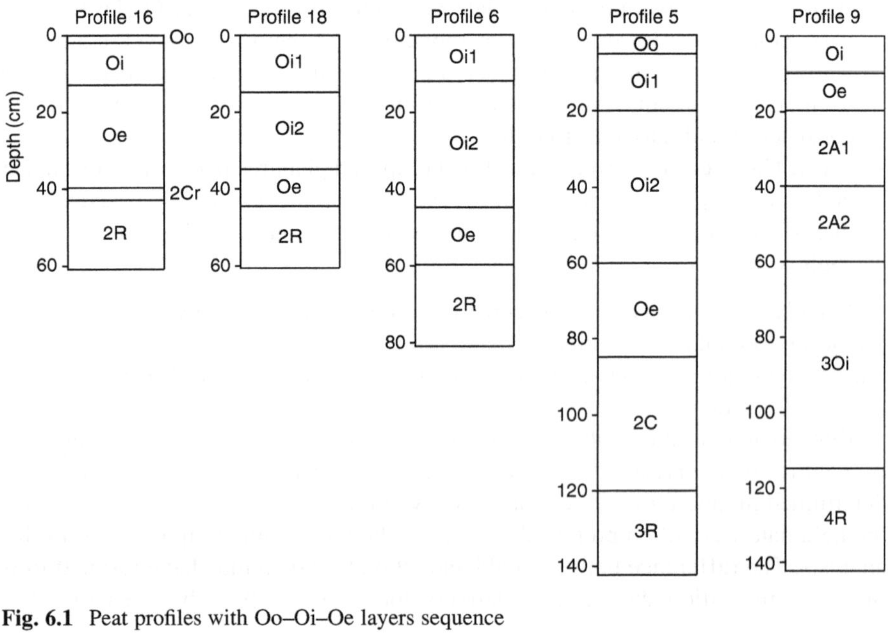

Fig. 6.1 Peat profiles with Oo–Oi–Oe layers sequence

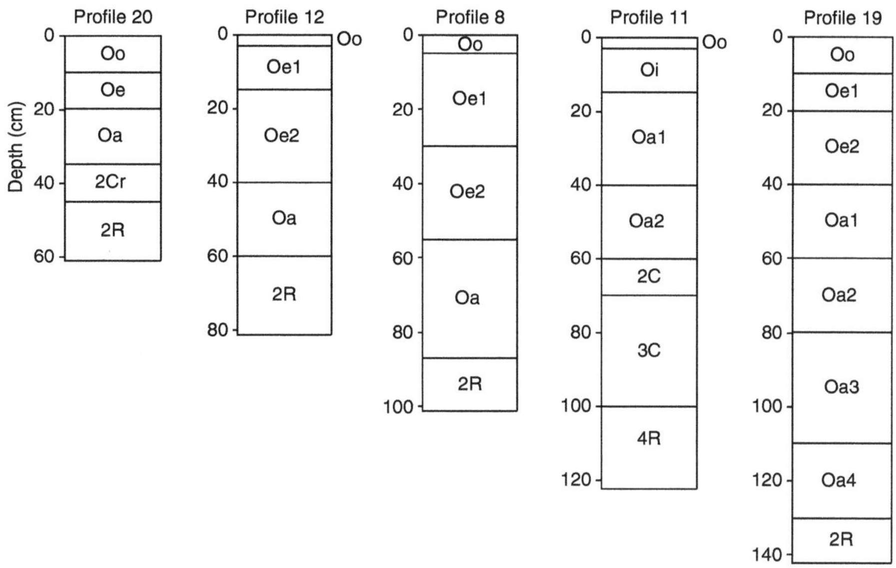

Fig. 6.2 Peat profiles with Oo–(Oi)Oe–Oa layers sequence

The upper peat layers in tepui environment are less decomposed than the deeper ones, in contrast to what happens usually in most tropical and subtropical lowland peats where the stages of decomposition decrease with depth (Zelazny and Carlisle 1974; Driessen and Rochimah 1976; Andriesse 1988). Variations in thickness per type of layer and horizon are shown in Table 6.1. In general, thickness variations of the organic layers tend to decrease from the surface Oi (CV of 74%) to the subsurface Oe and Oa layers (CV around 60%). The Oi layers are 25 cm thick on average and lie usually at the soil surface, but may also occur buried. They contain a dense mat of live and dead roots, from very coarse to very fine, and some leaves in different stages of decomposition. The Oe layers are 23 cm thick on average and start at 3–60 cm depth from the soil surface. Root content is fairly similar to that of the Oi layers; roots are medium to very fine and horizontally distributed. Structure is massive. Most Oi and Oe layers are nonplastic and nonsticky. The Oa layers are 45 cm thick on average. They start at 15–110 cm depth from the soil surface and extend to 80–150 cm depth in some profiles. In general, Oa layers are the deepest organic materials in the layer sequence, close to the bedrock or to sandy C horizons. They include very few roots, and are massive, slightly plastic, and slightly sticky.

Some profiles have mineralized umbric A horizons, 28–40 cm thick and starting at 0–20 cm depth from the soil surface. They have few to many, fine to medium, live and dead roots. Structure is blocky, and consistence is slightly sticky and slightly plastic. At some locations, the organic tier overlies sandy or finer-textured C horizons that start at 13–85 cm depth from the soil surface. The thickness of the

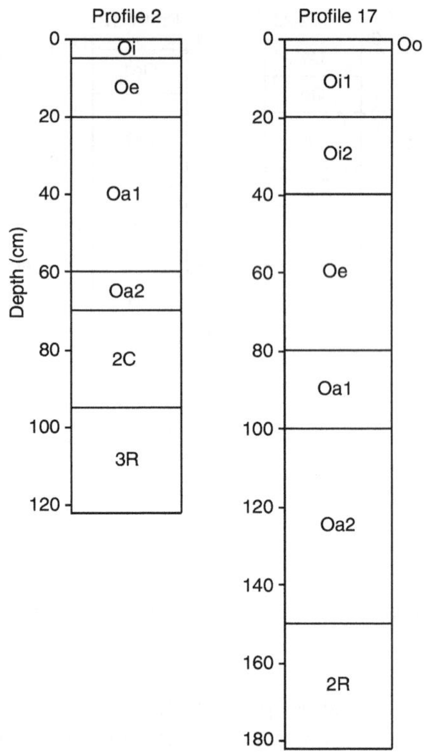

Fig. 6.3 Peat profiles with Oo–Oi–Oe–Oa layers sequence

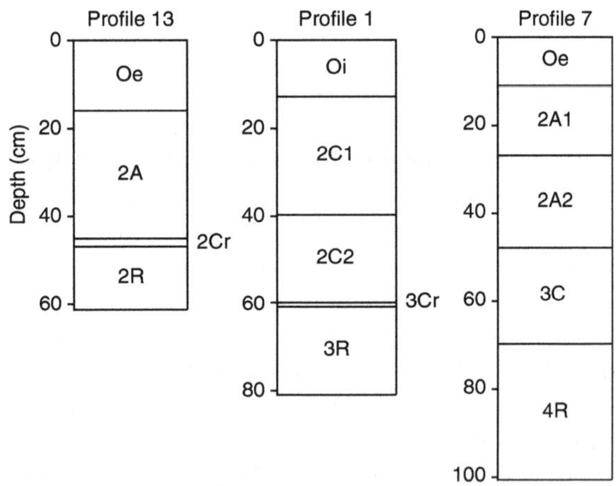

Fig. 6.4 Organomineral profiles

C horizons is 29 cm on average, but varies from 2 to 107 cm. Underlying the mineral C horizons is the bedrock, starting at 40–150 cm depth from the soil surface.

Table 6.1 Morphological features of organic and organomineral layers

Layer	Statistics	Thickness (cm)	Roots dry weight (%)	Munsell colors most common
Oi	*n*	9	6	10YR4/1; 10YR2/2; 5YR2.5/1; 2.5Y2/0
	Range	5–55	3.7–14.2	
	Mean	24.7	9.0	
	SD	18.3	4.0	
	CV (%)	74.1	44.4	
Oe	*n*	13	10	10YR2/2; 10YR2/1; 2.5Y2/0; 7.5YR4/2; 5YR2.5/1; 5YR3/1
	Range	10–50	1.6–16.8	
	Mean	22.7	8.4	
	SD	13.3	6.2	
	CV (%)	59.6	73.8	
Oa	*n*	7	5	10YR2/1; 10YR2/2; 2.5Y2/0; 5YR2.5/1
	Range	15–90	1.2–7.2	
	Mean	44.6	3.6	
	SD	28.1	2.9	
	CV (%)	63.0	80.5	
A	*n*	6	3	10 YR2/1; 2.5Y2/0; 10YR3/1
	Range	28–40	1.1–2.7	
	Mean	25.8	1.8	
	SD	12.8	–	
	CV (%)	49.6	–	

6.2.2 Root Content

Roots vary with soil depth in terms of quantity, size, distribution pattern (vertical or horizontal), and state (proportion of live and dead roots). As peats are dominantly covered by meadow vegetation and, to a minor extent, by grassland, living roots usually concentrate in a dense surficial mat (0–15/20 cm). Some roots go down to 50–60 cm; below this depth, only few shrub roots were found. A common vertical sequence of layers and root distribution based on visual estimation in the field is as follows:

- Oi layers: many roots, from fine (1–2 mm) to very coarse (10 mm or larger), vertically distributed, mainly live, and some dead slightly decomposed.
- Oe layers: moderately few to common and sometimes many roots, fine (1–2 mm) to medium (2–5 mm), vertically and horizontally distributed, commonly dead, and moderately decomposed.
- Oa layers: very few to few roots, fine and very fine (<1 mm), commonly dead, and strongly decomposed.

In general terms, root content decreases with soil depth (Fig. 6.5). There is no clear difference in average root content between Oi and Oe layers, but the amount of roots drops substantially in the Oa layers, being about 60% lower than in the overlying layers (Table 6.1). If the whole layer population is taken into account, the root content drops by 41–94% between the top of the Oi layers and the bottom

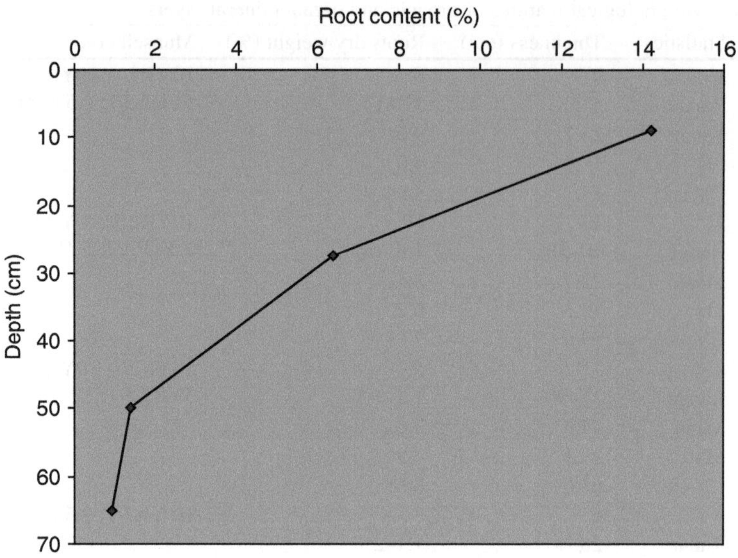

Fig. 6.5 Variation of root content with depth in peatsoil P11 (% moist roots in field moist material)

of the Oe layers. This reflects greater root growth in the surface layers and increasing decomposition of the organic residues with depth as a result of layer age. Relative variability increases from surface layers (CV of 44%) to subsurface tiers (CV of 74–81%) as a consequence of wider root content ranges. Low root content in some A horizons could indicate faster humification taking place in soils on better drained landforms (e.g., colluvial glacis and alveoli on stepped terrains).

Although live and little decomposed roots were removed prior to physical and chemical analyses, root content and organic matter content by ignition show moderate, positive correlation ($r = 0.39$). Most likely, only the coarser roots were removed and a significant proportion of the organic matter is accounted for by the remaining root mass. In some layers, the amount of dry matter removed with the weakly decomposed roots is quite high, close to 30% of the total dry matter.

The presence of thick surficial root mats in the Oo and upper portion of the Oi and Oe layers may result from waterlogging and oligotrophic conditions of the surface organic tier. This dense carpet of organic litter is an important feature of the Guayana peats, indicating that nutrient cycling basically occurs in the upper leaf and fibrous litter layers (Oo and Oi layers) which behave as the acrotelm stratum, rather than in the deeper permanently waterlogged peat, the catotelm stratum.

6.2.3 *Soil Color*

There are slight color differences between organic layers (Table 6.1). The color value tends to decrease from top to bottom of the soil profiles. The Oi layers vary between dark gray and very dark brown, although black layers can occur at the soil surface. In the Oe and Oa layers, the dominant colors are black and very dark brown. Hues of 10YR and 2.5Y are the most common in the soil matrix. Reddish hues (5YR and 7.5YR) occur in Oi and Oe layers of soils formed on granodiorite, reflecting the effect of the iron oxides released by the weathering of the mafic rock minerals. Iron oxides form quelate complexes with the organic matter that can be leached and deposited on the colloidal particles of the soil matrix. Hues of 10YR, 7.5YR, and 5YR predominate in the subsurface organic tier, indicating stronger chemical decomposition. Lighter colors in the Oi layers reflect a lower degree of decomposition and higher fiber content relative to deeper layers.

Surface A horizons are lighter colored (10YR 3/1) than subsurface ones (10YR 2/0, 10YR 2/1). Surficial A horizons form under better drainage conditions and with higher mineralization than the buried A horizons. The dark color of the latter is caused by colloidal substances percolating from the upper layers.

6.2.4 *Drainage Conditions*

Peats are extremely soft and unconsolidated surficial deposits and are water-saturated the largest part of the year. The water table is usually near or above the ground surface. Soils on colluvial fans and stepped slopes have deeper water table (80–90 cm depth) or have wet nonsatiated organic layers in response to better drainage conditions. Water surplus due to high rainfall and low evapotranspiration, low hydraulic conductivity, water retention by the organic materials, and sloping topography contribute to the concentration, accumulation, and ponding of surface water in depressions giving rise to peat bogs (Fig. 6.6).

6.3 Physical Properties

In Table 6.2, the physical property data are presented per type of organic layer and organomineral horizon. These data have been averaged to provide an overall characterization of all organic layers together in Table 6.3 and all organomineral horizons together in Table 6.4. Physical soil properties are moderately to highly variable. In general, the most significant and abrupt changes occur at 15–30 cm depth. Because of sampling and laboratory constraints, some layers lack physical data causing differences in sample size (Table A.4 in Appendix).

Fig. 6.6 Peatland drainage features. Peatlands are water-saturated over the year with patches of free water at the peat surface (**a**; *Brocchinia*, *Stegolepis, Drosera*, among others). Excess water is evacuated by small outlets (**b**) or by blackwater creeks (**c**) according to the size of the peat area (photos Zinck)

6.3.1 Dry Matter and Fiber Contents

Dry matter consists of both the organic and mineral materials after oven drying the field moist soil samples at 110°C. Average values and ranges of dry matter are as follows: 12% (5–30%), 12% (3–35%), and 16% (2–49%) in Oi, Oe, and Oa layers, respectively (Table 6.2). In most soils, dry matter increases and field water content decreases with soil depth, except in profile P19, which has very high field water

Table 6.2 Physical properties of organic and organomineral layers

Layer	Statistics	Dry matter (%)	NRF1 (%)	NRF2 (%)	NRF3 (%)	Rubbed fiber (%)	Mineral content (%)	Bulk density at field moisture (Mg m^{-3})	Bulk density of dry sample (Mg m^{-3})	Field water content (%)
Oi	*n*	10	9	9	6	9	10	9	1	12
	Range	5.3–30.2	20–60	10–50	5–40	5–50	0–77	0.56–0.70	–	231–1,782
	Mean	12.4	42.2	30	20.8	15.6	29.0	0.56	0.11	1,026
	SD	8.2	14.8	16.6	12.8	17.3	30.0	0.05	–	460
	CV (%)	66.1	35.1	55.3	61.5	110.9	103.4	8.9	–	45
Oe	*n*	16	13	13	12	13	17	10	2	15
	Range	2.6–35.3	5–60	5–60	5–40	5–40	0.86–83.6	0.54–0.78	0.10–0.23	183–3,710
	Mean	12.4	36.5	27.7	22.5	15.8	26.3	0.64	0.17	1,170
	SD	8.9	19.1	15.8	11.8	11.8	31.9	0.09	–	918
	CV (%)	71.7	52.3	57.0	52.4	74.7	121.3	14.1	–	78
Oa	*n*	10	8	8	7	8	9	5	3	10
	Range	1.9–48.7	5–50	5–50	5–40	5–20	7.3–84.8	0.56–0.76	0.14–0.23	106–5,229
	Mean	15.8	16.8	13.8	10.7	6.8	31.2	0.64	0.20	1,532
	SD	15.1	14.8	14.8	13.0	5.3	30.3	0.09	–	1,899
	CV (%)	95.6	88.1	107.2	121.5	78.0	97.1	14.1	–	124
A	*n*	6	3	3	3	3	7	3	2	7
	Range	35.2–76.2	50–60	30–50	30–40	5–30	71.3–97.1	0.66–0.76	0.78–1.34	17–184
	Mean	57	53.3	40.0	36.7	21.7	88.4	0.71	1.06	79
	SD	16.3	–	–	–	–	9.5	–	–	61
	CV (%)	28.6	–	–	–	–	10.7	–	–	77

Dry matter: organic and mineral materials after oven drying the field moist soil samples at 110°C; NRF1 to NRF3: nonrubbed fibers in three steps of increasing treatment intensity; Mineral content: mineral content after ignition at 110°C and 400°C; FWC: soil water content at field condition

Table 6.3 Variability of the physical properties in the organic layers

Statistics	Roots (%)	Dry matter (%)	NRF1 (%)	NRF2 (%)	NRF3 (%)	RF (%)	Bulk density at field moisture (Mg m^{-3})	Bulk density of dry sample (Mg m^{-3})	Mineral content (%)	Organic content (%)	Field water content (%)
n	21	36	30	30	25	30	25	6	35	38	36
Range	1–16.8	1.9–48.7	5.0–60	5.0–60	5.0–40	5.0–50	0.54–0.78	0.1–0.24	0–84.8	15.2–100	106–5,229
Mean	7.4	13.3	33	24.7	18.8	13.3	0.63	0.18	26.7	74.3	1,226
SD	5.3	10.6	19.1	16.6	13	12.7	0.07	0.06	29.1	28.5	1,171
CV (%)	71.6	79.7	57.8	67.2	69.1	95.5	11.1	33.3	108.9	38.4	96

NRF1 to NFR3: nonrubbed fibers in three steps of increasing treatment intensity; RF: rubbed fibers

Table 6.4 Variability of the physical properties in the organomineral horizons

Statistics	Roots (%)	Dry matter (%)	NRF1 (%)	NRF2 (%)	NRF3 (%)	RF (%)	Bulk density at field moisture (Mg m^{-3})	Bulk density of dry sample (Mg m^{-3})	Mineral content (%)	Organic content (%)	Field water content (%)
n	5	13	3	3	3	3	3	2	14	14	14
Range	0.9–2.7	20.3–76.2	50–60	30–50	30–40	5–30	0.66–0.76	0.78–1.34	42.7–97.8	2.2–57.3	17–392
Mean	1.44	52.4	53.3	40	36.7	21.7	0.71	–	85	15	117
SD	0.76	18.4	–	–	–	–	–	–	14.3	14.3	111
CV (%)	52.8	35.1	–	–	–	–	–	–	16.8	95.3	95

NRF1 to NFR3: nonrubbed fibers in three steps of increasing treatment intensity; RF: rubbed fibers

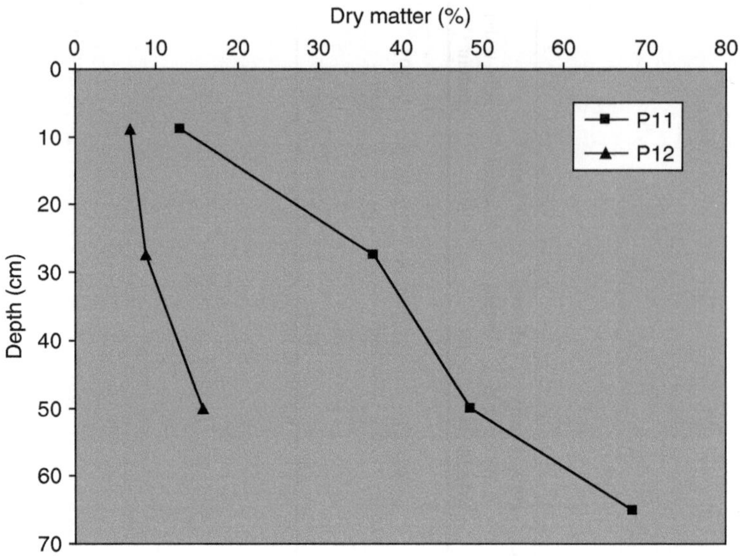

Fig. 6.7 Variation of dry matter content with depth in peatsoils on sandstone (P11) and granite (P12)

content even in the deeper, permanently saturated layers. Dry matter content increases also in the relatively better drained A horizons, especially toward the underlying C horizons. Pore space and field water content decrease markedly with increasing mineral matter, as reflected by wet and dry bulk densities. In older peat layers (e.g., P8, P12, and P17), dry matter content increases and field moisture content decreases as a consequence of the loss of pore space with age and depth. Figure 6.7 shows the vertical variation of dry matter content in peats on sandstone and granite, respectively.

Rubbed fiber content is an indicator of the physical decomposition of the organic materials and relates to physical and chemical processes occurring in the soil as a consequence of peat evolution with age (Boelter and Blake 1964; Boelter 1969; Nichols and Boelter 1984). Clear and precise layer delimitation and characterization based on fiber content were difficult due to high levels of water saturation and the presence of water tables. According to visual appraisal in the field, most peats show a physical decomposition of the fibers in the hemic range. In general, unrubbed and rubbed fibers decrease with soil depth, indicating that fiber decomposition is weak in the surface peat layers, but increases with depth close to the mineral C horizon or the lithic contact. These results agree with those reported by D'Amore and Lynn (2002), who noted increasing organic matter decomposition below the fibric surface layers in natural undrained Histosols.

Rubbed fiber amounts to about 16% in both Oi and Oe layers, but only 7% in Oa layers (Table 6.2). In contrast to this general trend, the rubbed fiber content increases with soil depth in a few young peat profiles, as for instance profile P18 that shows a marked fiber increase from 35 to 45 cm downward. In some profiles (e.g., P9, P10, P11, P12, P13, and P17), the values of rubbed and nonrubbed fiber are relatively low in the upper layer exposed to seasonal fluctuation of the water table that favors the decomposition of the organic materials.

Fiber determination strongly depends on the strength of the laboratory treatment applied. For instance, between treatment 1 and treatment 3, as described in Schargel et al. (2011), the unrubbed fiber estimates decrease by about 51, 38, and 36% for the Oi, Oe, and Oa layers, respectively (from NRF1 to NRF3 in Table 6.2). Although rubbed fiber shows the presence of quartz sand in most layers, the latter is really important and dominant in layers rich in mineral matter (e.g., 0–60 cm in P9, 0–5 cm in P10, and 15–60 cm in P11). Sand coarser than 0.15 mm that is retained on the 100-mesh sieve can contribute to the breakdown of fibers during the rubbing procedure and in treatment 3 of unrubbed fiber. Coarse sand particles are infrequent in most of the organic soils, but they are often an important component in the small and shallow peat deposits surrounding rock outcrops. Separating the coarser mineral matter by decantation is essential for fiber determination in these soils.

Although the vegetation of the peatsoils does not vary substantially in terms of taxonomic composition, differences were found in the proportions of herbaceous (tepuian grassland and meadows) and woody components (tepuian forests, bushes, and shrubs). This controls the age of the plant residues and the biochemical composition of their tissues in terms of being mainly composed by either more resistant ligneous debris or, on the contrary, readily decomposed litter. These differences affect the natural breakdown of organic materials and the proportion of rubbed fiber as an indicator of the degree of decomposition of the organic materials.

6.3.2 Mineral Content

The mineral content consists of the plant residue ash and the mineral particles that remain after organic matter has been removed on ignition at 400°C, expressed on oven-dry weight basis (Davies 1974). Mineral content increases globally with soil depth in most of the peatsoils. Average values decrease first slightly from Oi (29%) to Oe (26%) layers and then increase in the Oa layers (31%) (Table 6.2). Mineral matter content in the organic layers is highly variable, ranging from nothing to about 85%. It is much less variable in the organomineral A horizons and usually higher than 80%. Variations in mineral content are mostly related to the geomorphic processes that favored the accumulation of mineral matter in low-lying landscape positions, together with the accumulation and preservation of the organic residues. Terric materials can have different origins, including in situ rock weathering, admixture of colluvial sediments coming from surrounding rocky terrains, and alluvial deposits intercalated with organic layers in small

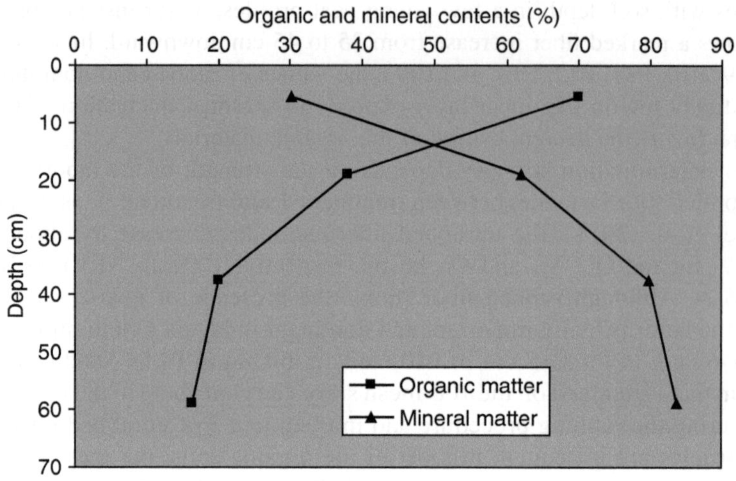

Fig. 6.8 Variation of organic matter and mineral matter contents in peatsoil P5 on sandstone

floodplains. The inclusion of mineral matter in peats is common in mountain peatlands exposed to receiving erosion debris from upslopes (Chimner and Karberg 2008). Mineral matter also results from organic matter decomposition but to a lesser extent. Higher amounts in Oa layers can be associated with increased decomposition of the organic material and the transition to the underlying mineral layers. Mineral contents and organic matter contents by Walkley–Black are closely and negatively correlated ($r = -0.59$). The pattern of mineral content increment with depth in an organic soil developed on sandstone is shown in Fig. 6.8.

6.3.3 Bulk Density

Soil bulk density was determined using two different procedures: from samples at field water content (i.e., water-saturated samples) and from oven-dry samples. Samples were taken in the center of the natural profile layers delineated in the field. Additionally, some samples were collected at 10 cm intervals at 0–40 cm depth in profiles P8, P9, and P12 to assess changes in bulk density at constant depth variation. These samples were oven dried, and bulk density was calculated on a volume base of 125 cm^3.

Wet bulk density is the physical parameter with the lowest variability. In most soils, bulk density at field moisture slightly increases with depth in the organic layers, with average values of 0.56 Mg m^{-3} in Oi layers and 0.64 Mg m^{-3} in Oe and Oa layers (Table 6.2). In some cases, the values decrease in deeper layers. Wet bulk

Table 6.5 Bulk density of dry samples

Soil profile	Layer	Depth (cm)	Bulk density (Mg m^{-3})
P8	Oo–Oe	0–10	0.04
	Oe	10–20	0.05
	Oe	20–30	0.07
	Oe	30–40	0.06
P12	Oo–Oe	0–10	0.08
	Oe	10–20	0.091
	Oe	20–30	0.091
	Oe	30–40	0.094
P9	Oi	0–10	0.23
	Oe	10–20	0.29
	A	20–30	0.63
	A	30–40	0.81

density values range from 0.54 to 0.78 Mg m^{-3}, without significant variability differences between types of organic layer. Bulk density values of the A horizons are similar to those of the organic layers in spite of larger mineral contents. The high proportion of water per soil volume obscures the differences in bulk density between organic and mineral materials.

Dry bulk density was determined on a few samples only. Average values vary with depth from 0.11 to 0.17 to 0.20 Mg m^{-3} in Oi, Oe, and Oa layers, respectively (Table 6.2). Dry bulk density variations in samples collected at 10 cm intervals reflect the proportion of organic matter in the samples (Table 6.5). In profiles P8 and P12 with more than 90% organic matter content, the values of dry bulk density range from 0.04 to 0.09 Mg m^{-3} with slight increases with depth, while in profile P9 bulk density varies from 0.23 to 0.81 Mg m^{-3} related with organic matter contents decreasing from 83 to 15% with depth. These values are in agreement with those reported by Andriesse (1988) and D'Amore and Lynn (2002). They are also partially in line with the data provided by Driessen and Rochimah (1976), and Tie and Kueh (1979) for lowland peats from Malaysia and Indonesia, showing bulk density values of less than 0.1 Mg m^{-3} in fibric peats and 0.13–0.2 Mg m^{-3} in sapric peats.

The low bulk density of the peat materials in our study area is related to the high pore space occupied by water. The permanently waterlogged layers are only slightly compacted with depth. In the surface layers, bulk density values are also low because of the abundance of large and irregular pores, many of them being probably root channels as already noted by Caldwell et al. (2007) in peatlands from southeastern USA. In contrast, in many tropical lowland peats, bulk density is higher in the surface than subsurface layers because of more advanced organic matter decomposition. Furthermore, bulk density determined at field water content increases with increasing mineral content ($r = 0.58$) because part of the total pore space is occupied by mineral matter. Large mineral contents in A horizons cause dry bulk density to increase (0.78–1.34 Mg m^{-3}).

Bulk density is related to organic matter content, with $r = -0.40$ for organic matter by Walkley–Black (OM1) and $r = -0.57$ for organic matter by ignition (OM2). The weight loss on ignition explains better the effect of organic matter on bulk density than the data provided by the other method. In fact, OM1 is not considered a good estimator of organic carbon in organic soils. The correlation of bulk density to the differences between OM2 and OM1 values gave a coefficient of −0.52. Smaller differences were found in soils with lower organic matter content by ignition.

6.3.4 Field Water Content

The field water content (FWC %) represents the difference in weight between the soil sample at (saturated) field conditions and the sample oven dried at 105°C. Average FWC values are 1,026, 1,170, and 1,532% for Oi, Oe, and Oa layers, respectively (Table 6.2). FWC is highly variable among individual organic layers and whole profiles, with extreme recorded values of 106 and 5,229%. The above-mentioned FWC averages obscure the fact that, in most soil profiles, field water contents actually decrease with depth from hemic to sapric layers, in agreement with similar results obtained by Katimon and Melling (2007).

Field water content is related to other physical and chemical properties. FWC decreases with increasing dry matter content ($r = -0.71$) and increasing mineral matter content ($r = -0.43$). FWC correlates with organic matter contents by Walkley–Black (OM1) and by loss on ignition (OM2) at $r = 0.53$ and $r = 0.43$,

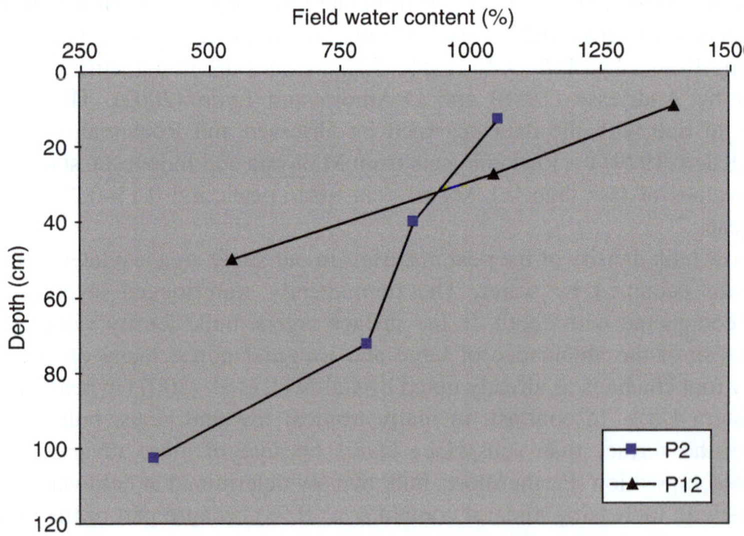

Fig. 6.9 Variation of field water content with depth in peatsoils on sandstone (P2) and granite (P12)

respectively. Water content increases as bulk density decreases ($r = -0.58$). Field water contents of peatsoils on sandstone and granite are compared in Fig. 6.9. Levels of FWC in organomineral soils are significantly lower than those in organic layers (Table 6.2).

6.4 Chemical Properties

In Tables 6.6 and 6.7, the chemical property data are presented per type of organic layer and organomineral horizon, respectively. These data have been averaged to provide an overall characterization of all organic layers together in Table 6.8 and all organomineral horizons together in Table 6.9. Some layers lack chemical determinations causing differences in sample size that can affect the statistical analysis. The full dataset used for describing and assessing the chemical properties is shown in Table A.5 in the Appendix. The peatsoils vary markedly in organic matter content from nearly pure organic material (e.g., P17) to layers borderline to mineral soils. Trend analysis with soil depth is based upon the different organic layers as identified in the field.

6.4.1 Organic Matter Content

Average organic matter content in organic layers is 74% by loss on ignition (OM2) and 28% using the Walkley–Black method (OM1) (Table 6.8). OM1 and OM2 are moderately correlated ($r = 0.47$), while OM1 shows a moderate negative correlation with mineral material ($r = -0.33$). In general, OM1 values are substantially lower than OM2 values. The average OM2:OM1 ratio for all sampled organic layers is 3.8. However, in a few layers with high mineral content in profiles P7, P9, P11, and P13, OM1 values resulted to be larger than OM2 values. This is an unlikely outcome, most probably due to insufficient ignition of samples containing mineral material that prevented the organic matter from full combustion.

In general, organic matter contents decrease with peat depth (Fig. 6.10). The decrease is more pronounced for OM2 than for OM1 in most soils high in organic matter (profiles P8, P12, P17, P18, and P19). Lee et al. (1988) found that the easily oxidizable organic carbon, obtained using the Walkley–Black procedure, remained relatively constant with depth in fibric, hemic, and some sapric materials, decreasing only in the most decomposed sapric material. In general, changes in organic matter content with depth are more related to the mixture of organic and mineral materials along the profiles than to the decomposition of the organic materials.

Peat materials vary spatially, with amounts of organic matter higher than 90% at some sites (P8, P12, P16, P17, P18, and P19). On average, OM2 values are spatially less variable than OM1 values. Organic matter contents vary with elevation in organic as well as in organomineral soils. Values of OM1 are higher at

Table 6.6 Chemical properties of the organic layers

Layer	Statist.	OM1	OM2	OM2: OM1	pH1	pH2	Ca	Mg	Na	K	Bases	Acd1	Acd2	CEC1	CEC2	BS1	BS2	P
Oi	*n*	12	12	12	9	9	13	13	13	13	13	12	9	11	8	11	8	9
	Range	6.7–55.2	23–100	1.1–4	4–5.2	3.5–3.8	0–2.1	0.1–0.8	0.2–1.1	0.1–1.3	0.7–3.3	29–185	1.6–9.0	30.7–186.7	3.2–10.7	1–7	16–67	0–7
	Mean	33.3	72.6	2.5	4.6	3.6	0.42	0.34	0.47	0.4	1.63	68.6	3.1	72.5	5	2.9	42.5	1.48
	SD	16.5	28.4	0.92	0.41	0.13	0.53	0.21	0.28	0.32	0.77	44.3	2.3	45.8	2.3	2.1	15.1	2.81
	CV (%)	49.5	39.1	36.8	8.9	3.6	126.2	61.7	59.6	80.0	47.2	64.6	74.2	63.2	46.0	72.4	35.5	189.8
Oe	*n*	13	15	11	13	13	14	14	14	14	15	16	8	14	8	14	8	12
	Range	12.1–45.7	20.2–99.1	1.1–5.7	3.2–5.3	3.2–5.2	0–3.4	0–2.1	0.2–1.1	0–1.6	0.6–7.4	19–136	0.6–5.2	21.2–141.3	2.4–8.2	1–15	16–79	0–24
	Mean	27.8	82.2	2.9	4.2	3.8	1.03	0.56	0.61	0.65	2.9	65	3.2	69.3	5.1	5.1	37.4	6.0
	SD	10.7	26.9	1.3	0.65	0.74	1.1	0.58	0.29	0.53	2.11	32.2	1.46	34.5	2.1	4.5	19.7	8.8
	CV (%)	38.5	32.7	44.8	15.4	19.5	106.7	103.6	47.5	81.5	72.7	49.5	45.3	49.7	41.2	88.2	52.6	146.7
Oa	*n*	7	10	5	8	8	10	10	10	10	10	10	5	10	5	10	5	9
	Range	3.4–30.9	15.2–92.7	2.9–27.1	3.2–5.0	3–4.7	0–2.1	0–0.7	0.2–1.1	0–0.3	0.5–3.4	16.5–117	2.6–10.2	19.9–118.3	3.1–11.5	3–17	9–29	0–18
	Mean	19.6	68.3	8.7	4.3	3.9	0.63	0.2	0.44	0.12	1.39	53.0	5.64	54.4	6.7	5.0	16.4	4.9
	SD	11.1	28.6	10.3	0.5	0.69	0.77	0.23	0.3	0.1	0.95	30.3	2.92	30	3.1	2.22	7.8	5.5
	CV (%)	56.6	41.8	118.4	11.6	17.7	122.2	115	68.2	83.3	68.3	57.2	51.7	55.1	46.3	44.4	47.5	112.4

OM1: % organic matter by Walkley–Black; OM2: % organic matter by loss on ignition (105–400°C); pH1: pH measured at field moisture; pH2: pH measured in 0.1 M $CaCl_2$;Ca, Mg, Na, K: exchangeable cations by neutral 1 N ammonium acetate; Acd1: acidity determined with trietanolamine; Acd2: acidity determined with KCl; CEC1: cation exchange capacity by sum of cations; CEC2: effective cation exchange capacity from KCl values; BS1: % base saturation from CEC1; BS2: % base saturation from CEC2; P: phosphorus in mg kg^{-1} using the Olsen method; exchangeable cations, total bases, acidity, and cation exchange capacity in cmol(+) kg^{-1}

Table 6.7 Chemical properties of the organomineral horizons

Horizon	Statist.	OM1	OM2	OM2:OM1	pH1	pH2	Ca	Mg	Na	K	Bases	Acd1	Acd2	CEC1	CEC2	BS1	BS2	P
A	*n*	8	7	7	8	8	7	7	7	7	7	7	7	7	7	7	7	7
	Range	1.2–30.9	2.9–28.7	0.47–1.2	3.4–5.0	3.1–4	0.1–0.7	0–0.1	0.2–0.7	0.1–0.4	0.6–1.5	16–110	1.2–9.8	16.7–111	2.2–10.8	1–7	9–54	–
	Mean	12.2	11.6	0.76	3.9	3.5	0.27	0.04	0.51	0.19	1.0	52.4	3.8	53.4	4.8	2.9	30.3	0
	SD	8.9	9.5	0.27	0.57	0.29	0.23	0.05	0.19	0.11	0.33	38.8	3.3	38.9	3.2	2.5	20.3	–
	CV (%)	72.9		81.9	14.6	8.3	85.2	125	37.3	57.9	33.0	74.0	86.8	72.8	66.6	86.2	66.9	–
C	*n*	7	5	–	8	8	10	10	10	10	10	10	3	10	3	7	6	4
	Range	0.34–8.6	2.2–57.3	–	3.4–5.5	3.1–4.6	0–0.3	0–0.1	0.07–0.6	0–0.1	0.07–0.1	0.2–35	1.4–4.2	0.27–36	2–4.6	1–7	3–30	0–4
	Mean	3.7	19.9	–	4.8	4.0	0.11	0.01	0.23	0.04	0.37	13.7	3.0	14.1	3.7	2.83	15.7	1.0
	SD	3.4	19.7	–	0.71	0.49	0.13	0.03	0.18	0.05	0.31	12.5	–	12.8	–	2.2	10.2	2.0
	CV (%)	91.9	98.9	–	14.8	12.3	118.2	300	78.3	125	97.8	91.2	–	9.1	–	77.7	64.5	200.0

OM1: % organic matter by Walkley–Black; OM2: % organic matter by loss on ignition (105–400°C); pH1: pH measured at field moisture; pH2: pH measured in 0.1 M $CaCl_2$; Ca, Mg, Na, K: exchangeable cations by neutral 1 N ammonium acetate; Acd1: acidity determined with trietanolamine; Acd2: acidity determined with KCl; CEC1: cation exchange capacity by sum of cations; CEC2: effective cation exchange capacity from KCl values; BS1: % base saturation from CEC1; BS2: % base saturation from CEC2; P: phosphorus in mg kg^{-1} using the Olsen method; exchangeable cations, total bases, acidity, and cation exchange capacity in cmol(+) kg^{-1}

Table 6.8 Variability of the chemical properties in the organic layers

Statistics	OM1	OM2	OM2:OM1	pH1	pH2	Ca	Mg	Na	K	Bases	Acd1	Acd2	CEC1	CEC2	BS1	BS2	P
n	30	32	28	30	30	37	37	37	37	37	38	22	35	21	35	20	29
Range	3.4–55.2	15.2–100	1.1–27.1	3.2–5.3	3.1–5.2	0–3.4	0–2.1	0.2–1.1	0–1.6	0.5–7.4	16.5–185	0.6–10.2	19.9–186.7	2.4–11.5	1.0–17.0	9.0–79.0	0–24.0
Mean	28.1	74.3	3.8	4.3	3.7	0.71	0.38	0.52	0.42	2.0	63	3.7	66.1	5.4	4	33.4	4.3
SD	13.8	27.7	4.6	0.57	0.6	0.87	0.42	0.29	0.43	1.6	35.6	2.3	36.9	2.4	4	18.3	6.8
CV (%)	49.1	37.3	121.1	12.3	16.2	122.5	110.5	55.7	104.8	80.0	56.5	62.2	55.8	44.4	100.0	54.8	158.1

OM1: % organic matter by Walkley–Black; OM2: % organic matter by loss on ignition (105–400°C); pH1: pH measured at field moisture; pH2: pH measured in 0.1 M $CaCl_2$; Ca, Mg, Na, K: exchangeable cations by neutral 1 N ammonium acetate; Acd1: acidity determined with trietanolamine; Acd2: acidity determined with KCl; CEC1: cation exchange capacity by sum of cations; CEC2: effective cation exchange capacity from KCl values; BS1: % base saturation from CEC1; BS2: % base saturation from CEC2; P: phosphorus in mg kg^{-1} using the Olsen method; exchangeable cations, total bases, acidity, and cation exchange capacity in cmol(+) kg^{-1}

Table 6.9 Variability of the chemical properties in the organomineral horizons

Statistics	OM1	OM2	pH1	pH2	Ca	Mg	Na	K	Bases	Acd1	Acd2	CEC1	CEC2	BS1	BS2	P
n	14	14	16	16	18	18	18	19	18	18	10	18	10	15	13	4
Range	0.34–30.9	2.2–57.3	3.4–5.5	3.1–4.6	0–0.7	0–0.1	0.07–0.7	0–0.4	0.07–1	0.2–45.8	1.2–9.8	0.27–111	2–10.8	1.0–7.0	3.0–54.0	0–4
Mean	8.5	15	4.5	3.8	0.17	0.02	0.35	0.09	0.36	16.6	3.5	31.2	4.5	2.8	24.5	1.0
SD	7.6	14.3	0.78	0.48	0.19	0.04	0.24	0.1	0.44	31.5	2.8	31.8	2.7	2.2	17.1	2.0
CV (%)	89.4	95.3	17.3	12.6	111.7	200.0	68.6	111.1	122.2	189.8	80.0	101.9	60.0	78.6	69.8	200.0

OM1: % organic matter by Walkley–Black; OM2: % organic matter by loss on ignition (105–400°C); pH1: pH measured at field moisture; pH2: pH measured in 0.1 M $CaCl_2$; Ca, Mg, Na, K: exchangeable cations by neutral 1 N ammonium acetate; Acd1: acidity determined with trietanolamine; Acd2: acidity determined with KCl; CEC1: cation exchange capacity by sum of cations; CEC2: effective cation exchange capacity from KCl values; BS1: % base saturation from CEC1; BS2: % base saturation from CEC2; P: phosphorus in mg kg^{-1} using the Olsen method; exchangeable cations, total bases, acidity, and cation exchange capacity in cmol(+) kg^{-1}

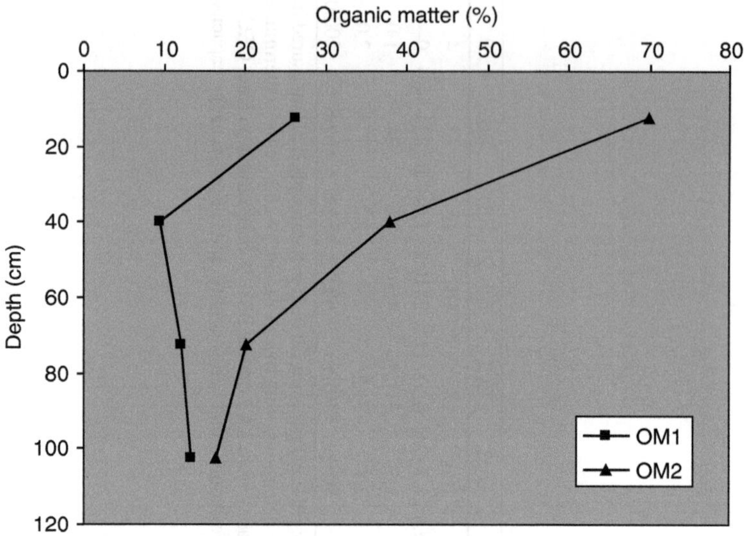

Fig. 6.10 Variation of organic matter with depth in a full organic peatsoil (P5) on sandstone. OM1 determined by Walkley–Black and OM2 by loss on ignition

650–1,800 m a.s.l. (14–55%) than above 1,800 m a.s.l. (6–26%). This trend can be explained by the higher soil temperature at lower elevation that increases the mineralization rate of organic materials in the less anoxic surface layers.

Leaves that accumulate at the soil surface constitute a large proportion of the folic layers (Oo). Temporary aeration of the topsoil during the short dry season in January–March promotes fairly rapid decomposition of this surface litter. Roots and wooden plant fragments such as branches and stems from bushes and shrubs are the main constituents of subsurface and deeper organic tiers. They decompose slowly under anaerobic conditions because of high lignin content and high C:N ratio. They are the most important source of carbon for peat formation.

In organomineral soils, organic matter contents of the A and C horizons are significantly lower than those of peat layers (Tables 6.7 and 6.9). OM1 and OM2 values are alike in A horizons (around 12% on average) except in the uppermost topsoil (Fig. 6.11), while OM2 values are several times higher than OM1 values in the C horizons (20% vs. 3%, respectively). In the mineral and organomineral horizons, organic matter is strongly mineralized and transformed into humic substances. The OM2:OM1 ratio in the A horizons varies between 0.5 and 1.2 (average 0.76). Values equal to or lower than 1 indicate higher degree of organic matter decomposition and humification. The two methods used for organic carbon determination generate similar results when dealing with organic materials evolving in better drained conditions and with higher mineralization rate.

On the basis of organic material and organic matter contents, the peat profiles can be clustered into three groups:

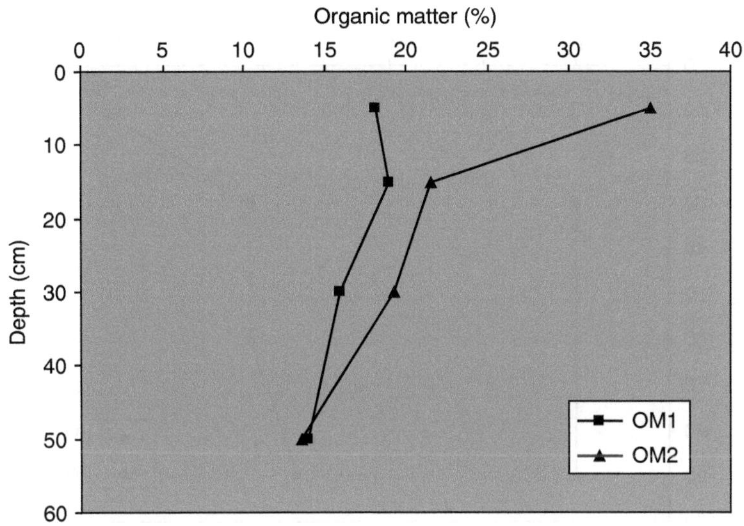

Fig. 6.11 Variation of organic matter with depth in an organomineral peatsoil (P9) on sandstone. OM1 determined by Walkley–Black and OM2 by loss on ignition

- Group A: soils having 40 cm or more organic material within 80 cm of the soil surface, and an average of more than 80% organic matter by loss on ignition (profiles P8, P12, P16, P17, P18, and P19). In most profiles, the organic material overlies the rock substratum.
- Group B: soils having 40 cm or more organic material within 80 cm of the soil surface, and an average of less than 80% organic matter (profiles P2, P5, P6, P9, P11, and P20). In general, an unconsolidated mineral layer of variable origin (weathering, colluvial, or alluvial products) is found between the organic material and the underlying bedrock. In a few cases, the organic layers overlie directly the rock substratum.
- Group C: soils having less than 40 cm organic material (profiles P1, P7, P10, and P13). Thin organic layers overlie alluvium, colluvium, or weathered rock.

6.4.2 Soil Reaction and Acidity

All organic soils in the Guayana Highlands are very acid. The pH values measured at field moisture (pH1) and in $CaCl_2$ solution (pH2) vary from ultra acid (pH 3.1–3.5) to strongly acid (pH 5.2–5.3). Values of pH1 and pH2 are closely correlated ($r = 0.57$). There are no significant differences between pH1 and pH2 values in both organic layers and mineral horizons. On average, pH1 values are higher by 0.2–1.4 units than pH2 values in surface layers, but differences are minimal in subsurface layers. In a few soils (e.g., P8 and P19), the opposite occurs.

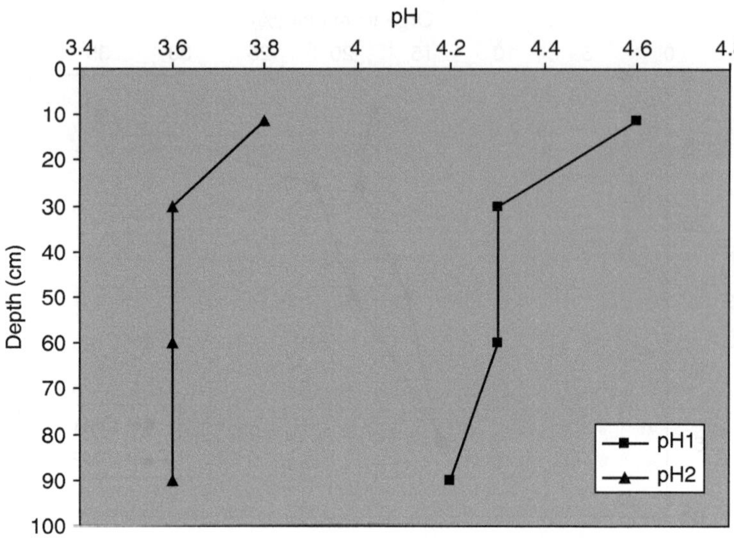

Fig. 6.12 Variation of soil reaction with depth in peatsoil P17 on sandstone. Values of pH1 measured at field moisture and pH2 in $CaCl_2$

In most soils, pH1 and pH2 values decrease gradually with depth or sometimes do not vary substantially, while increasing in the deeper mineral horizons (Fig. 6.12). Increases of 0.2–0.6 pH units were recorded in subsurface organic layers around 35–60 cm depth. In most soils, Oi layers have higher pH values than Oe and Oa layers and show more significant differences between pH1 and pH2 probably because of lesser organic matter decomposition and lower content of H^+. Generally speaking, pH is the least variable property of the organic layers. Electric conductivity determined on a few profiles (P19 and P20) was very low (0.1 dS m^{-1} on average).

Exchangeable acidity is strong in peatsoils. Both total acidity (Acd1) and salt-replaceable acidity (Acd2) show high variability within and among soils. Total acidity is 3–43 times greater than the active salt-replaceable acidity. In general, total acidity decreases with soil depth from Oi to Oa layers (from 69 to 53 cmol(+) kg^{-1}), while salt-replaceable acidity increases from Oi to Oa layers (from 3.1 to 5.6 cmol(+) kg^{-1}) (Table 6.6). Topsoils including Oi layers, shallow Oe layers, and A horizons have the lowest values of salt-replaceable acidity (3.1–3.8 cmol(+) kg^{-1}). In some soils, the decrease in total acidity with depth is related to the inclusion of mineral matter, while in others the decrease occurs in layers with similar organic matter content. In some cases, total acidity changes little with depth in profiles with high organic matter content (P17 and P19) or with organic matter decrease (P5 and P13). Higher total acidity in most of the less decomposed surface layers indicates that decomposition does not increase the density of functional groups; in most cases, a decrease occurs with greater decomposition of the organic material.

Total acidity is closely related to the content of organic matter by Walkley–Black ($r = 0.48$), but not so to the organic matter content by loss on ignition ($r = 0.24$). A large proportion of the organic matter is not chemically active at the low pH of these soils. High values of total acidity are related to the pH-dependent surface charge of the humus substances, with H^+ ions strongly tied up in the carboxyl and hydroxyl functional groups which behave as weak acids. Entrained Fe^{3+} and Al^{3+} ions or hydroxy-Al and -Fe ions are responsible for the weak-acid behavior of the organic matter (Coleman and Thomas 1967).

The salt-replaceable acidity could be associated with the presence of carboxyl groups behaving as strong acids in layers with low pH. The pH at field moisture is often 4.2 or less in layers where the salt-replaceable acidity is mostly composed of H^+ or layers that have very high organic matter content. Al^{3+} is a significant part of the salt-replaceable acidity in some layers relatively high in mineral matter, as well as in some layers with very high organic matter content. Al^{3+} saturation changes with soil depth, decreasing from Oi (12%) to Oe layers (6%) and strongly increasing in the Oa layers (43%). Also A horizons have fairly high exchangeable Al^{3+} (35%). Mineral content and exchangeable Al^{3+} show moderate-to-high correlation ($r = 0.59$). Salt-replaceable acidity and age of the organic layers are moderately related ($r = 0.45$), although variations in this relationship among profiles with similar organic matter contents suggest that factors other than age may be more important.

6.4.3 Cation Exchange Capacity and Exchangeable Cations

The cation exchange capacity (CEC) of the peatsoils is highly variable between layers (Table 6.6) and among profiles (Table A.5 in Appendix). The cation exchange capacity by sum of bases plus total acidity (CEC1) ranges between 20 and 187 cmol(+) kg^{-1}, with an average of 66 cmol(+) kg^{-1}. Mean values of CEC1 are 73, 69, and 54 cmol(+) kg^{-1} for Oi, Oe, and Oa layers, respectively, with coefficients of variation ranging between 50 and 63%. The effective cation exchange capacity determined by the sum of bases plus salt-replaceable acidity (CEC2) varies between 2 and 12 cmol(+) kg^{-1}, with an average of 5 cmol(+) kg^{-1}. Mean values of CEC2 are 5, 5, and 7 cmol(+) kg^{-1} for Oi, Oe, and Oa layers, respectively, with coefficients of variation ranging between 41 and 46%. In most profiles, CEC1, CEC2, and total acidity decrease with soil depth (Fig. 6.13). As the total bases are higher in the upper layers, their decrease with depth is more pronounced than the decrease of salt-replaceable acidity. Soils with similar organic matter contents differ significantly in CEC values. In some cases, soils with lower organic matter contents have higher CEC1 and CEC2 values than those with higher organic matter contents. Kamprath and Welch (1962) report cation exchange capacity values at pH 7 for soil organic matter to vary between 62 and 279 cmol(+) kg^{-1} and relate these considerable fluctuations to the stage of

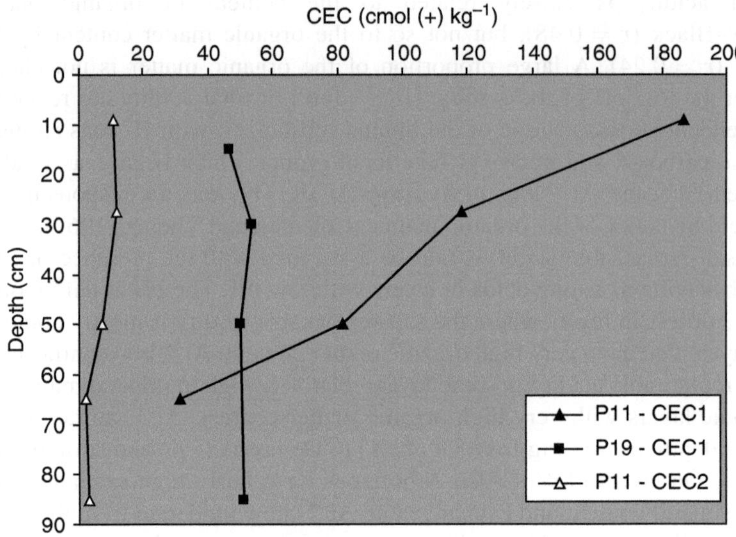

Fig. 6.13 Variation of cation exchange capacity with depth in peatsoils on sandstone (P11) and granite (P19). CEC1 determined by sum of cations and CEC2 from KCl

decomposition of the organic residues and differences in the original source of the organic materials.

In most organic soils, the relationship between cation exchange capacity and organic matter content by Walkley–Black is stronger for CEC1 ($r = 0.59$) than for CEC2 ($r = 0.31$). The relationship between CEC and organic matter content by ignition is in general weak ($r = 0.25$), indicating that the easily oxidizable organic matter is a better estimator of the cation exchange capacity in these soils. Exchangeable bases and effective cation exchange capacity (CEC2) are low in these very acid organic soils. Only strongly ionized sites are involved in the exchange reactions. The much higher values of CEC1 include weakly ionized sites which are inactive in acid soils.

In the organomineral soils, the values of CEC1 and CEC2 are significantly lower and vary from 0.3 to 111 cmol(+) kg^{-1} and from 2 to 11 cmol(+) kg^{-1}, respectively (Table 6.7). Commonly, CEC values decrease with soil depth. However, in cases where organomineral layers overlie organic layers (e.g., Oe layers), CEC increases as a response to higher contents of organic matter (Table A.5 in Appendix).

Levels of exchangeable bases are low in general, but highly variable among and within soil profiles, with coefficients of variation above 100% in the case of Ca and Mg (Table 6.6). For most soils, Ca and Na are the least scarce bases. For all organic layers together, the relative abundance of exchangeable bases is approximately $Ca^{+2} > Na^{+1} > K^{+1} > Mg^{+2}$, with average levels of 0.71, 0.52, 0.42, and 0.38 cmol(+) kg^{-1}, respectively. Exchangeable bases are mainly stored in the surface layers through nutrient cycling, while often reaching trace values in deeper layers (Fig. 6.14). This pattern is common in peats on sandstone, while peats on

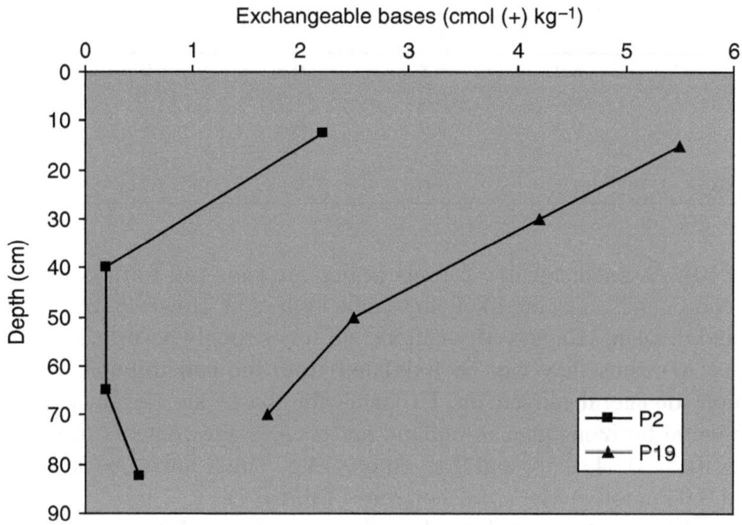

Fig. 6.14 Variation of the sum of exchangeable bases with depth in peatsoils on sandstone (P2) and granite (P19)

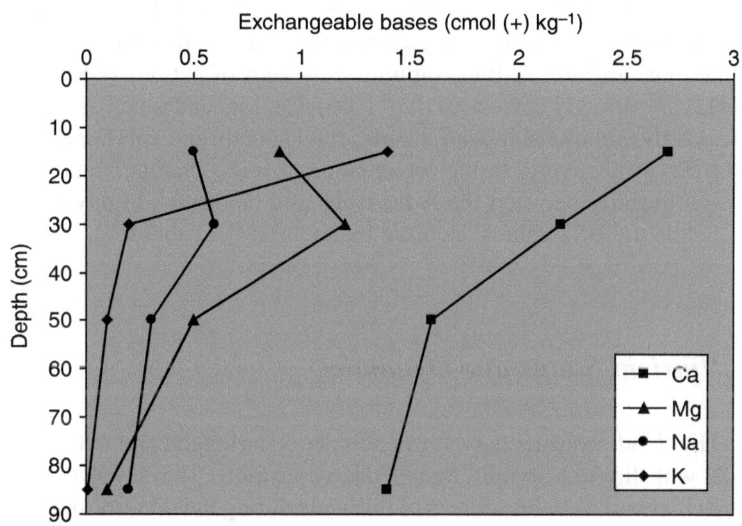

Fig. 6.15 Variation of exchangeable bases with depth in peatsoil P19 on granite

granite show a slight increase in exchangeable bases in the layers close to the rock weathering front rich in base-bearing minerals. Average values of the four macroelements Ca, Mg, Na, and K increase from Oi to Oe layers, but decrease substantially in the Oa layers (Table 6.6 and Fig. 6.15). Exchangeable K and Mg are the cations with the lowest levels of saturation in the deeper organic layers

Table 6.10 Exchangeable base saturation (%) in organic and organomineral layers

Layers	Ca^{2+}	Mg^{2+}	Na^{1+}	K^{1+}	Sum of bases
Oi	8.0	6.8	9.4	8.0	32
Oe	20.2	10.9	11.9	12.7	56
Oa	9.4	2.9	6.6	1.8	21
A	5.6	0.8	10.6	3.9	21
Organic layers	13.1	7.0	9.6	7.7	37

(Table 6.10). As exchangeable cations in organic soils can form complexes with organic compounds, they are likely to remain in the soil. However, monovalent (Na and K) and divalent (Ca and Mg) cations are less strongly adsorbed than trivalent ones (i.e., Al); thus they can be lixiviated from the soil through water runoff, percolation, or lateral movement. Exchangeable bases are significantly lower in organomineral horizons than in organic layers. Average contents of Ca, Mg, Na, and K are 0.27, 0.04, 0.51, and 0.19 cmol(+) kg^{-1} in A horizons, and 0.11, 0.01, 0.23, and 0.04 cmol(+) kg^{-1} in C horizons (Table 6.7).

Values of base saturation depend on the method used for determination of the cation exchange capacity. In the organic soils, average values of BS1 (from CEC1) and BS2 (from CEC2) are 4 and 33%, respectively. The BS2 values decrease by about 61% from the surface Oi layers to the bottom Oa layers. The BS1 values increase slightly from Oi to Oe layers but remain similar between the Oe and Oa layers (Table 6.6). Organic soils on granitic rocks have higher base saturation than soils formed on sandstones. Base saturation by sum of cations (BS1) and pH by $CaCl_2$ (pH2) are strongly correlated ($r = 0.66$). Ca, Mg, and total bases have low to moderate positive relationship with the pH at field conditions (pH1), with $r = 0.33$, 0.41 and 0.33, respectively. In the organomineral soils, average values of BS1 do not vary significantly between the A horizons and the sandy, highly dystrophic C horizons, while the BS2 values decrease by about 48% (Table 6.7).

6.4.4 *Nutrient Status and Dynamics*

Guayana Highland peatsoils have formed under waterlogged and nutrient-deficient conditions with limited organic matter decomposition. They evolve in a pluvial environment where the main sources of nutrients during the minerotrophic stage are rainfall and a tiny provision of minerals from sandstones, granites, and very locally from intrusive basic rocks (diabases). During the ombrotrophic stage, these nutrients are fixed by the vegetation successions, then released and incorporated into the peat material during biomass decomposition, and finally recycled again by the vegetation.

During peat evolution, most of the nutrients are washed out by runoff, deep percolation, seasonal fluctuation of the water table, and lateral water movement. The intensity of nutrient depletion depends on the nature and permeability of the bedrock, sandstones being more permeable than granitic rocks. Although organic materials have high total exchange capacity, the effective cation exchange capacity is low at the very acid pH of these soils. Only the strongly acid functional groups of the organic matter dissociate. High acidity is a consequence of the environmental conditions in which these organic soils evolve.

The presence of a thick surficial root mat in the Oo layers and shallower portions thereof in some Oi and Oe layers is an important feature for nutrient cycling in the Guayana peats. The concentration of exchangeable bases in the topsoil (0–30 cm) indicates that nutrient transfer and recycling occur within the upper peat layers, where organic matter decomposition increases temporarily, releasing nutrients during the short dry season. In a similar manner, Jackson et al. (2008) show that the activity of bacterial enzymes associated with carbon, phosphorus, and nitrogen cycling is highest at the peat surface and declines markedly with depth.

Conventional methods for determining available phosphorus are of limited use in acid organic soils under native vegetation, where organic phosphorus compounds are relevant. Phosphorus was determined using the Bray–Kurtz1 method for P19 and P20 and the Olsen method for the other soils. In general, available phosphorus is low, often no more than traces (Tables 6.6 and 6.7). Only a few layers in the upper 40 cm of profiles P19 and P20 have moderate levels of available P (Table A.5 in Appendix). Phosphorus is related to the organic matter content by ignition ($r = 0.40$) and to total acidity ($r = 0.65$), reflecting the relationship between available P and the functional groups of the organic compounds. These results are consistent with the findings of Andriesse (1988), who states that oligotrophic tropical peats in pristine conditions have usually very low phosphorus contents and that the net primary productivity of most natural peatlands, particularly freshwater wetlands, is limited by P deficiency.

6.5 Spatial Variations

6.5.1 Relationships with Landscape Position

Data obtained from transects at several locations show differences in peatland length, width, and depth (Table 6.11; see also Figs. 4.17–4.21 and Fig. 4.24). In general, peatland configuration is elongated. The average length-to-width ratio is 3.5:1, but can be as large as 14:1. The length of peat areas varies from 140 to 325 m, while the width ranges from 10 to 50 m. Peat areas are longest in the Maigualida massif (280–325 m) compared to 140–166 m in the other study areas. Length and width are controlled by factors such as rock hardness, joints and fractures of the bedrock, intensity of rock weathering, and (pseudo-)karst formation.

Thickness of the organic deposits varies among and within peatland areas (Table 6.12). A maximum thickness of 170 cm was observed in the Maigualida massif, with mean values of 60 cm on longitudinal transects and 113 cm on

Table 6.11 Morphometric variations of peatlands

Area	Lithology	Slope (%)	Transect	Peat length (m)	Peat width (m)	Peat depth (cm)
Cuao massif (C)	Sandstones	0.3–1	CL-1	166	20–45	20–90
			CT-1	–	28	25–120
			CT-2	–	50	40–80
			CT-3	–	30	10–50
Sipapo massif (S)	Granitic rocks	0.5–1	SL-1	140	10–40	5–90
			ST-1	–	10–15	10–50
			ST-2	–	40	30–95
		3–5	ST-3	–	38–40	15–70
Maigualida massif (M)	Granitic rocks	3–5	ML-1	325	50	30–170
		3–16	ML-2	280	30–50	10–40
		2–3	MT-1	–	50	50–140

L longitudinal transect, *T* transversal transect

Table 6.12 Thickness variability of the organic deposits (cm)

Area	Lithology	Statistics	Longitudinal transect	Transversal transect
Cuao massif	Sandstones	Total transects	1	3
		N° observations	14	9
		Range	20–90	10–120
		Mean	53.6	48.9
		SD	22.1	33.8
		CV (%)	41.2	69.1
Sipapo massif	Granitic rocks	Total transects	1	3
		N° observations	14	18
		Range	5–90	10–95
		Mean	53.6	47.6
		SD	22.1	23.5
		CV (%)	41.2	49.4
Maigualida massif	Granitic rocks	Total transects	2	1
		N° observations	62	9
		Range	10–170	50–140
		Mean	60.1	113.3
		SD	38.7	29.2
		CV (%)	64.4	25.8

transversal transects. In the Cuao-Sipapo massif, the mean depth to bedrock is 74 cm and the range is 45–150 cm, with no significant differences in peat thickness between longitudinal and transversal transects.

Soils developed on stepped slopes, colluvial fans, small floodplains, and narrow valleys are mixtures of organic and mineral layers (sand, silt loam, and silty clay loam). Mineral layers, often of allochthonous origin, occur at variable depth (13–70 cm) and with variable thickness (3–100 cm). Surficial organomineral horizons up to 20 cm thick occur locally. In (pseudo-)karstic depressions, sandy basal layers originate from rock weathering.

The lower content of roots in some A horizons, in comparison to Oi and Oe layers, indicates that a large proportion of the dead roots has been mineralized. This occurs in soils on landforms with better drainage conditions such as colluvial glacis and alveoli on stepped terrains.

There is good correlation ($r = 0.58$) between elevation and thickness of the moderately to strongly decomposed Oe and Oa organic layers. This shows the effect of elevation on temperature and evaporation, which control water surplus, and the influence of these environmental factors on the rate of accumulation and decomposition of the organic materials.

6.5.2 Relationships with Rock Substratum

Hues of 10YR and 2.5Y are the dominant colors in the soil matrix. However, soils on granite show reddish hues of 5YR and 7.5YR in the Oi and Oe layers, which could be related to the presence of iron oxides resulting from the weathering of the granitic rocks.

Root contents in the surface layers of organic soils formed on sloping landforms (karstic and pseudokarstic alveoli and colluvial glacis) are larger on granitic rocks than on sandstones (Table 6.13). Lower root content in peats on sandstones could be related to this mineralogically poorer and more permeable substratum and sandy basal layers, where nutrient depletion is favored by better drainage in sloping conditions. Vegetation differences could be an additional factor affecting the volume of roots in the surface layers.

Dry matter contents in organic soils on sandstones are larger (almost twice as much) than those of organic soils on granitic rocks, resulting in lower field water contents in the former than in the latter (Table 6.13). This could be related to higher proportions of sand, accounted as a part of the dry matter content, resulting from

Table 6.13 Variability of physical properties in peats on sandstones and granitic rocks

Bedrock	Statistics	Roots (%)	Dry matter (%)	Bulk density at field moisture (Mg m^{-3})	Bulk density of dry sample (Mg m^{-3})	Mineral content (%)	Field water content (%)
Sandstones	n	11	23	17	3	22	24
	Range	1.2–14.2	5.3–48.7	0.54–0.76	0.11–0.24	0–84.8	106–1,782
	Mean	6.6	18.0	0.63	0.19	37.5	790
	CV (%)	75.8	64.4	11.1	–	84.8	58
Granitic rocks	n	9	12	6	3	12	12
	Range	1.6–16.8	1.9–15.6	0.48–0.78	0.10–0.23	0.60–26	542–5,229
	Mean	9.1	7.2	0.60	0.15	8.23	2,081
	CV (%)	59.3	61.2	15.0	–	81.4	80

Table 6.14 Variability of % base saturation in peats on sandstones and granitic rocks

Bedrock	Ca^{2+}	Mg^{2+}	Na^{1+}	K^{1+}	Sum bases
Sandstones	7.4	5.4	9.3	5.9	28
Granitic rocks	24.4	10.5	9.9	11.4	56

weathering of the sandstones and admixture of colluvial materials. Also peats on aggradational surfaces such as narrow floodplains and glacis with large mineral contents occur mainly in sandstone areas.

Values of pH at field conditions are slightly higher in organic soils on granitic rocks (4.0–5.3) than in organic soils on sandstones (3.2–5.2). The ranges are 3.2–5.2 and 3.0–3.8, respectively, for the pH values determined with $CaCl_2$. Average levels of exchangeable Mg, K, and total bases in peats on granitic rocks are about twice as high as those in peats on sandstones, with Ca being even three times higher but no difference in Na (Table 6.14). The highest levels of exchangeable bases, total bases, and pH were obtained in peatsoils on granite in the Maigualida massif. These variations could be related to (1) lower lixiviation of bases and nutrients on massive granitic substratum, which constraints the water flow through the soil, (2) higher provision of cations from weathering of the substratum together with the inflow of cations from surrounding areas, and (3) local differences in the chemical composition of the granitic rocks, which could affect the uptake of nutrients by the vegetation and their accumulation in the plant tissues during the minerotrophic stage of peat formation, with residual effect in the ombrotrophic stage.

In conclusion, spatial variations in peatsoils are related to a set of factors that may interact at specific sites, including type and dynamics of the landforms, drainage conditions of the peatland, permeability and mineralogy of the rock substratum, age and history of the organic deposits, and type and composition of the vegetation cover, among others.

6.6 Taxonomic Classification

The soils of the Guayana Highlands are poorly to very poorly drained. They have layers composed of organic soil material of variable thickness overlying layers of mineral soil material or hard bedrock including sandstones and granites. The average content of organic carbon varies from nearly pure peat material to borderline to mineral soil material (i.e., 12% organic carbon if the mineral fraction contains no clay; up to 18% organic carbon if the mineral fraction contains 60% or more clay). Most organic soil materials in the study area contain little clay (CVG 1991).

The soils of the study area were classified according to the Soil Taxonomy (Soil Survey Staff 2006) and the World Reference Base for Soil Resources or WRB (FAO 2006), following the criteria used for the definition of organic and mineral soil materials by both systems. Table 6.15 shows the classification of the organic

Table 6.15 Classification of the organic soils

Profile	USDA classification (2006)	WRB classification (2006)
P2	Lithic Haplosaprist	Endoleptic Rheic Sapric Histosol (Hyperdystric)
P5	Lithic *Terric* Haplofibrist	Rheic Fibric Histosol (Hyperdystric)
P6	Lithic Haplofibrist	Endoleptic Rheic Fibric Histosol (Hyperdystric)
P8	Lithic Haplohemist	Endoleptic Rheic Hemic Histosol (Hyperdystric)
P9	Lithic *Terric* Haplofibrist	Rheic Fibric Histosol (Hyperdystric)
P11	Lithic *Terric* Haplosaprist	Ombric Sapric Histosol (Hyperdystric)
P12	Lithic Haplosaprist	Endoleptic Ombric Hemic Histosol (Hyperdystric)
P16	Lithic Haplohemist	Epileptic Ombric Hemic Histosol (Hyperdystric)
P17	Typic Haplosaprist	Ombric Hemic Histosol (Hyperdystric, *Thaptosapric*)
P18	Lithic Haplohemist	Epileptic Rheic Fibric Histosol (Hyperdystric)
P19	Typic Haplosaprist	Ombric Sapric Histosol (Hyperdystric)
P20	Lithic Haplosaprist	Epileptic Rheic Sapric Histosol (Hyperdystric)

Proposed: *Terric, Thaptosapric*

soils in the study area. All organic soil profiles support the requirements of the Histosol order in both classification systems (Jiménez 1995). Complementary morphological properties of the organic layers such as color and unrubbed fiber content used in the Soil Taxonomy (Soil Survey Staff 1999) were evaluated.

6.6.1 Organic Soils

6.6.1.1 Classification According to the Soil Taxonomy System

The Soil Taxonomy provides two alternatives for organic soils to be classified as Histosols: (1) the soils are saturated for 30 days or more per year in normal years, have organic soil material within 40 cm of the soil surface, and have a total thickness of either 60 cm if the bulk density moist is less than 0.1 Mg m^{-3} or 40 cm if the bulk density is higher; and (2) organic soil materials constitute two thirds or more of the total thickness of the soil to a densic, lithic, or paralithic contact, and have no mineral horizons or have mineral horizons with a total thickness of 10 cm or less. Alternative (1) is applicable to all organic soils of the study area, with profile P9 that has 12.5% organic carbon at 10–20 cm depth being a borderline case.

The suborder level in the Soil Taxonomy is defined by the degree of decomposition of the organic soil material, which is determined by the amount of rubbed fiber and pyrophosphate color in the organic part of the subsurface tier (usually 60–120 cm) or in the combined thickness of the organic parts of the surface and subsurface tiers if a mineral layer 40 cm or more thick has its upper boundary within the subsurface tier. Field determination of the degree of decomposition of the organic soil material was considered decisive, as the pyrophosphate measurement was not performed. Rubbed fiber was determined on selected layers of a few soils only. The transportation, storage, and interaction with quartz sand may have decreased the rubbed fiber content. On the basis of these criteria, most soils classify as Saprists. However, the field water content in many layers is very high when the mineral matter is low. The decomposed organic matter seems to float in water and is not compacted with age. Soil P17 is a borderline case; although the 40–80 cm layer was described in the field as moderately decomposed, the soil is placed with the Saprists because the rubbed and unrubbed fiber content is very low and the color is black. Soils P8, P16, and P18 classify as Hemists. They have relatively high field water contents and the bulk density is below 0.1 Mg m^{-3} in moderately decomposed layers of very high organic matter content. Soils P5, P6, and P9 classify as Fibrist. The latter has a slightly decomposed organic layer at 60–115 cm depth, underlying a very young mineral soil layer with relatively high organic matter content.

All the soils classified belong to the Haplo-great groups because they lack features such as a sulfuric horizon, sulfidic materials, or a cryic temperature regime, among others. At the subgroup level, most soils are Lithic extragrades because they have a lithic contact within the control section (0–130 cm), except soils P17 and P19 that classify as Typic Haplosaprists. We propose the creation of Lithic Terric subgroups to fit the soils P5, P9, and P11 because they have mineral soil layers 30 cm or more thick within the control section. In the Soil Taxonomy system, Terric subgroups key out after Lithic, and so far both criteria have not been used together, probably because Lithic Haplohemists and Lithic Haplosaprists occupy small extents in the United States, and the Lithic Haplofibrists are not known to occur there (Soil Survey Staff 1999).

At the family level, the particle-size and mineralogy classes are required for the terric layers. Histosols without terric layers do not use mineralogy classes unless the soils have ferrihumic or limnic materials. With the information available, the terric layers are labeled as sandy siliceous in P5 and P11, and as loamy siliceous in P9. Reaction classes are used for all Histosol families based upon the pH in 0.01 M $CaCl_2$ determined on undried soil. All soils evaluated by this procedure are dysic (pH < 4.5), except P19 and P20, which are euic because the pH in 0.01 M $CaCl_2$ is equal to or greater than 4.5 in one or more organic layers within the control section. P19 and P20 have formed on granitic rock and have a higher total base content, relatively low total acidity, and higher base saturation than peats high in organic matter on sandstone. It is probable that most of the strong acidic functional groups of the organic matter in these two soils are occupied by exchangeable bases. The soil temperature regime is approximated from elevation and temperature data recorded under bare soil in Venezuela by Comerma and Sánchez (1980). In organic soils saturated mostly with rainwater and under natural vegetation, the soil temperature is lower than the values obtained on bare mineral soils. On the basis of these considerations, an isomesic soil temperature is proposed for P2, P5, and probably P19, isohyperthermic for P16, and isothermic for the remaining soils. In addition, soils P16, P18, and P20 belong to the shallow soil depth class because they have a root-limiting layer between 18 and 50 cm.

Figures 6.16–6.18 show three representative peatsoils with fibric (P6), hemic (P8), and sapric (P2) materials as main profile components, respectively.

Fig. 6.16 (**a**) FIBRIC PEATSOIL on a sandstone tepui at 1,770 m a.s.l. in the Huachamakari massif. The soil is a 60 cm deep, slightly over moderately decomposed Lithic Haplofibrist with ^{14}C age 1,080 BP in the lower tier (45–60 cm) (Profile 6). (**b**) Shows a gently undulating sandstone meseta covered by shrubby meadow vegetation where the peatsoil of (**a**) is located. *Heliamphora* is a typical component of the tepui meadow (**c**) (photos Zinck)

6.6.1.2 Classification According to the WRB Framework

In the first tier of categorical detail in the WRB classification framework, the Histosol reference soil group presents two alternatives as follows: (1) organic material 10 cm or more thick starting at the soil surface and immediately overlying

Fig. 6.17 (**a**) HEMIC PEATSOIL on a sandstone meseta at 1,800 m a.s.l. in the Cuao-Sipapo massif. The soil is a 87 cm deep, moderately to strongly decomposed Lithic Haplohemist with ^{14}C age 1,665 BP in the middle tier (30–55 cm) and 3,415 BP in the lower tier (55–87 cm) (Profile 8). The summit area in (**b**) shows a mosaic of rock outcrops and small karstic depressions covered by shrubby meadow vegetation where the peatsoil of (**a**) is located. Typical plants of the meadow cover in (**c**) include *Brocchinia, Kunhardtia* and *Stegolepis* (photos Zinck)

continuous rock; and (2) organic material cumulatively within 100 cm of the soil surface either 60 cm or more thick if 75% (by volume) or more of the material consists of moss fibers or 40 cm or more thick and starting within 40 cm of the soil surface. Alternative (2) is applicable to all organic soils, and P9 is not a borderline case in the WRB classification framework.

At the second tier of categorical detail, the Histosol reference soil group is characterized by prefix qualifiers that refer to associate and intergrade attributes, and by less important suffix qualifiers that are written between brackets. The degree of fiber decomposition is not specified for the subsurface layer (below 30 cm) as it is for deep Histosols in the Soil Taxonomy; consequently, the less decomposed surface layers often determine the selection of the appropriate prefix. Field decomposition of the organic material is used in the WRB classification framework,

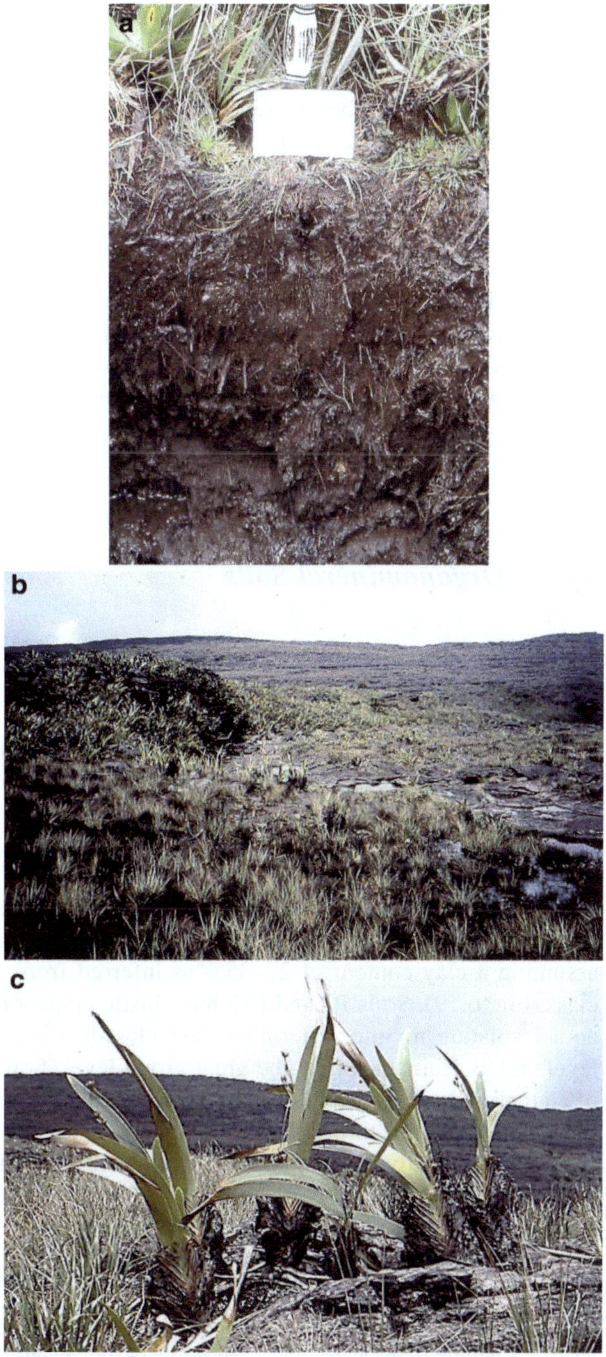

Fig. 6.18 (**a**) SAPRIC PEATSOIL on a sandstone tepui summit at 2,580 m a.s.l. in the Marahuaka massif. The soil is a 95 cm deep, strongly decomposed Lithic Haplosaprist with ^{14}C age 5,880 BP in the middle tier (20–60 cm) and 7,490 BP in the lower tier (70–95 cm) (Profile 2). (**b**) Shows a gently undulating plateau summit where the peatsoil of (**a**) has formed in a karstic depression (center of the picture) surrounded by rock outcrops. Meadow cover dominated by *Everardia* and *Stegolepis* (**c**) (photos Zinck)

because no laboratory criteria are provided for the determination of fiber content (FAO 2006). The fibric prefix indicates that an average of more than two thirds of the volume of organic material to a depth of 1 m is fiber after rubbing. In sapric material, the rubbed fiber averages less than one sixth of the volume, and hemic material is intermediate. Endoleptic means that the soil rests on continuous rock starting between 50 and 100 cm from the soil surface, and epileptic means that the soil lies on continuous rock within 50 cm. Ombric means that the Histosol is saturated predominantly with rainwater, and rheic means that the saturation is mostly with groundwater or flowing surface water. All soils are Hyperdystric at the suffix level, which means that the base saturation (by 1 M NH_4OAc) is less than 50% between 20 and 100 cm from the soil surface, and less than 20% in some layer within 100 cm of the soil surface. The Thaptosapric suffix indicates that the soil has a buried sapric layer, which is about 4,000 years older in this specific case than the overlying hemic layer.

6.6.2 Organomineral Soils

Table 6.16 shows the classification of the organomineral soils in the study area. They comprise a set of soils that do not fulfill the thickness requirements of the organic soils. They have a thin surface layer of organic soil material that is usually less than 20 cm thick. They occur in association with the Histosols and classify as Inceptisols and Entisols (Soil Survey Staff 2006), and Gleysols and Fluvisols (FAO 2006).

Applying the rules of the Soil Taxonomy, soil P1 lacks diagnostic horizons except a surface ochric horizon, which does not meet the minimum thickness of 20 cm required to fit the histic epipedon. This soil is affected by endosaturation and classifies as Typic Endoaquent, fine-silty, siliceous, semiactive, acid, and isomesic, assuming a clay content of 27–35% as inferred from the silty clay loam textural class (Fig. 6.19). Soils P7 and P13 have histic epipedons, which, together with the aquic moisture regime, determine their classification as Aquepts at the suborder level and as Humaquepts at the great group level. Soil P7 meets the requirements of an umbric horizon until the underlying rock at 70 cm. It has an irregular decrease of organic matter with depth and classifies as Cumulic Humaquept, coarse-loamy, siliceous, superactive, acid, and isothermic. Soil P13 classifies as Typic Humaquept,

Table 6.16 Classification of the organomineral soils

Profile	USDA classification (2006)	WRB classification (2006)
P1	Typic Endoaquent	*Endoleptic* Gleysol (Hyperhumic, Hyperdystric, Episiltic)
P7	Cumulic Humaquept	Histic *Endoleptic* Fluvisol (Hyperdystric)
P13	Typic Humaquept	Histic *Epileptic* Fluvisol (Hyperdystric, Epiarenic)

Proposed: *Endoleptic, Epileptic*

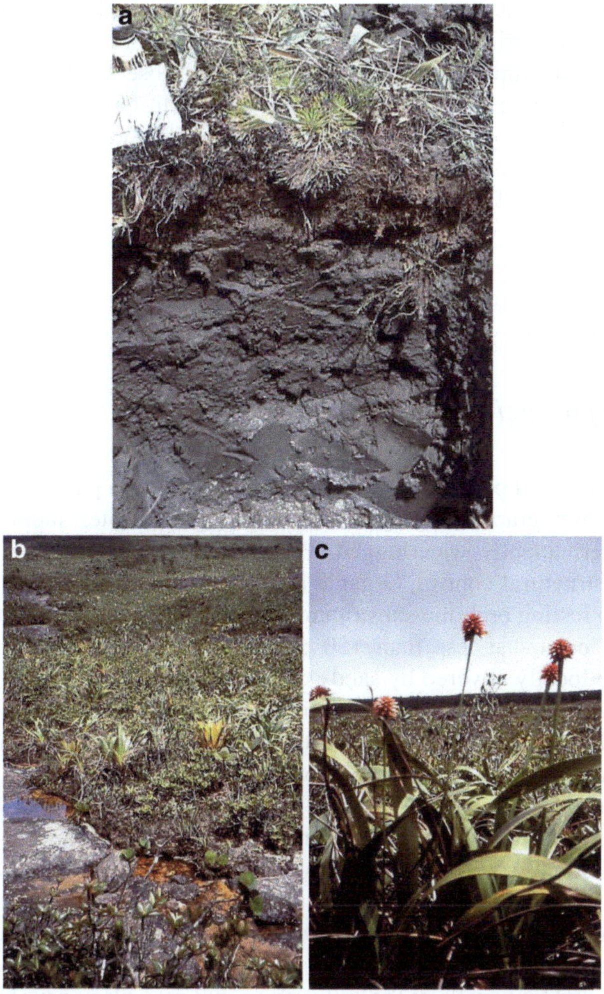

Fig. 6.19 (**a**) ORGANOMINERAL SOIL on a sandstone plateau at 2,570 m a.s.l. in the Marahuaka massif. The soil is a 60 cm deep, poorly drained Typic Endoaquent that shows a fibric surface root mat atop a colluvio-alluvial deposit containing allochthonous organic material with ^{14}C age 7,100 BP in the bottom layer (40–60 cm) (Profile 1). Narrow floodplains are frequent on tepui summits with small creeks running on bare rock and draining excess water from peatlands. Creeks also collect, transport and redeposit mineral debris provided by the weathering of rock outcrops surrounding the peatlands. The composite organomineral profile shown in (**a**) is located in the floodplain of (**b**). Mixed vegetation cover includes herbs and grasses. *Stegolepis* is a typical component (**c**) (photos Zinck)

sandy, siliceous, acid, isothermic, and shallow; the latter due to a root-limiting layer within 50 cm depth.

According to the WRB classification framework, soils P7 and P13 classify as Fluvisols because they have fluvic material characterized by stratification or by an organic carbon content decreasing irregularly with depth. The prefixes reflect the presence of a histic horizon and the depth to continuous rock. The suffixes show low base saturation and the presence of sandy layers close to the surface. Soil P1 classifies as Endoleptic Gleysoil due to reducing conditions, the lack of diagnostic horizons, and depth to continuous rock between 50 and 100 cm. The suffixes indicate high organic matter content, low base saturation, and silty texture in the upper 50 cm of the soil profile.

6.7 Conclusions

6.7.1 Morphological Features

- Guayana Highland peats show a fragmented distribution pattern and are located on specific topographic sites provided with regular water supply that ensures they are permanently waterlogged. Peatlands are generally small because of strong geostructural control. Organic and organomineral soils are associated in space, both resting on lithic substratum (mainly sandstone or granitic rocks) that commonly occurs at less than 150 cm depth. Surrounding stony and rocky terrains are locally covered by sandy colluvial mantles.
- Organic layers, including folic Oo, fibric Oi, hemic Oe, and sapric Oa, were recognized using field criteria such as color, amount of roots and other plant residues, and degree of decomposition of the organic materials. Variations in layer arrangement are largely controlled by profile position on the landscape. Peatsoils on (pseudo-)karstic landforms show commonly an Oo–Oi–Oe layer sequence. Some soil profiles exhibit Oi–Oe or, more locally, Oi–Oe–Oa sequences resting on sandy or rocky substratum.
- The thickness of the soil profiles above the lithic substratum, including organic and mineral materials, varies from 40 to 150 cm. The Oi layers are 25 cm thick on average and lie usually at the soil surface, but may also occur buried. The Oe layers are 23 cm thick on average and start at 3–60 cm depth from the soil surface. The Oa layers are 45 cm thick on average. They start at 15–110 cm depth from the soil surface and extend to 80–150 cm depth in some profiles. In general, Oa layers are the deepest organic materials in the layer sequence, close to the bedrock or to sandy C horizons. Some profiles have umbric A horizons, 28–40 cm thick and starting at 0–20 cm depth from the soil surface. At some locations, the organic tier overlies sandy or finer-textured C horizons that start at 13–85 cm depth from the soil surface. The thickness of the C horizons is 29 cm on average, but varies from 2 to 107 cm. Underlying the mineral C horizons is the bedrock, starting at 40–150 cm depth from the soil surface.
- In general, root content decreases with soil depth. There is no clear difference in average root content between Oi and Oe layers, but the amount of roots drops substantially in the Oa layers, being about 60% lower than in the overlying

layers. The presence of thick surface root mats in the Oo and upper portion of the Oi and Oe layers may result from waterlogging and oligotrophic conditions of the surface organic tier. This dense carpet of organic litter is an important feature of the Guayana Highland peats, indicating that nutrient cycling basically takes place in the upper leaf and fibrous litter layers (Oo and Oi layers).

- There are slight color differences between organic layers. The color value tends to decrease from top to bottom of the soil profiles. The Oi layers vary between dark gray and very dark brown, although black layers can occur at the soil surface. In the Oe and Oa layers, the dominant colors are black and very dark brown. Hues of 10YR and 2.5Y are most common in the soil matrix. Reddish hues (5YR and 7.5YR) occur in Oi and Oe layers of soils formed on granodiorite, reflecting the effect of iron oxides that result from the weathering of the mafic rock minerals.
- Peats are soft and unconsolidated deposits and are water-saturated the largest part of the year. The water table is usually near or above the ground surface. Water surplus due to high rainfall and low evapotranspiration, low hydraulic conductivity, water retention by the organic materials, and sloping topography contribute to the concentration, accumulation, and ponding of surface water in depressions giving rise to peat bogs.

6.7.2 Physical Properties

- Dry matter content increases with depth from 35 to 47% in Oi layers and from 43 to 87% in Oe layers. Also in the A horizons, relatively better drained, dry matter content increases especially toward the underlying C horizons. Pore space and field water content decrease markedly with increasing mineral matter.
- In general, unrubbed and rubbed fibers decrease with soil depth, indicating that physical fiber decomposition is weak in the surface peat layers, but increases with depth close to the mineral C horizon or the lithic contact. Rubbed fiber amounts to about 16% in both Oi and Oe layers, but only 7% in Oa layers. Fiber determination strongly depends on the strength of the laboratory treatment applied and the presence of sand grains in the organic layers that contribute to the breakdown of fibers during the treatments.
- Mineral matter content in organic layers is highly variable, ranging from nothing to about 80%. The average mineral content decreases first slightly from Oi (29%) to Oe (26%) layers and then increases in the Oa layers (31%). Variations in mineral content are mostly related to the geomorphic processes that favored the accumulation of mineral matter in low-lying landscape positions, together with the accumulation and preservation of the organic residues.
- Wet bulk density is the physical parameter with lowest variability. In most soils, bulk density at field conditions slightly increases with depth in the organic layers, while in some others it decreases in deeper layers. Wet bulk density values range from 0.54 to 0.78 Mg m^{-3}, without significant differences between organic layer types. Dry bulk density varies with depth from 0.11 to 0.16 to

0.20 Mg m^{-3} in Oi, Oe, and Oa layers, respectively. The low bulk density of the peat materials is related to the high pore space occupied by water. The permanently waterlogged layers are only slightly compacted with depth. In contrast, in many tropical lowland peats, bulk density is higher in the surface than subsurface layers because of more advanced organic matter decomposition.

- Average values of field water content are 1,026, 1,170, and 1,532% for Oi, Oe, and Oa layers, respectively. This physical parameter is highly variable among individual organic layers and whole profiles, with extreme values of 106 and 5,229%.

6.7.3 *Chemical Properties*

- The Guayana Highland peatsoils vary markedly in organic matter content from nearly pure organic material to layers borderline to mineral soils. Average organic matter content is 74% by loss on ignition and 28% using the Walkley–Black method. In general, organic matter contents decrease with peat depth. Changes in organic matter content are often more related to the mixture of organic and mineral materials along the profiles than to the degree of decomposition of the organic materials. Peat materials vary spatially, with amounts of organic matter higher than 90% at some sites, and according to elevation with the highest amounts of organic matter occurring at 650–1,800 m a.s.l. In organomineral soils, organic matter contents of the A and C horizons are significantly lower than in peat materials.
- Soil pH is the least variable chemical property of the organic layers. The pH values vary from ultra acid (pH 3.1–3.5) to strongly acid (pH 5.2–5.3). The pH at field moisture is usually higher than the pH in $CaCl_2$ solution. The Oi layers have higher pH values than the other organic layer types. In most soils, lower pH values are associated with increase in organic matter because the latter is a source of H^+ ions and contributes to soil acidification. Total acidity is around 3–43 times higher than salt-replaceable acidity, which indicates the high pH-dependent surface charge of the organic compounds.
- The cation exchange capacity (CEC) of the peatsoils is very variable among profiles and between layers. The cation exchange capacity by sum of bases plus total acidity ranges between 20 and 187 cmol(+) kg^{-1}, with an average of 66 cmol(+) kg^{-1}. The effective cation exchange capacity determined by the sum of bases plus salt-replaceable acidity varies between 2 and 12 cmol(+) kg^{-1}, with an average of 5 cmol(+) kg^{-1}. Although organic materials have high total exchange capacity, the effective cation exchange capacity is low at the very acid pH of these soils. In most profiles, acidity decreases with soil depth.
- Levels of exchangeable bases are low in general, but highly variable among and within soil profiles. For all organic layers together, the relative abundance of exchangeable bases is $Ca^{+2} > Na^{+1} > K^{+1} > Mg^{+2}$, with average levels of

0.71, 0.52, 0.42, and 0.38 cmol(+) kg^{-1}, respectively. Exchangeable bases are mainly stored in the surface layers through nutrient cycling, while often reaching trace values in deeper layers. Base saturation by sum of cations and by $CaCl_2$ is 4 and 33%, respectively. Organic soils on granitic rocks have higher base saturation than those formed on sandstones.

- Peatsoils have formed under waterlogged and nutrient-deficient conditions with limited organic matter decomposition. During the minerotrophic stage, the main sources of nutrients are rainfall and a tiny provision of minerals from bedrocks. During the ombrotrophic stage, some of these nutrients are fixed by the vegetation successions, then released and incorporated into the peat material during biomass decomposition, and finally recycled again by the vegetation. Most of the nutrients are washed out by runoff and deep percolation. However, the concentration of exchangeable bases in the topsoil (0–30 cm) indicates that nutrient transfer and recycling occur within the upper peat layers, where organic matter decomposition increases temporarily, releasing nutrients during the short dry season. In general, available phosphorus is low, often no more than traces.

6.7.4 Soil Variability

- Spatial variations in peatsoils are related to a set of factors that may interact at specific sites, including type and dynamics of the landforms, drainage conditions of the peatland, permeability and mineralogy of the rock substratum, age and history of the organic deposits, and type and composition of the vegetation cover, among others. In general terms, landscape position and the nature of the bedrock mainly control the spatial variations of peats and peatlands.
- In general, peatland configuration is elongated. The average length-to-width ratio is 3.5:1, but can be as large as 14:1. The length of peat areas varies from 140 to 325 m, while the width ranges from 10 to 50 m. Thickness of the organic deposits varies among and within peatland areas. A maximum thickness of 170 cm was observed, but the most common range is 45–150 cm.
- Peats formed on sandstones and on granitic rocks, respectively, differ in several aspects including color, root content, dry matter content, pH, and available macronutrients, among others. Peats are slightly redder on granitic rocks than on sandstones. Root contents in the surface organic layers are larger on granitic rocks than on sandstones, the latter being a mineralogically poorer and more permeable substratum. Dry matter contents in organic soils on sandstones are larger (almost twice as much) than those of organic soils on granitic rocks, resulting in lower field water content in the former than in the latter. Average levels of exchangeable bases in peats on granitic rocks are almost twice as high as those in peats on sandstones, with Ca being even three times higher.

6.7.5 Soil Classification

- Soils were classified according to the Soil Taxonomy and the World Reference Base for Soil Resources (WRB), following the criteria used for the definition of organic and mineral soil materials by both systems. All organic soil profiles support the requirements of the Histosol order in both classification systems.
- At the suborder level of the Soil Taxonomy, most soils classify as Saprists (50%), followed by Hemists (25%), and Fibrists (25%), on the basis of the degree of decomposition of the organic soil material. All soils belong to the Haplo-great groups and most soils are Lithic extragrades because they have a lithic contact within the control section (0–130 cm).
- According to the WRB framework, the organic soils classify as Sapric, Hemic, and Fibric Histosols in equal proportions. Rheic Histosols are more frequent (60%) than Ombric Histosols (40%). All soils are hyperdystric.
- The organomineral soils classify as Endoaquents and Humaquepts according to the Soil Taxonomy, and as Gleysols and Fluvisols according to the WRB framework.

References

Andriesse JP (1988) Nature and management of tropical peat soils. FAO Soils Bulletin 59, Food and Agriculture Organization of the United Nations, Rome

Boelter DH (1969) Physical properties of peats as related to degree of decomposition. Soil Sci Soc Am Proc 33:606–609

Boelter DH, Blake GR (1964) Importance of volumetric expressions of water contents of organic soils. Soil Sci Soc Am Proc 28:176–178

Bord na Mona (1984) Fuel peat in developing countries. Study Report for the World Bank, Dublin

Caldwell PV, Vepraskas MJ, Gregory JD (2007) Physical properties of natural organic soils in Carolina Bays of the Southeastern United States. Soil Sci Soc Am J 71:1051–1057

Chimner RA, Karberg JM (2008) Long-term carbon accumulation in two tropical mountain peatlands, Andes Mountains, Ecuador. Mires and Peat 3: art. 4. http://www.mires-and-peat.net/

Coleman NT, Thomas GW (1967) The basic chemistry of soil acidity, chapter 3. In: Soil acidity and liming. Agronomy, vol 12. American Society of Agronomy, Madison, WI, pp 1–41

Comerma JA, Sánchez J (1980) Consideraciones sobre el régimen de temperatura del suelo en Venezuela. Actas VII Congreso Latinoamericano de la Ciencia del Suelo. San José

CVG Técnica Minera CA (1991) Informe de avance NC20-11-12-15-16: clima, geología, geomorfología, suelos, vegetación. Proyecto Inventario de los Recursos Naturales de la Región Guayana. Corporación Venezolana de Guayana. Ciudad Bolívar

D'Amore DV, Lynn WC (2002) Classification of forested Histosols in Southeast Alaska. Soil Sci Soc Am J 66:554–562

Davies BE (1974) Loss-on-ignition as an estimate of soil organic matter. Soil Sci Soc Am Proc 38: 150–151

Driessen PM, Rochimah L (1976) The physical properties of lowland peats from Kalimantan and their significance for land suitability appraisal. In: Research on peat and podzolic soils in Indonesia and their potential for agriculture. Soil Research Institute, Bogor

EKONO (1981) Report on energy use of peat. Contribution to U.N. Conference on New and Renewable Sources of Energy. Nairobi

FAO (1990) Guidelines for soil description, 3rd edn (revised). Soil Resources, Management and Conservation Service; Land and Water Development Division. FAO, Rome

FAO (2006) World reference base for soil resources. A framework for international classification, correlation and communication. World Soil Resources Reports 103. IUSS, ISRIC, FAO, Rome

Farnham RS, Finney HR (1965) Classification and properties of organic soils. Adv Agron 17: 115–162

Jackson CR, Liew KC, Yule CM (2008) Structural and functional changes with depth in microbial communities in tropical peat swamp forest sediments. Microb Ecol http://dx.doi.org/10.1007/s00248-008-9409-4

Jiménez D (1995) Aplicación de la taxonomía de suelos en la caracterización de cuatro suelos orgánicos venezolanos. MSc Thesis, UNELLEZ, Guanare

Kamprath EJ, Welch CD (1962) Retention and cation exchange properties of organic matter in coastal plain soils. Soil Sci Soc Am Proc 26:263–268

Katimon A, Melling L (2007) Moisture retention curve of tropical sapric and hemic peat. Malays J Civ Eng 19(1):84–90

Lee GB, Bulligton SW, Madison FW (1988) Characteristics of histic materials in Wisconsin as arrayed in four classes. Soil Sci Soc Am Proc 52:1753–1758

Lucas RE (1982) Organic soils (Histosols). Formation, distribution, physical and chemical properties, and management for crop production. Research Report 435 (Farm Science), Michigan State University, East Lansing, MI

Mohamed M, Padmanabhan E, Mei BLH, Siong WB (2002) The peat soils of Sarawak. STRAPEAT Status Report. Universiti Malaysia Sarawak, Malaysia

Nichols DS, Boelter DH (1984) Fiber size distribution, bulk density, and ash content of peats in Minnesota, Wisconsin and Michigan. Soil Sci Soc Am J 48:1320–1328

Nogué S, Rull V, Montoya E, Huber O, Vegas-Vilarrúbia T (2009) Paleoecology of the Guayana Highlands (northern South America): Holocene pollen record from the Eruoda-tepui, in the Chimantá massif. Palaeogeogr Palaeoclimatol Palaeoecol 281:165–173

Ramos B (1997) Los suelos en las cumbres de las Sierras de Maigualida y Uasadi-Jidi, Guayana venezolana. In: Huber O, Rosales J (eds) Ecología de la cuenca del río Caura, Venezuela. II. Estudios especiales. Scientia Guaianae 7. Caracas, pp 423–440

Rull V (1991) Contribución a la paleoecología de Pantepui y la Gran Sabana (Guayana Venezolana): clima, biogeografía y ecología. Scientia Guaianae 2, CVG-EDELCA, Caracas

Rull V (2004) An evaluation of the Lost World and the vertical displacement hypotheses in the Chimantá Massif, Venezuelan Guayana. Glob Ecol Biogeogr 13:141–148

Rull V (2005) Vegetation and environmental constancy in the Neotropical Guayana Highlands during the last 6000 years? Rev Palaeobot Palynol 135:205–222

Satrio AE, Gandaseca S, Ahmed OH, Majid NMA (2009) Effect of precipitation fluctuation on soil carbon storage of a tropical peat swamp forest. Am J Appl Sci 6(8):1484–1488

Schargel R, García P, Jiménez D (2011) Laboratory methods for characterization of peat materials. In: Zinck JA, Huber O (eds) Peatlands of the Western Guayana Highlands, Venezuela, Chap. 5. Springer, Heidelberg

Schubert C, Fritz P (1985) Radiocarbon ages of peat, Guayana Highlands (Venezuela). Some paleoclimatic implications. Naturwissenschaften 72:427–429

Schubert C, Fritz P, Aravena R (1994) Late Quaternary paleoenvironmental studies in the Gran Sabana (Venezuelan Guayana Shield). Quatern Int 21:81–90

Soil Survey Staff (1993) Soil survey manual. USDA-SCS Agriculture Handbook 18. U.S. Gov. Print. Office, Washington, DC

Soil Survey Staff (1999) Soil taxonomy, a basic system of soil classification for making and interpreting soil surveys. Agriculture Handbook 436, 2nd edn. Natural Resources Conservation Service, U.S. Department of Agriculture. Washington, DC

Soil Survey Staff (2006) Keys to soil taxonomy, 10th edn. Natural Resources Conservation Service, U.S. Department of Agriculture. Washington, DC

Tie YL, Kueh HS (1979) A review of lowland organic soils of Sarawak. Department of Agriculture, Technical Paper 4, Research Branch, Sarawak, Malaysia

Zelazny LW, Carlisle VW (1974) Physical, chemical, elemental, and oxygen-containing functional group analysis of selected Florida Histosols. In: Histosols: their characteristics, classification and use. SSSA Special Publication 6, Soil Science Society of America. Madison, WI, pp 63–78

Chapter 7
Tepui Peatlands: Age Record and Environmental Changes

J.A. Zinck, P. García, and J. van der Plicht

7.1 Introduction

Peat deposits in tepui environment have been described in the eastern part of the Venezuelan Guayana Highlands, focusing on peat dating (Schubert and Fritz 1985; Rull 1991; Schubert et al. 1994; Nogué et al. 2009) and pollen analysis (Rull 1991, 2004, 2005; Nogué et al. 2009). The present study deals with the description, characterization, and dating of peat deposits on sandstone–quartzite tepui mesetas and, to a lesser extent, on granitic domes in the western Guayana Highlands, in contact with the Venezuelan Amazon Basin. Three exploration missions were carried out between 1992 and 1996, covering parts of three high-plateau and mountain massifs in the western Amazon state: the Marahuaka-Huachamakari massif (including the northern edge of the adjoining Duida massif), the Cuao-Sipapo massif, and the western border of the Maigualida massif (see location maps in Huber and García 2011). No new peat data have been collected in these areas since our exploration missions in the early 1990s.

This chapter focuses on the description of the sampling sites, the carbon-14 dating of selected peat layers, and the interpretation of the peat age record with respect to peat formation and environmental changes during the Holocene.

7.2 Sampling and Dating Method

7.2.1 Peat Sampling

Peat samples for carbon-14 determination were collected in 1992 (24–26 March), 1993 (10–18 February), and 1996 (08–09 March), in the framework of multidisciplinary exploration missions to the Venezuelan Amazon region. Helicopter was

J.A. Zinck and O. Huber (eds.), *Peatlands of the Western Guayana Highlands, Venezuela*, Ecological Studies 217, DOI 10.1007/978-3-642-20138-7_7,

used to access the top of selected tepuis and mountain ranges. So far, our samples are the only ones ever collected in this region of the western Guayana Highlands.

Preliminary cartographic sketches were obtained from visual interpretation of satellite images to guide the location of fieldwork transects and observation points. In the case of the Cuao-Sipapo massif, the most extensive and diverse of the three study areas, a photo-geomorphic map was prepared at the scale of 1:250,000 (see Fig. 4.1). Observation points were distributed as evenly as possible to cope with the geographical variations within this wide territory in terms of flora, relief, and peat deposits. At each site, the general landscape features were described, including vegetation, relief form, topography, drainage, and geology. Pits were opened for peat and soil description and sampling. Profile descriptions are reported together with laboratory data in the Appendix of this volume.

Peat sampling was made difficult because of the looseness of the organic material and high groundwater level. In a few places where peat was dense and consistent enough, a Wardenaar peat profile sampler (Eijkelkamp 1989) was used. In general, this apparatus was not appropriate for sampling the most commonly occurring organic material in the area: loose, water-saturated fibric and hemic deposits. Therefore, peat layers were described and sampled mostly in pits, some of them opened by deepening natural exposures. As vertical differentiation of the peat mantle was in general very gradual, sampling was done in many cases at regular intervals of 20 cm (in some cases, 40 cm). Sampling proceeded from top downwards while the pit was deepened to restrict as much as possible the contamination of lower layers by material moving down from upper ones. Material was collected in several places within each layer, mainly from its central section, so that the age of the compound sample can be considered to be the average age of the layer. All discernible organic layers and mineral horizons were sampled for laboratory analyses. Organic materials needed the application of special procedures and tests that are described and discussed in Schargel et al. (2011).

In most of the sites, the middle and lower layers of the peat mantle were water-saturated by the time of sample collection. It was thus unavoidable that the bottom layers be somewhat contaminated by admixture of modern organic substances proceeding from the overlying layers. Field samples were submitted to a pretreatment to control contamination by modern plant material. Roots and rootlets were removed by sieving, and the remaining contamination was removed by the AAA procedure, a 3-step chemical treatment using HCl and NaOH solutions successively as described in Mook and Streurman (1983) and in Mook and Waterbolk (1985).

7.2.2 Radiocarbon Dating and Calibration

The ^{14}C dates are indicated as ^{14}C BP. BP is a defined unit, which means that it is calculated from a radioactivity measurement relative to a standard, using a conventional half-life, and corrected for isotopic fractionation (Van der Plicht and Hogg 2006). The ^{14}C dates have to be calibrated into calendar ages using calibration curves, which are obtained from paired ^{14}C/dendrochronological wood samples. The presently recommended curve is IntCal04 (Reimer et al. 2004). Calibrated ^{14}C dates are reported in calBC (years *before Christ*), calAD (*anno Domini*, years after Christ), or calBP, where calBP = 1950-calAD. Calibration is the transformation of years BP into years calBP, using calibration curves with the help of one of the existing calibration programs. In the following text, calibrated and noncalibrated carbon-14 dates are used. Although dates should be preferably expressed in calibrated calendar years, in many cases they are given in conventional ^{14}C years BP to allow for comparison with published noncalibrated dates.

The calibration resolution of conventional ^{14}C dates, with conversion into calendar years, has steadily improved over the last two decades (Van der Plicht 2004). Separate tree-ring data sets based on dendrochronologically dated samples have been cross-checked and integrated in a common data base to cover the tree-ring period back to 12.4 kyr calBP. To stretch further back in the past, marine data including coral and foraminifera records were used to cover the period of 12.4–26.0 kyr calBP, upon their conversion to the atmospheric equivalent with site-specific marine reservoir corrections (Reimer et al. 2004). This is the basis of the internationally accepted calibration curve IntCal04, which improves the anterior IntCal98 curve by extending the period covered from 24 to 26 kyr calBP and by providing higher resolution beyond 11.4 kyr calBP. Calibration takes into account the natural variations of the carbon-14 content in the atmosphere. As the calibration curve shows irregularities (wiggles) due to these medium-term variations of the atmospheric carbon-14 content, the transformation of the Gaussian probability distribution representing the measured ^{14}C ages (BP $\pm$ σ) needs statistical adjustments (Van der Plicht 1993). The definition of the new calibration curve was enhanced by using a random-walk statistical model to cope with the uncertainty affecting calendar and ^{14}C ages (Reimer et al. 2004). The temporal resolution of the curve is five calendar years.

The radiocarbon age of the selected peat samples was determined at the Center for Isotope Research (CIO) of the University of Groningen, the Netherlands, using acid/alkali/acid extraction and dating of the extraction residues (Mook and Waterbolk 1985). In a few cases, the alkali extract was dated separately. The resulting conventional radiocarbon ages were calibrated into calendar ages using IntCal04 (Table 7.1).

7.2.3 Calibration Graphics

Calibration results are displayed graphically using the program WinCal25 (Van der Plicht 2007). This is an updated version of the original Groningen calibration program developed about 20 years ago (Van der Plicht and Mook 1987, 1989). Figure 7.1 shows the graphical computer output of the sample Marahuaka I.1

Table 7.1 Site characteristics, carbon-14 data, and calibrated calendar ages using IntCal04

Profile	Site	Elevation (m a.s.l.)	Substratum	Depth (cm)	Dated material	Layer	$\delta^{13}C$ (‰)	^{14}C yr BP (±σ)	Cal yr BC/*AD	Cal yr BP
P1	Marahuaka I.1	2,570	Siltstone	40–60	OM	C2	−26.2	7100 ± 60	6065–5845	8015–7795
P2	Marahuaka I.2	2,580		20–60	Sapric peat	Oa1	−26.4	5880 ± 50	4890–4615	6840–6565
P2	Marahuaka I.2		Sandstone	70–95	OM	2C	−27.1	7490 ± 50	6435–6245	8385–8195
P5	Marahuaka II.5	2,600		20–60	Hemic peat	Oi2	−25.0	170 ± 40	*1650–*1880	300–70
P5	Marahuaka II.5		Sandstone	85–120	Hemic peat	2C	−25.5	3400 ± 50	1875–1535	3825–3485
P6	Huachama I.6	1,770		45–60	Hemic peat	Oe	−26.3	1080 ± 40	*890–*1020	1060–930
P6	Huachama I.6		Sandstone	50	Stemwood	Oe	−25.2	4840 ± 40	3700–3525	5650–5475
P7	Cuao 1	1,800		48–70	Hemic peat R	3C	−25.3	2105 ± 45	350–10	2300–1960
P7	Cuao 1		Sandstone	48–70	Hemic peat E	3C	−26.1	2730 ± 50	995–805	2945–2755
P8	Cuao 2	1,800		30–55	Hemic peat	Oe2	−26.1	1665 ± 35	*260–*530	1690–1420
P8	Cuao 2		Sandstone	55–87	Sapric peat	Oa	−27.3	3415 ± 30	1865–1630	3815–3580
P9	Cuao 3	1,600	Sandstone	40–60	Hemic peat	2A2	−24.6	165 ± 30	*1665–*1880	285–70
P11	Cuao 6	1,200		40–60	Sapric peat R	Oa2	−25.8	5380 ± 50	4335–4050	6285–6000
P11	Cuao 6		Quartzite	40–60	Sapric peat E	Oa2	−24.8	6745 ± 45	5720–5570	7670–7520
P12	Cuao 7	1,370		15–40	Hemic peat	Oe2	−26.5	850 ± 35	*1050–*1260	900–690
P12	Cuao 7		Granodiorite	40–60	Sapric peat	Oa	−27.1	3110 ± 35	1445–1300	3395–3250
P16	Cuao 11	650	Granite	13–40	Hemic peat	Oe	−28.6	185 ± 30	*1650–*1810	300–140
P17	Cuao 12	1,500		40–80	Hemic peat	Oe	−26.5	2990 ± 40	1380–1090	3330–3040
P17	Cuao 12		Sandstone	80–100	Sapric peat	Oa1	−27.1	7030 ± 50	6005–5805	7955–7755
P18	Cuao 13	700	Sandstone	35–45	Hemic peat	Oe	−25.9	105.3 ± 0.4%		
P19	Maigualida 1	2,100		60–80	Sapric peat	Oa2	−27.2	5190 ± 40	4220–3850	6170–5800
P19	Maigualida 1		Granite	110–130	Sapric peat	Oa4	−26.4	6580 ± 50	5615–5475	7565–7425
P20	Maigualida 2	1,650		20–35	Sapric peat	Oa	−25.7	2530 ± 40	795–535	2745–2485
P20	Maigualida 2			35–45	OM	2Cr	−26.3	3610 ± 40	2125–1880	4075–3830
P20	Maigualida 2		Granite	30–35	Rootwood	Oa/2Cr	−23.4	2350 ± 30	510–380	2460–2330

^{14}C yr BP = conventional radiocarbon age before 1950; Cal yr BC/*AD = calendar age BC/*AD using IntCal04 (calibrated age ranges at 2-sigma probability; ages rounded off to the nearest 5); Cal yr BP = calendar age before 1950; OM = organic matter in mineral soil; R = carbon-14 determined on residue; E = carbon-14 determined on extract; Huachama stands for Huachamakari

(40–60 cm) dated 7100 ± 60 BP. Figure 7.1a represents (1) the selected part of the calibration curve going through the calibration data points, plotted together with their respective error bars, (2) the Gaussian probability distribution of the ^{14}C ages along the *y*-axis, and (3) the resulting calibrated calendar age probability distribution along the *x*-axis. The segment of the calibration curve plotted along the *x*-axis covers the span of the ^{14}C ages BP ± 3σ represented on the *y*-axis (i.e., a 99.7% probability interval). Because of the presence of wiggles in the calibration curve, the calibrated age distribution of the Marahuaka I.1 sample shows two peaks, centered roughly around 5985 and 5935 calBC.

In Fig. 7.1b, the upper and lower horizontal bars correspond to the 68.3% (1σ) and 95.4% (2σ) confidence level on the normalized probability distribution scale, respectively. The intercepts of the horizontal bars with the *x*-axis determine calibrated calendar age ranges within which a conventional ^{14}C age BP has a given probability to fit at the 1σ or 2σ confidence level (Van der Plicht and Mook 1987, 1989; Van der Plicht 1993; Dehling and Van der Plicht 1993). In the case of the Marahuaka I.1 (40–60) sample, the calibrated age ranges are 6029–5969 and 5952–5909 calBC (1σ), and 6067–5872 and 5861–5845 calBC (2σ). Converted into calBP (absolute years relative to AD 1950), the calibrated ranges are 7979–7919 and 7902–7859 calBP (1σ), and 8017–7822 and 7811–7795 calBP (2σ). Thus, the real age of the dated material is in fact older than the conventional ^{14}C age. If the calendar age probability distribution at the 2σ confidence level is taken as reference, then the ^{14}C age of 7100 ± 60 BP calibrates into 6065–5845 calBC (=8015–7795 calBP).

Figure 7.1c shows the normalized cumulative probability distribution. In the case of a sample yielding a cumulative curve with a nice sigmoid shape, it would be possible to read a central value of calendar age BC at the probability of 0.5 (the median). In general, probability curves depart substantially from the ideal S-shaped curve because of the irregular distribution, with peaks and troughs, of the calibrated ages BC along the *x*-axis. In the example of Fig. 7.1c, the median calibrated age BC is not a representative central value of the probability distribution as it actually corresponds to the trough between the two peaks. Therefore, the full extent of the calibrated probability distribution at the 2σ confidence level was taken as reference to assign calendar year ranges to the dated samples.

7.2.4 Conventional and Calendar Ages

Radiocarbon dating based on the decay of the carbon-14 isotope provides apparent absolute ages that were corrected to account for the effect of the variations of ^{14}C in the atmosphere. In the time span in which lie many of our ^{14}C dates, the IntCal04 calibration curve shows a wiggly shape that increases the level of uncertainty in the calibrated probability distribution. In most cases, calibration generated several calendar age ranges at both the 1σ (68.3%) and 2σ (95.4%) confidence levels. To compare calendar ages and conventional ^{14}C ages, we

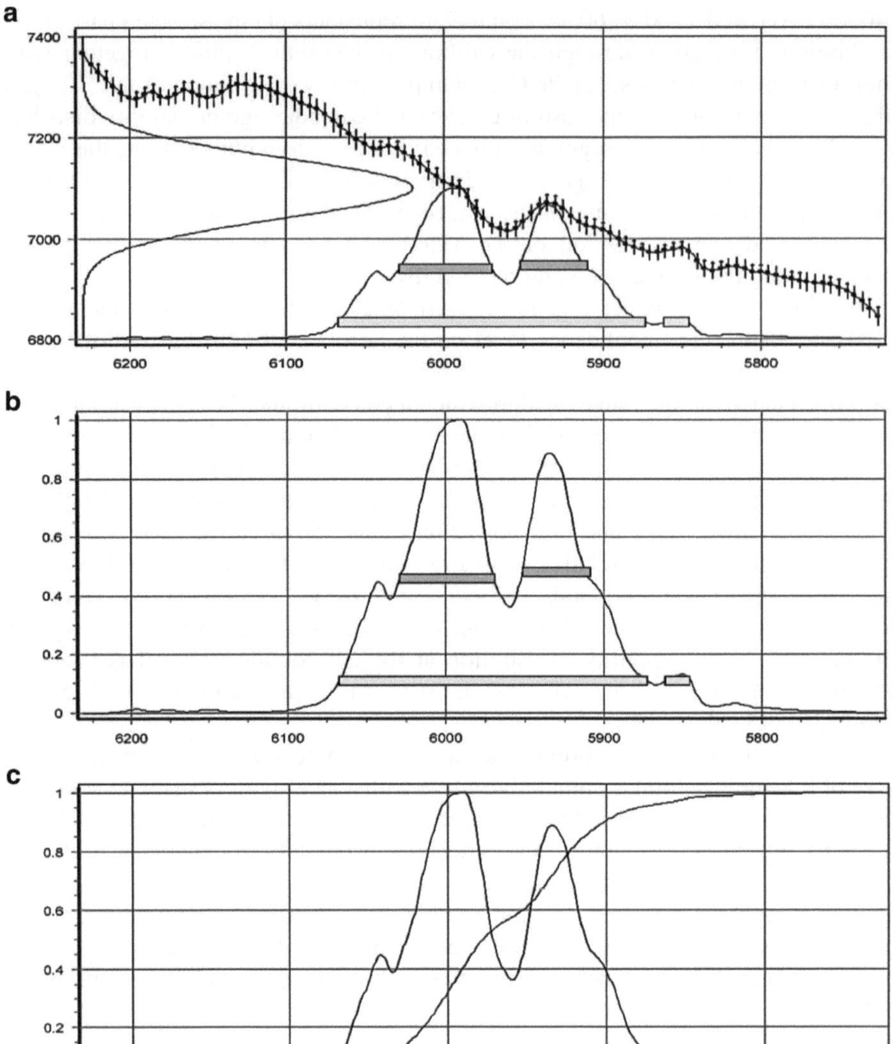

Fig. 7.1 Calibration graphics obtained applying IntCal04 high-resolution (Reimer et al. 2004) to the ^{14}C data of the Marahuaka I.1 (40–60) site. *Y*-axis: ^{14}C years BP in (**a**) and probability scale in (**b**, **c**); *X*-axis: calendar years BC in all three figures (see comments in Sect. 7.2.3)

selected the calibrated probability distribution range at the 2σ confidence level in which the real age of a calibrated sample had presumably the highest chance to fit. In general, the calibrated calendar data were not normally distributed, thus estimating a central value (i.e., mean or median) for the selected calibrated ranges is questionable, and comparisons were therefore based on the ranges. Calibrated BC/AD ranges were converted into calibrated BP ranges, and the deviations of the latter with respect to the conventional ^{14}C ages were expressed in percentages of increase (calendar ages older than ^{14}C ages) or decrease (calendar ages younger than ^{14}C ages).

In general, calibrated calendar ages are older than conventional ^{14}C ages: by 11–15% on average for the ^{14}C ages older than 4840 BP, and by 4–11% on average for the range of 3610–2530 BP. Calibration of ^{14}C ages within the range of 2350–850 BP generates calendar ages that vary on average from +6% (= calendar age older than ^{14}C age) to −11% (= calendar age younger than ^{14}C age). Very recent ages (185–165 BP) show large deviations of calendar ages vs. conventional ages, varying from +70% to −47%. Calibrated probability distribution curves of these subrecent ^{14}C dates are the most irregular among all the samples, with many peaks and calendar age ranges. In the case of one sample at the Maigualida 2 site (2350 BP calibrating into 2460–2330 calBP), calendar and conventional ages are almost equivalent.

Calibrated calendar ages stretch the initiation of peat accumulation back to ca 8400 calBP, instead of the conventional ca 7500 BP. Above 4840 BP, calibration generated calendar ages that are on an average 699–922 calendar years older than the corresponding ^{14}C ages, whereas in the range of 3610–2530 BP calendar ages are on an average only 114–335 calendar years older than their corresponding ^{14}C ages. In the range of 2350–850 BP and in the case of very recent ages (185–165 BP), calibration resulted in calendar ages that are either older or younger by several tens of calendar years than the related ^{14}C ages (Table 7.1).

7.3 Site Description and Age Record

In this section, each visited peatland site is characterized in terms of geomorphic setting, drainage condition, and origin of the peat material. Out of 12 full peat profiles, eight lie on sandstone–quartzite mesetas and the remaining four lie on granitic substratum on domes and in mountain ranges. Additional four profiles have organic layers interstratified with mineral material. The layers sampled for radiocarbon determination are shown in Figs. 7.2–7.4, with their respective ^{14}C ages (23 radiocarbon dates at 14 different sites). The properties of the dated peat materials are briefly described hereafter and, for each visited site, reference is made to the full profile description reported in the Appendix. Radiocarbon ages are expressed in conventional ^{14}C years BP and in calendar year ranges (calBC/calAD/calBP) at the 2σ confidence level of the calibrated age probability distribution. Peat ages are interpreted in the context of the site characteristics and history. The location of the three study areas and individual study sites is shown in Figs. 3.8, 3.9, 3.13, and 3.16.

7.3.1 The Marahuaka-Huachamakari Area (Fig. 7.2)

The general landscape is a table-shaped highland or high-plateau (2,000–2,800 m a.s.l.), developed on horizontally lying to slightly folded or tilted sandstone–siltstone and quartzite formations of the Roraima Group (Precambrian).

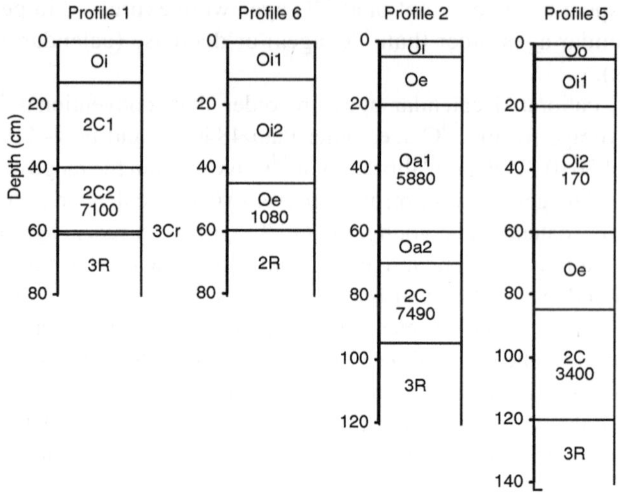

Fig. 7.2 Peat profiles from Marahuaka-Huachamakari massif. *O* organic layer, *o* folic, *i* fibric, *e* hemic, *a* sapric, *C* mineral horizon, *r* sandy and fine gravelly material derived from rock weathering, *R* bedrock. In selected layers: peat age in ^{14}C yr BP

The most common relief type corresponds to gently undulating (3–5% slope) tepui mesetas, with rugged ruiniform rock outcrops and deep crevices across the summit areas, and vertical rock walls several hundreds of meters high on the edges.

7.3.1.1 Marahuaka I.1 (Profile 1)

The Marahuaka I.1 site is located on an almost flat, narrow floodplain along a small creek running on bare rock at 2,570 m elevation. The area is covered by high-tepui grassland with occasional shrub islands. A shallow soil has developed from water-transported sediments of local origin, lying atop the sandstone substratum. The sample was taken from a dark gray (2.5Y 4/1) mineral C horizon at 40–60 cm depth, composed of massive, plastic, humus-containing silty clay loam. The ^{14}C age of the material is 7100 ± 60 BP that calibrates into 6065–5845 calBC (=8015–7795 calBP) (see Fig. 7.1). This is one of the three oldest ^{14}C ages (i.e., older than 7000 BP) determined in the study area.

Narrow floodplains in tepui environment are unstable landscape units. The feeding creeks run usually on bare bedrock and therefore tend to move laterally, reworking periodically their own deposits. From the context of the site, it can thus be assumed that the dated organic matter has not formed in situ. The site lies at the foot of a slope-peat mantle and is enriched in old allochthonous organic acids. These were transported by lateral groundwater flowing atop the rock substratum toward the narrow floodplain, where they impregnated the lower mineral horizon. This would support the hypothesis that old decomposed organic material, dating back to the early times of peat formation in the area, was depleted from upslope basal peat layers and moved within the groundwater to lower positions on the landscape, or was exported through joints, crevices, and fractures to karstic underground pipes, ultimately feeding the blackwater rivers in the lowlands.

7.3.1.2 Marahuaka I.2 (Profile 2)

The Marahuaka I.2 site is a karstic depression at 2,700 m elevation, 30 m × 20 m in size, surrounded by rock outcrops that dominate the depression by 1–5 m height. Local topography is flat to slightly concave (0.3–0.5% slope), and micro-topography is slightly hummocky. The site is very poorly drained and water-saturated during the largest part of the year, with patches of stagnating water at the peat surface and water table at 10 cm depth. The area is covered by high-tepui meadow vegetation.

The depression is filled with very dark brown (10YR 2/2) to black (10YR 2/1), moderately decomposed hemic material over more decomposed sapric material. Between the base of the peat and the rock substratum, a dark gray (10YR 4/1) sandy loam horizon containing 4% organic matter was found. The mineral material of this horizon can be considered allochthonous, eroded from the surrounding slopes, and filling the bottom of the depression before peat accumulation started. The humus mixed with the mineral material can have been either washed in from the surrounding slopes by lateral groundwater flow or washed down from the basal peat layers and trapped in the mineral horizon. This humus-rich mineral horizon (70–95 cm) was dated together with the central layer (20–60 cm) of the peat deposit.

Marahuaka I.2 (20–60), a decomposed sapric material with 5–10% fibers, provided a ^{14}C age of 5880 ± 50 BP that calibrates into 4890–4615 calBC (=6840–6565 calBP). Marahuaka I.2 (70–95), a humus-containing mineral C horizon of allochthonous origin that lies on the rock substratum, provided a ^{14}C age of 7490 ± 50 BP that calibrates into 6435–6245 calBC (=8385–8195 calBP). This is the oldest date recorded in the study area.

The time gap of 1610 ^{14}C years (7490–5880 BP) between the two dated samples corresponds to an intermediate sapric layer at 60–70 cm depth. The magnitude of the gap compared with the shallowness of this 10-cm layer underscores the absence of genetic relationship between the peat cover and the humus in the underlying mineral bottom horizon. This supports the hypothesis that the mineral horizon has been enriched by organic matter washed in from the surrounding slopes.

7.3.1.3 Marahuaka II.5 (Profile 5)

The Marahuaka II.5 site is a large karstic, amphitheater-like depression (500 m diameter) at 2,600 m elevation. It is dominated by a rocky crest and subdivided into smaller bogs (60 m × 30 m) that are surrounded by rock outcrops carved with micro-channels (karren field). Local topography is flat to slightly concave with less than 0.5% slope. The site is very poorly drained and water-saturated during the largest part of the year, with patches of stagnating water at the peat surface.

The vegetation is a very dense high-tepui meadow, with shrubs covering rocky islands close to the peat site.

The depression is filled with 120 cm peat resting on sandstone. The upper half (0–60 cm) of the deposit is mainly a dense fibric root mat, with coarse tubular roots (*Brocchinia*), mixed with very dark brown (10YR 2/2), moderately decomposed hemic material (40–50%). The dating of the hemic material in a layer at 20–60 cm depth provided a ^{14}C age of 170 ± 40 BP that calibrates into 1650–1880 calAD (=300–70 calBP).

The lower part (60–120 cm) of the deposit is a very dark gray (10YR 3/1) to black (10YR 2/1) moderately decomposed hemic material, with 10–20% root fibers. A sample taken from a layer at 85–120 cm depth, just above the rock substratum, gave a ^{14}C age of 3400 ± 50 BP that calibrates into 1875–1535 calBC (=3825–3485 calBP). This is the youngest peat date at comparable depth (±1 m) recorded in the study area.

Between the upper and lower tiers of this peat profile there is a time gap of 3230 ^{14}C years (3400–170 BP), which indicates a long depositional hiatus. Taking into account the presence of an intermediate hemic layer at 60–85 cm depth, it can be hypothesized that organic matter stopped accumulating somewhen after 3400 BP and resumed depositing a long time later around the seventeenth to nineteenth centuries.

7.3.1.4 Huachamakari I.6 (Profile 6)

The Huachamakari I.6 site is located at 1,770 m elevation. The relief is formed by a sequence of shelf-like sandstone slabs, 20–30 m wide, flat to slightly sloping (2–3%), arranged in consecutive stairway steps separated by small scarps 1–2 m high. The site is very poorly drained and water-saturated during the largest part of the year, with patches of stagnating water at the peat surface. In general, the peat cover is shallow, with frequent interspersed rock outcrops. Peat started accumulating along the intersection between the horizontal rock surfaces and the vertical scarps, and from there expanded over the sandstone slabs. The peat cover in this kind of sloping position is unstable and tends to slide on the rock substratum. There is evidence of past peat sliding events, causing the fragmentation of the peat mantle and the occurrence of bare rock patches in places where peat has been ripped apart and stripped away. This causes the peat accumulation to restart locally at variable times. Vegetation is a shrubby tepui meadow. Small *Bonnetia* colonies grow on rock outcrops, especially along the scarps separating consecutive sandstone slabs.

At the observation site, the peat mantle is 60 cm thick on sandstone substratum. It includes 45 cm of fibric material, followed by 15 cm of hemic material. The sample, taken at 45–60 cm depth from a very dark gray (10YR 3/1), moderately decomposed hemic material with some roots, gave a ^{14}C age of 1080 ± 40 BP that calibrates into 890–1020 calAD (=1060–930 calBP).

The dating of this sample located at the bottom of the peat mantle indicates that peat accumulation in this site started in historical times, about one millennium ago.

This is a much younger age than the peat ages obtained at similar depths in other places, especially from bog peats. The particular geodynamics that controls peat formation in slope situations, with frequent sliding causing partial destruction followed by local renewal of the peat mantle, can explain the difference.

In the same layer (45–60 cm), a *Bonnetia* stem fragment was found lying horizontally at 50 cm depth. The dating of this wood sample gave a ^{14}C age of 4840 ± 40 BP that calibrates into 3700–3525 calBC (=5650–5475 calBP). The wood sample is thus much older than the ^{14}C age (1080 BP) of the wrapping peat material. The *Bonnetia* tree of which this sample is a remnant died long before the stem fragment was incorporated in the current peat mantle about one millennium ago. It is a remainder of an earlier peat deposit. Possibly, the relict *Bonnetia* fragment survived thanks to the strong acidity and fairly anaerobic conditions of the peat environment.

7.3.2 The Cuao-Sipapo Area (Fig. 7.3)

The general landscape is a table-shaped mountain massif or high-plateau (600–2,000 m a.s.l.), developed on horizontally lying to slightly folded or tilted sandstone–siltstone and quartzite formations of the Roraima Group (Precambrian). Individual relief units are gently undulating (3–5% slope) tepui mesetas, with rugged ruiniform rock outcrops and deep crevices across the summit areas, and vertical rock walls several hundreds of meters high on the edges. Within the plateau landscape, there are enclaves of hilly dome-shaped relief corresponding to the igneous-metamorphic basement rocks of the Guayana Shield that dominate eastwards (see Fig. 4.1).

7.3.2.1 Cuao 1 (Profile 7)

The site is a karstic depression within an almost flat area (1–2% slope) at 1,800 m elevation, on a sandstone tepui meseta. The depression is surrounded by steep rocky slopes (30%) and periodically flooded by a small creek running on bare rock. The area is covered by a low, dense riparian forest.

The site is very poorly drained and water-saturated during the largest part of the year, with patches of stagnating water at the peat surface. The depression is filled with water-transported loam to sandy loam sediments of local origin. This mineral material buries a peat layer at 48–70 cm depth that rests on the sandstone substratum. The peat is a black (10YR 2/1), fairly well-decomposed hemic material with loamy touch, slightly plastic and slightly sticky. It includes common fine and medium roots (1–3 mm) and few coarse roots (up to 10 mm).

At this site, residue and extract ages of the peat layer at 48–70 cm depth were determined separately. The ^{14}C age of the residue is 2105 ± 45 BP that calibrates into 350–10 calBC (=2300–1960 calBP); and the ^{14}C age of the extract is

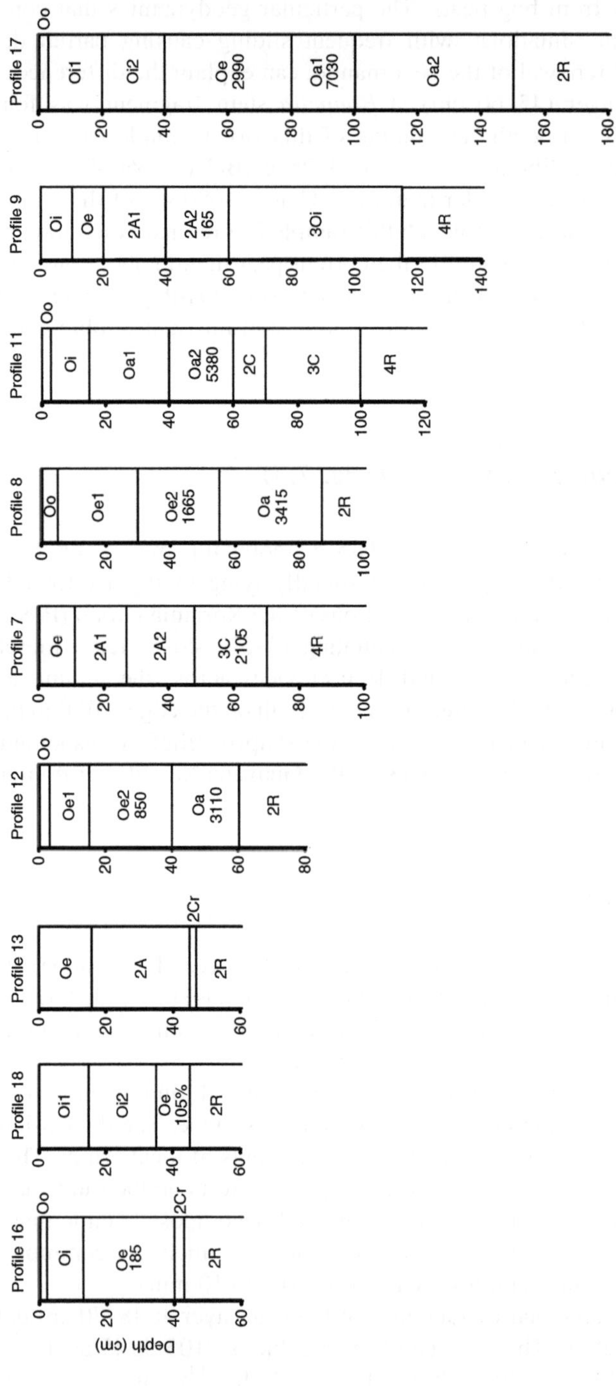

Fig. 7.3 Peat profiles from Cuao-Sipapo massif. *O* organic layer, *o* folic, *i* fibric, *e* hemic, *a* sapric, *A* organo-mineral horizon, *C* mineral horizon, *r* sandy and fine gravelly material derived from rock weathering, *R* bedrock. In selected layers: peat age in ^{14}C yr BP

2730 ± 50 BP that calibrates into 995–805 calBC (=2945–2755 calBP). Extract ages are usually smaller than the residue ages (Mook and Waterbolk 1985). Here, the opposite occurs, the extract being older than the residue. This could be either due to the penetration of living roots that rejuvenate the residue or the presence of organic substances in the extract that reflect a peat formation phase older than that of the residue. Considering the depth (48–70 cm) of the dated material and the fact that the latter is buried under a younger mineral cover, the second hypothesis might be more realistic. In this case, the peat layer would be polygenetic. Younger organic material would have been superimposed on the older one in the same layer after a depositional interruption of 625 ^{14}C years (2730–2105 BP).

The dated material corresponds to a peat layer lying just above the rock substratum. The site is a closed depression where loss of surface peat material is unlikely to take place. The relatively younger age of this peat material, when compared with older peats occurring at comparable depths, suggests that peat started accumulating at different dates during the Holocene throughout the area. Furthermore, the peat deposit at this site has been fossilized by transported mineral material. Thus, the age of the peat (2105/2730 BP) is also the time at which peat stopped accumulating at this place. It could be hypothesized that the age of the extract (2730 BP) indicates the beginning of peat accumulation and the age of the residue (2105 BP) reflects the end thereof. The history of this site is different from that of other sites where only organic material deposited.

7.3.2.2 Cuao 2 (Profile 8)

The landform is a narrow, almost flat (1–2% slope) interfluve between two small creeks running on bare rock at 1,800 m elevation, on a sandstone tepui meseta. The selected site is one of the many small poorly drained karstic alveoli (30–50 m^2) that are scattered across the interfluve. Vegetation is a high-tepui meadow with *Bonnetia* bushes close to rock outcrops.

An 87-cm-thick peat deposit has accumulated in the depression above the sandstone substratum. It is composed of a sequence of organic material increasingly decomposed with depth, including folic (0–5 cm), hemic (5–55 cm), and sapric (55–87 cm) layers. Two layers were sampled for carbon-14 determination.

The first layer (30–55 cm) is a black (10YR 2/1), moderately decomposed hemic material, with massive structure, slightly plastic and slightly sticky. The layer contains many roots, mainly coarse ones (up to 5 cm), with reddish surface, often horizontally lying. The material has a ^{14}C age of 1665 ± 35 BP that calibrates into 260–530 calAD (=1690–1420 calBP). This indicates that deposition took place in historical times.

The second layer (55–87 cm) is a black (10YR 2/1), fairly well-decomposed sapric material, with massive structure, loamy touch, slightly plastic and slightly sticky, containing only a few roots. This is the basal layer of the peat mantle that lies directly on the sandstone substratum. It has a ^{14}C age of 3415 ± 30 BP that calibrates into 1865–1630 calBC (=3815–3580 calBP).

There is a considerable time gap between these consecutive layers (3415–1665 BP = 1750 ^{14}C years). Considering that there is a clear difference in degree of decomposition between the hemic and sapric materials, the gap reflects a real peat sedimentation hiatus between the two consecutive layers. As the loss of material from the bog through sliding is unlikely, a climate change could be invoked.

This site is located close to the Cuao 1 site. The hemic material of the bottom layer (48–70 cm) at site Cuao 1 has a ^{14}C age of 2105 BP, somehow comparable to the 1665 BP age of the hemic material (30–55 cm) at site Cuao 2. Thus, the time around these ages could mark the starting of a new phase of peat formation, although not at similar depths.

7.3.2.3 Cuao 3 (Profile 9)

The site is a karstic depression close to a small creek running on bare rock, on a tepui meseta at 1,600 m elevation. Local topography is flat to slightly concave (1–2% slope). The site is very poorly drained and water-saturated during the largest part of the year, with patches of stagnating water at the peat surface and free water table at 40 cm depth. Vegetation is a tubiform high-tepui meadow with isolated shrubs.

A 115-cm-thick peat deposit has accumulated in the depression above the sandstone substratum. It is composed of a sequence of folic/fibric (0–10 cm), hemic (10–60 cm), and again fibric (60–115 cm) materials. Peat decomposition is in a very initial stage.

A sample was taken at 40–60 cm depth from a black (10YR 2/1), slightly-to-moderately decomposed hemic material, with many medium roots (2–4 mm) essentially from *Brocchinia*. This material, the most decomposed one of the peat sequence, gave a very recent ^{14}C age of 165 ± 30 BP that calibrates into 1665–1880 calAD (=285–70 calBP). Thus peat deposition took place in historical times, around the seventeenth to nineteenth centuries.

The layer underlying the dated one is a loose, floating fibric bottom mat. This suggests that a mantle of raw peat material about one meter thick may have accumulated at this place in a matter of two to three centuries.

7.3.2.4 Cuao 6 (Profiles 10 and 11)

This site is located in a unique piece of landscape, the Cerro Autana tepui, a sacred tepui meaning "tree-of-life" in Piaroa language, which has been declared national monument in 1978. Autana is an isolated quartzite meseta at 1,200–1,240 m elevation, with a summit area of only a few hectares. It rises from the surrounding lowlands like a natural tower, limited by vertical walls about 700–800 m high (see Fig. 4.6b). The Autana tepui is a remnant of one of the oldest erosion surfaces recognized by Schubert et al. (1994) in the eastern Guayana Highlands. It shows caves that were once galleries of an underground karstic drainage system (see Fig. 4.10). At one edge of the tepui, a slightly rounded hill formed by sandstone–quartzite layers dominates the summit surface by about 100 m height. Sandy material eroded from the rocky slopes of this hill has accumulated at the foot forming a colluvial glacis with 3–5% slope (Profile 10). On the almost flat distal part of the glacis, peat has deposited fossilizing the sandy colluvium. This is the place where the Cuao 6 site is located, showing a 60-cm-thick peat cover atop the sandy substratum (Profile 11). Vegetation is a tubiform tepui meadow with isolated shrubs.

A sample was taken from the basal layer (40–60 cm) of the peat cover, recording thus the starting time of peat accumulation. The layer corresponds to a black (2.5Y 2/0), decomposed sapric material, massive, slightly plastic and slightly sticky, with frequent fine and very fine roots. At this site, residue and extract ages were determined separately. The ^{14}C age of the residue is 5380 ± 50 BP that calibrates into 4335–4050 calBC (=6285–6000 calBP), whereas the ^{14}C age of the extract is 6745 ± 45 BP that calibrates into 5720–5570 calBC (=7670–7520 calBP). The cumulative curve of the extract approximates a reasonably good S-shape, providing a median age (at probability 0.5) of 5650 calBC.

Similarly to site Cuao 1, the dating of the extract gave an age older than that of the residue, in contrast to what usually occurs. This could be either due to the penetration of living roots that rejuvenate the residue or the presence of organic substances in the extract that reflect a peat forming phase older than that of the residue. In the case of the second hypothesis, younger organic material would have been superimposed on the older one in the same layer after a depositional interruption of 1365 ^{14}C years (6745–5380 BP).

There is no evidence that the mechanical erosion–deposition process that originated the colluvial glacis is operating in the current moist environment of the region. It possibly reflects a former drier climate. Subsequent peat deposition fossilizing the sandy colluvium may indicate a return to moister environmental conditions.

7.3.2.5 Cuao 7 (Profile 12)

The site is located at 1,370 m elevation, on the narrow summit of a bare rocky dome, with backslopes above 100%, that belongs to an incised hilland surrounded by higher mountains. At the slightly convex top of the dome (3–5% slope), small pseudokarstic depressions have been excavated from granodiorite bedrock, rich in feldspar phenocrysts (rapakivi texture). Vegetation is a shrubby broadleaved meadow with isolated shrubs. Samples were taken from two peat layers that accumulated in a poorly drained pseudokarstic depression.

The first layer (15–40 cm) is a very dark gray (5YR 3/1), moderately decomposed hemic material, massive, slightly plastic and slightly sticky, with abundant medium and few large dead and live roots. The ^{14}C age of the material is 850 ± 35 BP that calibrates into 1050–1260 calAD (=900–690 calBP). Thus peat deposition took place in historical times, around the eleventh to thirteenth centuries.

The second layer (40–60 cm), lying just above the bedrock, is a black (5YR 2/1), slightly more decomposed sapric material. The ^{14}C age of the material is 3110 ± 35 BP that calibrates into 1445–1300 calBC (=3395–3250 calBP).

Between the two consecutive layers, there is a time gap of 2260 ^{14}C years (3110–850 BP), which suggests a depositional hiatus in the peat accumulation history of the site. As the peat mantle is confined in a closed depression, climate change might be a more likely cause explaining the pause in peat accumulation than the loss of material through sliding.

7.3.2.6 Cuao 11 (Profile 16)

This is the lowest visited site, located at 650 m elevation, at the bottom of an almost flat (1% slope), narrow, elongated valley, approximately one kilometer wide. The valley is deeply entrenched between alignments of granite–granodiorite domes with nearly vertical rocky walls and convex summits at 1,200–1,300 m a.s.l. Vegetation is an open forest with scattered shrubby meadow glades. The site is located in one of these open spaces surrounded by woodland. It is very poorly drained and water-saturated during the largest part of the year, with patches of stagnating water at the peat surface and free water table at 30 cm depth.

The peat mantle consists of fibric (2–13 cm) and hemic (13–40 cm) materials atop the granite–granodiorite substratum. A sample was taken from the hemic basal layer (13–40 cm): a dark reddish brown (5YR 2.5/2), moderately decomposed organic material with loamy, slightly sticky touch, containing fine to coarse (0.5–5 mm) live tubular roots, mainly horizontally lying (40–50%). Carbone-14 dating gave an age of 185 ± 30 BP that calibrates into 1650–1810 calAD (=300–140 calBP). Thus peat deposition took place in historical times, around the seventeenth to nineteenth centuries. This correlates well with the calendar age of the peat layer dated at the site Cuao 3 (1665–1880 calAD).

7.3.2.7 Cuao 12 (Profile 17)

The site is a karstic depression on a tepui meseta at 1,500 m elevation. Local topography is flat to slightly concave (2–3% slope). The site is very poorly drained and water-saturated during the largest part of the year, with patches of stagnating water at the peat surface and free water table at 30 cm depth. Vegetation is a tubiform tepui meadow with isolated shrubs.

A 150-cm-thick peat deposit has accumulated in the depression above the bedrock. It is composed of a sequence of fibric (3–40 cm), hemic (40–80 cm), and sapric (80–150 cm) layers atop the sandstone substratum. Two layers were sampled for carbone-14 determination.

The first layer (40–80 cm) is a black (5YR 2.5/1), moderately decomposed hemic material with loamy, slightly sticky touch, including common roots. The material has a ^{14}C age of 2990 ± 40 BP that calibrates into 1380–1090 calBC (=3330–3040 calBP).

The second layer (80–100 cm) is a black (10YR 2/1), fairly well-decomposed sapric material with loamy, slightly sticky touch, including very few roots. This material has a ^{14}C age of 7030 ± 50 BP that calibrates into 6005–5805 calBC (=7955–7755 calBP).

There is a large time gap (7030–2990 BP = 4040 ^{14}C years) between these two consecutive layers, which suggests an important depositional hiatus in the peat accumulation history of the site. As the peat mantle is confined to a closed depression, the loss of material through sliding might not be a suitable reason to explain the pause in peat accumulation. Likewise, the climate change hypothesis is improbable as the gap is too long in comparison with the total length of the whole Holocene period.

The ^{14}C age of 7030 BP is the oldest peat age recorded in the study area. Ages older than 7000 BP were determined at two other sites, but they concern humus mixed with mineral sediments of local origin: 7100 BP at the Marahuaka I.1 (40–60) site and 7490 BP at the Marahuaka I.2 (70–95) site.

7.3.2.8 Cuao 13 (Profile 18)

The relief unit is a rocky glacis covered by slope peat (3–5% slope), at the summit of a low-lying tepui meseta (700 m a.s.l.). The peat mantle is discontinuous and shows evidence of fragmentation resulting from differential sliding (see Fig. 7.5). The sliding process is activated at the front of the peat mantle by the lateral migration and cutting of a small creek that runs on the sandstone substratum. Buried karren micro-topography, with many grooves and small alveoli a few square meters large, is being exhumed in patches as the peat cover moves and fragments. This is a typical slope-peat situation, unstable and exposed to periodic renewal of the peat accumulation. The profile is located in an alveolus at the lower edge of the peat mantle, close to a creek. The site is poorly drained and water-saturated during the largest part of the year, with patches of stagnating water at the peat surface and free groundwater at 35 cm depth. Water in the alveolus is partly renewed at each important rainfall event because of the slope position. Open shrub with herbaceous layer covers the area.

The peat consists of fibric (0–35 cm) and hemic (35–45 cm) materials atop the sandstone substratum. A sample was taken from the black (5YR 2.5/1), moderately decomposed hemic material that forms 50–60% of the basal layer (35–45 cm), the rest being roots. The dated material has a ^{14}C relative activity of 105.3 ± 0.4%, which corresponds to an age of ca 1962 AD. This is thus a contemporaneous peat deposit that started accumulating in the early 1960s. In this place, 45 cm of peat have presumably formed in about 30 years (1962–1993).

7.3.3 The Maigualida Area (Fig. 7.4)

The general landscape (1,000–2,400 m a.s.l.) comprises elongated mountain ranges and dome-shaped hills developed from igneous-metamorphic basement of the Guayana Shield. Rocks are mainly coarse-grained rapakivi granite, rich in feldspar phenocrysts (about 50%). There are inclusions of tepui mesetas developed on horizontally lying to slightly folded sandstone and quartzite formations of the Roraima Group (Precambrian).

7.3.3.1 Maigualida 1 (Profile 19)

The site is located at 2,100 m elevation, in a narrow elongated vale (325 m × 60 m) that receives smaller lateral vales, gently undulating (1–3% slope) and flanked by granite outcrops with 20–30% slope. Local topography is flat to slightly concave, with small peat mounds 20–30 cm high that alternate with small water-filled alveoli colonized by *Drosera rotundifolia.* The sampled peat deposit consists of a small hummock 5 m × 10 m large, 20–30 cm above the surrounding water pools. The site is very poorly drained and water-saturated during the largest part of the year, with patches of stagnating water at the peat surface. There is free water at 0–60 cm depth, but peat is only moist without free water at 60–130 cm depth. Vegetation is a high-tepui grassland with short grasses and forbs and isolated shrubs. Pioneer vegetation develops along cracks in the granite.

The profile consists of a sequence of folic (0–10 cm), hemic (10–40 cm), and sapric (40–130 cm) layers atop the rapakivi granite substratum. This profile has the thickest sapric tier (90 cm) identified in the study area. Two layers belonging to this material were sampled for carbon-14 determination.

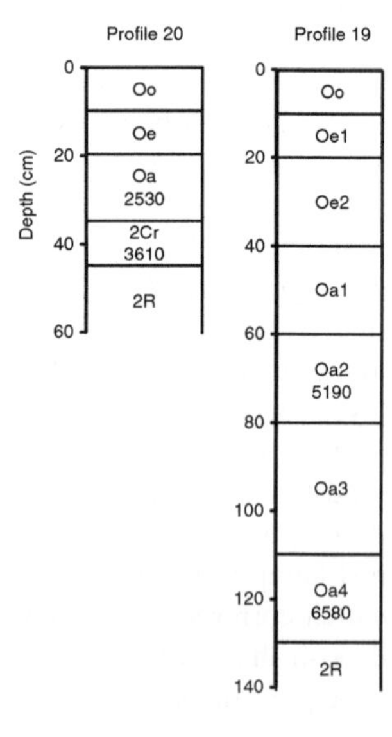

Fig. 7.4 Peat profiles from Maigualida massif. *O* organic layer, *o* folic, *e* hemic, *a* sapric, *C* mineral horizon, *r* sandy and fine gravelly material derived from rock weathering, *R* bedrock. In selected layers: peat age in ^{14}C yr BP

The first layer (60–80 cm) is a black (10YR 2/1), decomposed sapric material with common roots and some mineral inclusions. The material has a ^{14}C age of 5190 ± 40 BP that calibrates into 4220–3850 calBC (=6170–5800 calBP).

The second layer (110–130 cm) is the basal layer of the peat deposit at this site, with characteristics similar to those of the first one but without roots. The material has a ^{14}C age of 6580 ± 50 BP that calibrates into 5615–5475 calBC (=7565–7425 calBP).

The time gap between the two layers (6580–5190 BP = 1390 ^{14}C years) shall correspond to the formation time of the intermediate layer (80–110 cm). Thus 30 cm peat has accumulated in 1390 ^{14}C years, giving a deposition rate of 0.22 mm per ^{14}C year. If the center of the two compared layers is taken as reference thickness (50 cm), then the deposition rate is 0.36 mm per ^{14}C year (see Sect. 7.4.4).

7.3.3.2 Maigualida 2 (Profile 20)

The relief unit is a 300-m long rocky glacis covered by slope peat (3–5% slope) at the foot of a granite hill. The site is located at 1,650 m elevation, in the contact area (1–3% slope) between the distal sector of the glacis and the left bank of a blackwater creek. The site is very poorly drained and water-saturated during the largest part of the year, with patches of stagnating water at the peat surface and free water table at 35 cm depth. Vegetation is a very dense high-tepui broadleaved meadow mixed with shrubs.

The profile consists of folic (0–10 cm), hemic (10–20 cm), and sapric (20–35 cm) materials, followed by a mineral horizon mixed with decomposed organic matter (35–40/45 cm) atop the rapakivi granite substratum. Two peat layers and a root fragment were sampled for carbon-14 determination.

The first layer (20–35 cm) is a very dark brown (10YR 2/2) mixture of many fine and medium roots (1–3 mm) and decomposed organic material (40%). The dated material has a ^{14}C age of 2530 ± 40 BP that calibrates into 795–535 calBC (=2745–2485 calBP).

The second layer (35–40/45 cm) consists of coarse quartz–sand and fine gravel (1–5 mm), mixed with decomposed black (10YR 2/1) organic material (40%). The mineral material is most probably a residuum resulting from the weathering of the rapakivi granite substratum. The organic material has a ^{14}C age of 3610 ± 40 BP that calibrates into 2125–1880 calBC (=4075–3830 calBP).

Horizontally lying at the bottom of the first layer, at 30–35 cm depth, a flat, 100 cm long, 10 cm wide and 1–3 cm thick, red wooden root fragment yielded a ^{14}C age of 2350 ± 30 BP that calibrates into 510–380 calBC (=2460–2330 calBP). The root fragment is ca 180 ^{14}C years younger than the wrapping peat material (2530–2350 BP). In the case of this sample, the calibrated probability distribution curve yielded a very narrow high peak that concentrates the largest proportion of the normalized cumulative probability curve. From the whole data collection, this sample shows the best fit of the conventional ^{14}C age within the calendar range (±1%, considering solely the calibrated peak range).

As there is a relatively large time gap between the two layers (3610–2530 BP = 1080 ^{14}C years), it is unlikely that the humus impregnating the mineral material of the bottom horizon came from the overlying peat layer. Humus might have been brought in via the groundwater flowing on the glacis slope along the interface between the base of the peat mantle and the rock substratum.

7.4 Interpretations and Correlations: Peat Formation and Environmental Changes

In addition to the effect of ^{14}C variations in the atmosphere and the nature of the calibration curve implemented to correct for these variations, other factors have effect on the age of the peat materials at the studied sites, namely the water regime and dynamics, the surface geodynamics, and past climate conditions. Major peat age variations are related to the depth of the peat layers and the elevation of the peat sites. Changes in the composition of the vegetation cover over time influence peat nature. These relationships are analyzed hereafter.

The calibration of the ^{14}C dates generates one or more calendar year ranges for each ^{14}C date at the 1σ and 2σ confidence levels of the calendar age probability distribution (see Fig. 7.1), but fails to provide a reliable calendar central value (i.e., mean or median). Therefore, in the following section, we use mainly the conventional ^{14}C dates to construct graphics showing relationships between peat ages and environmental factors, and to compare our data with published radiocarbon data that are in general noncalibrated.

7.4.1 The Effect of Topography and Water Dynamics on Peat Formation

The water regime in peat deposits is controlled by the rainfall regime and the topographic configuration of the peat sites. In the Guayana Highlands, average monthly rainfall is usually above 100 mm except in February and March, with peaks above 400 mm per month in the period of May to August (Galán 1992). Annual rainfall estimates for the Chimantá massif in the eastern Guayana Highlands are on the order of 3,000–3,500 mm for high tepuis (above 2,000 m a.s.l.) and around 2,500 mm for mid-elevation tepuis (1,000–2,000 m a.s.l.) (Galán 1992).

Organic material accumulation and water concentration in the early stages of peat formation were guided by the landforms and the terrain irregularities inherited from the pre-Holocene relief evolution. Primary peat accumulation sites that vary from small alveoli of a few square meters to pans (dolines) of 40–50 m^2 are mainly fed by the concentration of water flowing on the bare surrounding rock surfaces (topogenous–rheophilous peats). In larger depressions (poljes), rainfall is the main source of direct water inflow (ombrophilous peats). Obviously, mixed situations are frequent where water is provided simultaneously by surface runoff and rainfall. As (pseudo-)karst formation is a dynamic ongoing process, peat deposits evolve with time from rheophilous to ombrophilous as depressions enlarge and deepen through rock weathering.

Bog peats are in general poorly to very poorly drained and water-saturated during the largest part of the year, with patches of stagnating water at the peat surface. During the period of lower rainfall in February–March, i.e., the period in which field observations and peat sampling took place, the water table is usually depressed to depths varying between 10 cm and 40 cm, mainly 30–40 cm. However, rainfall is enough to keep the upper peat layer always moist, a condition that allows but slows down the decomposition of the organic material while also diminishing fire hazard. These hydrologic conditions control the water movement in the peat mantle and the mobilization of organic substances in different ways.

- The seasonal fluctuations of the groundwater level move dissolved or suspended contemporary organic substances up and down in the upper layers of the peat, mixing materials of variable ages. In dense sapric layers, quelation of iron oxides by organic substances may also contribute to the leaching of these compounds. Iron oxides are provided by bedrock weathering, especially that of the mafic minerals contained in granites, and are washed into the peat pools by rainwater runoff on the bare rock slabs.
- During the peak of the rainy season, the water level in the bogs tends to overflow and pour blackwater into small creeks, causing the exportation of decomposed, presumably older, organic substances from the sites of formation.
- The joints and fractures of the bedrocks underlying the bogs, especially in sandstone–quartzite formations, allow leaching and deep drainage toward endokarstic pipes and galleries. This promotes sustained water percolation through the peat mantle and underlying rock fissures. Labile organic compounds are extracted from the bottom of the peat reservoirs, especially from the sapric layers, and dragged downwards; they end up feeding blackwater rivers in the lowlands. This process is more effective in sandstone formations than in granite.
- Because of the steady rainwater supply and deep water percolation, the classical acrotelm–catotelm model that distinguishes between an upper periodically aerated layer and a lower saturated layer in a closed hydrological system (Holden 2006) does not apply satisfactorily to the tepui peatland hydrology. In fact, basal peat layers at tepui and dome summits are generally more decomposed (sapric) than surface (fibric) and subsurface (hemic) layers.

In the case of slope peats, the water movement is less confined than in bog environment. Part of the rainwater is trapped in innumerable (pseudo-)karstic alveoli, but the rest moves freely along the interface between the base of the peat cover and the rock substratum toward small creeks. Usually, creeks run on bare bedrock but are often bordered by narrow floodplains where mineral sediments from upslopes are deposited. The bottom layers of these mineral deposits are frequently impregnated by decomposed organic material (humus). There is evidence that this organic material is allochthonous, as floodplains are unstable, ever changing landforms on tepui summits. It is actually the oldest organic material dated in the study area (7490 BP at Marahuaka I.2 and 7100 BP at Marahuaka I.1 sites). We assume that it has been extracted from basal layers of earlier peat mantles on upper slopes and drained to the narrow floodplains that border the distal front of the peat glacis. This process could contribute to rejuvenate depleted peats upslope and enrich recent mineral horizons downslope with older organic substances. Similar accumulation of soluble (sapric) organic matter by water movement down steep slopes has been observed in forested Histosols in Southeast Alaska (D'Amore and Lynn 2002).

7.4.2 The Effect of Geodynamics on Peat Formation

Geodynamic processes taking place at the terrain surface influence peat formation and age. These include peat sliding, the effect of creek migration on peat cutting, and the steady renewal of the micro-topography through (pseudo-)karstic alveolization.

There is evidence, in some places, that the peat mantle has been fragmented by differential sliding on rocky glacis slopes. For instance, fresh detachment scars were observed at the lower edge of a blanket peat with 3–5% slope at site Cuao 13 (Profile 18). The sliding process was activated at the front of the peat mantle by the lateral migration and cutting of a small creek running on bare bedrock. Buried karstic lapies micro-topography (karren field), with many grooves and small alveoli a few square meters large, was exhumed in patches as the peat cover moved and fragmented (Fig. 7.5). Sliding events, triggered by heavy rainfalls, are frequent in unstable slope-peat situations that are thus exposed to periodic renewal of peat accumulation, even at relatively gentle slope gradients. They result in variable peat ages in one single landscape unit. In blanket peats, the present basal layer at a given site does not always reflect the time of primary peat initiation at that site, as the original basal layer may have been removed by sliding upon shear failure at the peat-mineral interface. Peat sliding is recognized as an increasing environmental hazard (Dykes and Kirk 2006). Peat failures have been related to climate change (Dykes and Warburton 2007) or to slowly changing internal thresholds within the peat material (Dykes et al. 2008). A more contrasted rainfall regime at the tepui summits, resulting from ongoing climate trends, could generate short-term desiccation periods that have been recognized elsewhere as trigger for peat mass movements (Beven et al. 1978).

Peat accumulates preferentially in (pseudo-)karstic depressions that result from the weathering of the bedrocks, including processes such as cement and mineral dissolution, physical fragmentation, and exfoliation. The maintenance of warm moist conditions during most of the Holocene favored rock weathering, leading to slow but constant renewal of (pseudo-)karstic micro-topography. It can thus be assumed that new alveoli on rocky glacis and larger depressions in flat areas continued appearing on the landscape. This resulted in diachronic rather than cyclic peat inception, with starting dates of organic material accumulation spread over time.

Fig. 7.5 Peat sliding. (**a**) Shows karren microtopography of mounds with bare rock and depressions with sandy weathering products, uncovered upon sliding of the peat mantle (Cerro Autana in the left background). Sliding is promoted by water saturation at the peat–rock interface (**b**, Lithic Haplohemist of Profile 18). Close to the detachment scarp, the fragmentation of the peat mantle through differential sliding leaves behind isolated *Brocchinia* bunches as remnants of the original meadow cover (**c**) (photos Zinck)

In slope conditions, the ongoing creation of (pseudo-)karstic micro-topography might explain local differences in ^{14}C dates better than climate change does.

7.4.3 Peat Age–Depth Relationship

Peat ^{14}C ages vary from current and modern accumulation of organic material to deposits several thousands of years old (Fig. 7.6). In the Marahuaka-Huachamakari

Fig. 7.6 Old and young peat layers. (**a**) A 150 cm deep, strongly decomposed peat on sandstone (Profile 17) with ^{14}C age 7030 BP in the lower sapric layer (80–100 cm) under tubiform meadow cover (**b**, *Brocchinia* and *Kunhardtia*). (**c**) A 40 cm deep, slightly-to-moderately decomposed peat on granodiorite (Profile 16) with ^{14}C age 185 BP in the lower fibric layer (13–40 cm) under shrub cover (center of **d**) (photos Zinck)

area, peat ages range from 170 to 7490 BP. The oldest organic material (7490 BP) occurs at 70–95 cm depth, but similar ages are also found at 40–60 cm depth, while the deepest peat layer dated (85–120 cm) is only as old as 3400 BP. In the Cuao-Sipapo area, peat ages vary in a similar range: from 165 to 7030 BP. In this case, the oldest peat layer dated (7030 BP) is also the deepest one (80–110 cm). In the Maigualida area, the range of variation is narrower, stretching from 2530 to

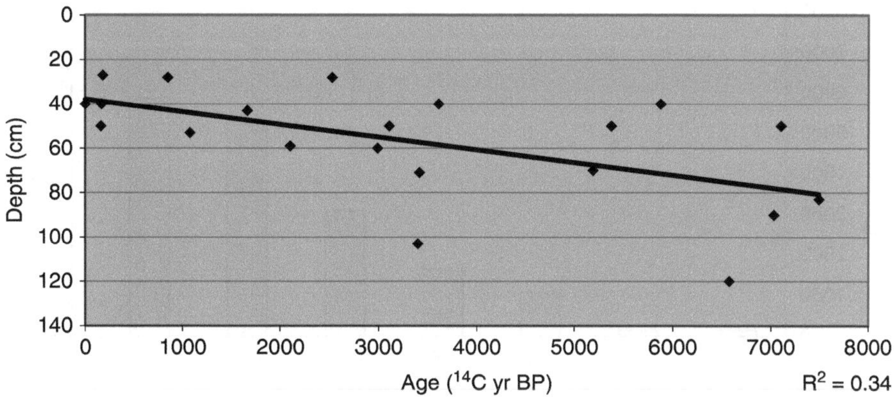

Fig. 7.7 Peat age–depth relationship in the western Guayana Highlands (peat residue samples; $n = 21$)

6580 BP. Like in the Cuao-Sipapo area, the oldest peat layer dated (6580 BP) is here also the deepest one (110–130 cm).

Ideally, assuming a constant cumulization rate of organic material, a kind of linear relationship between peat age and peat depth could be expected to occur, perhaps slightly skewed by the effect of compaction in the lower layers. To test this hypothesis, ^{14}C ages BP were plotted against peat depths in Fig. 7.7. A weak trend of increasing age with increasing depth can be observed ($R^2 = 0.34$). If two conspicuous outliers (3400 BP at 103 cm and 6580 BP at 120 cm) are disregarded, the relationship slightly improves ($R^2 = 0.38$). In fact, variable ages from 7490 BP to recent times were determined at similar depth ranges between ca 20 cm and 90 cm depth.

At mid-profile depth (20–60 cm), the age of subsurface peat layers and humus-rich mineral horizons ranges from young to old (Fig. 7.8). In the Marahuaka-Huachamakari area, the ^{14}C ages vary between 170 BP and 7100 BP at depths of 20/45–60 cm, while they vary between 165 BP and 5380 BP at depths of 30/40–60 cm in the Cuao-Sipapo area (Table 7.1). Thus, at less than 60 cm depth (roughly between 40 cm and 50 cm), peat ages vary from 165 to 7100 BP (Fig. 7.9).

At the bottom of the depressions or close thereto, peat ages also vary over a large range, from contemporary accumulation of organic material to 7030 BP for peat layers sensu stricto and to 7100–7490 BP for bottom horizons of mineral material mixed with allochthonous humus derived from earlier peat (Fig. 7.10). Thus, there is no common inception date of peat accumulation among the sampled sites. Organic material started depositing at variable dates during the Holocene, with an important gap in our data set between approximately 6600 BP and 3600 BP.

This suggests that the visited sites have undergone different evolution trends and have thus different peat formation histories. In some places, the peat mantle remained stable over time and was not disrupted by environmental hazards. In other places, the peat cover has been exposed to changes, with renewal of the organic material after truncation or even complete removal through peat sliding on slopes.

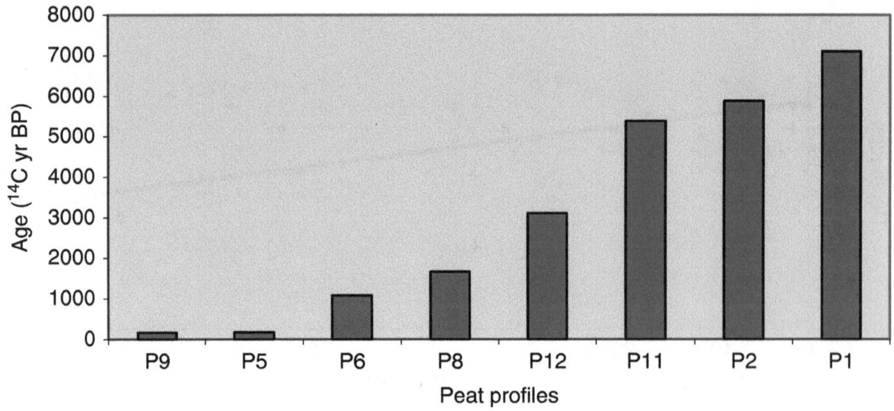

Fig. 7.8 Age of peat layers above 60 cm depth in the western Guayana Highlands

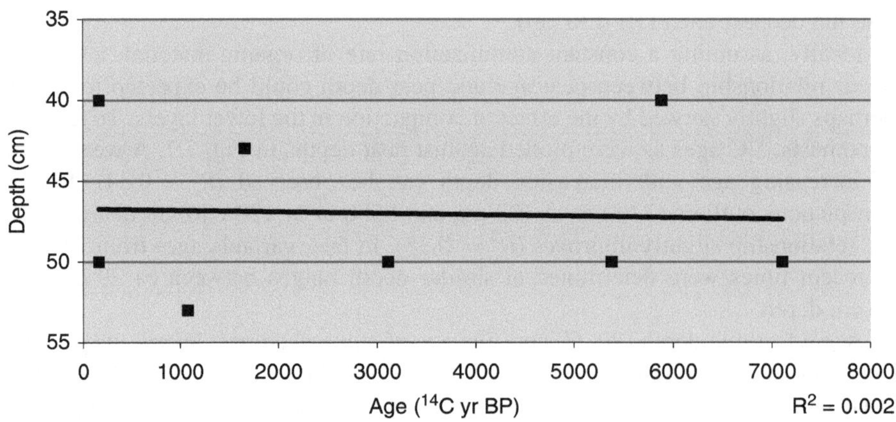

Fig. 7.9 Age–depth relationship in peat layers above 60 cm depth in the western Guayana Highlands

There is also evidence that organic material did not start accumulating everywhere at the same time, because sites favorable to peat deposition were created by (pseudo-)karstic morphogenesis all over the Holocene. This explains why there is no clear regional time trend in peat formation, and why there is no strong relationship between peat age and depth. However, the earliest organic material accumulations started at fairly similar ^{14}C dates (ca 7500–7000 BP) that calibrate into ca 8400–7800 calBP in both the Marahuaka-Huachamakari and Cuao-Sipapo areas. For the Maigualida area, there is less dispersion among the few available data, suggesting a more stable environment. In fact, on granitic domes, backslopes are too steep for peat to form. Peat develops mainly in closed depressions on dome summits, less exposed to renewal of material upon sliding.

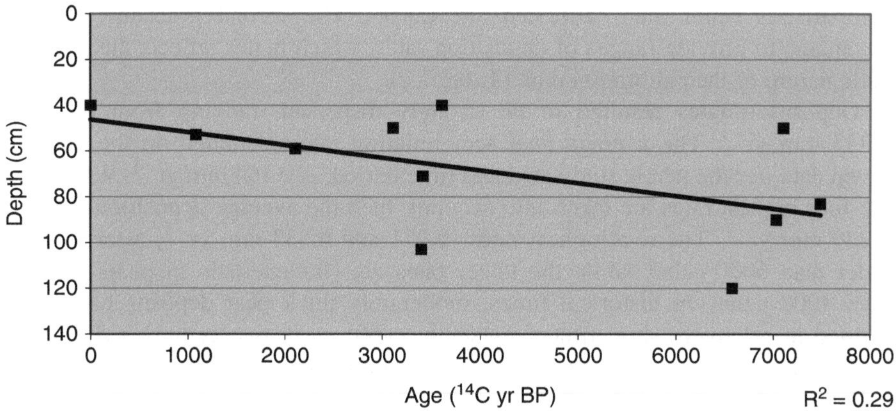

Fig. 7.10 Peat age–depth relationship in basal strata, including peat layers proper and mineral horizons with humus, in the western Guayana Highlands

7.4.4 Rates of Peat Deposition and Depositional Gaps

Budget limitations did not allow the dating of full layer sequences at all sites, an ideal condition to estimate peat deposition rates and rate changes over time. Deposition rates can be estimated from the relationship between the depth to the lowest layer and its age, assuming that peat accumulation was constant throughout the history of the peat cover formation. This approach seems unrealistic because peat deposition was not constant over time, as shown by the time gaps between layers, and the lower layers are likely to have been compacted under the weight of the overlying peat pack. So we preferred to estimate peat deposition rates from pairs of dated layers, taking the distance between consecutive layer centers as the peat reference thickness. Layer centers were considered more appropriate than layer boundaries because peat material for carbon-14 determination was mainly sampled in the central section of the layers to avoid contamination from neighboring layers. There were consecutive dated layers in four cases while, in other cases, a thin (10–30 cm) intermediate undated layer was present. However, such estimates may be less meaningful in the cases where there are large age gaps between layers that presumably indicate some discontinuity in peat deposition. In total, seven layer pairs were available, some of them with important age gaps. In all cases, the lower layer of the pair was the basal layer lying just above the rock substratum and corresponding thus to the inception of a peat accumulation phase.

Deposition rates are expressed in terms of calendar years. They can be estimated either from the means of the calibration ranges or from the time gaps between consecutive peat layers. Both procedures generate similar values. The first procedure is statistically questionable because the mean value may fall between wiggles on the calibration curve and may therefore not be representative of the calibrated probability distribution, but it is a convenient proxy central value that allows for comparisons better than calibrated age ranges. The second procedure has the advantage to provide ranges of deposition rates, which better reflects the probabilistic nature of the calibration data (Table 7.2).

Deposition rates resulted to be strongly dispersed, ranging from 0.065 to 0.333 mm yr^{-1}. The average peat accumulation rate calculated on the basis of seven data, for the whole study area and time period, is 0.164 mm yr^{-1}. When only the four highest rates are taken into account, then the average depositional rate is 0.229 mm yr^{-1}. The two highest rates (0.271 and 0.333 mm yr^{-1}) refer to peats older than 6000 calBP, while the lower rates are characteristic of peats younger than 4000 calBP. In historical times, moderately thick peat deposits have been formed in relatively short time lapses: up to 60 cm in two centuries (P9 at the Cuao 3 site); up to 45 cm in contemporaneous times (P18 at the Cuao 13 site). Peat continues forming nowadays.

Three deposition rates are particularly low (0.065 mm yr^{-1} for P17, 0.083 mm yr^{-1} for P20, and 0.087 mm yr^{-1} for P12), although they have been calculated on the basis of consecutive layers. This could indicate that peat at these sites has not accumulated at constant rate, including possible depositional hiatuses between consecutive layers, and that peat profiles might be polycyclic. In some cases, there are large time gaps between consecutive peat layers: for example, a gap of 2528 calendar years in P12, from 795 calBP (15–40 cm) to 3323 calBP (40–60 cm); a gap of 4670 calendar years in P17, from 3185 calBP (40–80 cm) to 7855 calBP (80–100 cm). At an average peat deposition rate of 0.229 mm yr^{-1}, a peat layer of 107 cm (4,670 years × 0.229 mm yr^{-1}) would have been formed at site P17, instead of the 30 cm peat actually accumulated, resulting thus in a deficit of 77 cm peat thickness. Likewise, at the same deposition rate, a peat layer of 58 cm (2,528 years × 0.229 mm yr^{-1}) would have been formed at site P12, instead of the 22 cm actually accumulated, resulting in a deficit of 36 cm peat thickness. This could be related to the following: (1) rejuvenation of the original peat profiles by truncation (peat sliding?), followed by new vegetation colonization and more recent peat deposition burying the truncated profiles; or (2) a depositional hiatus between consecutive layers because of unfavorable climate conditions. The first hypothesis is supported by the fact that peat layers at similar shallow depths (20–60 cm) have ages varying from a few hundred to a few thousand years. Peat truncation through sliding might have played an important role because gaps occur at different depths. The second hypothesis suggests a climate change, and that is addressed in Sect. 7.4.9.

In a previous study carried out on similar peat deposits in the eastern Guayana Highlands (Chimantá, Guaiquinima, Auyantepui, and Gran Sabana), Rull (1991) found that most peats had accumulated at rates of 0.2–0.4 mm yr^{-1} and at an overall rate of 0.296 mm yr^{-1}, with only two significant lower and upper outliers of 0.06 mm yr^{-1} and 1.53 mm yr^{-1} in a set of 15 data. The average rate of 0.296 mm yr^{-1} found in this area is comparable to our rate of 0.229 mm yr^{-1} but higher than our regional average rate of 0.164 mm yr^{-1}. A recent study carried out in the same region, at the Eruoda-tepui summit in the Chimantá massif, reports much higher peat accumulation rates (1.840 mm yr^{-1}) in the interval of 4.3–4.0 kyr calBP, lying between two periods of lower peat formation (0.006 mm yr^{-1} between

Table 7.2 Deposition rates (in calendar years)

Profile	Site	Depth (cm)	Dated material	Layer	Cal yr BP 2σ	Cal yr BP mean	Cal yr BP time gaps	Reference thickness (cm)	Deposition rate range (mm yr^{-1})	Deposition rate average (mm yr^{-1})
P2	Marahuaka I.2	20–60	Sapric peat	Oa1	6840–6565	6703	1545–1630	43	0.278–0.264	0.271
P2	Marahuaka I.2	70–95	OM	2C	8385–8195	8290				
P5	Marahuaka II.5	20–60	Hemic peat	Oi2	300–70	185	3525–3415	63	0.179–0.185	0.182
P5	Marahuaka II.5	85–120	Hemic peat	2C	3825–3485	3655				
P8	Cuao 2	30–55	Hemic peat	Oe2	1690–1420	1555	2125–2160	28	0.132–0.130	0.131
P8	Cuao 2	55–87	Sapric peat	Oa	3815–3580	3698				
P12	Cuao 7	15–40	Hemic peat	Oe2	900–690	795	2495–2560	22	0.088–0.086	0.087
P12	Cuao 7	40–60	Sapric peat	Oa	3395–3250	3323				
P17	Cuao 12	40–80	Hemic peat	Oe	3330–3040	3185	4625–4715	30	0.065–0.064	0.065
P17	Cuao 12	80–100	Sapric peat	Oa1	7955–7755	7855				
P19	Maigualida 1	60–80	Sapric peat	Oa2	6170–5800	5985	1395–1625	50	0.358–0.308	0.333
P19	Maigualida 1	110–130	Sapric peat	Oa4	7565–7425	7495				
P20	Maigualida 2	20–35	Sapric peat	Oa	2745–2485	2615	1330–1345	11	0.083–0.082	0.083
P20	Maigualida 2	35–45	OM	2Cr	4075–3830	3953				

OM = organic matter in mineral soil

12.7 and 4.3 kyr calBP, and 0.174 mm yr^{-1} between 4.0 kyr calBP and nowadays) (Nogué et al. 2009). High long-term average depth increments (1.3 mm yr^{-1}) were also found at Cayambre-Coca, 3,968 m elevation, in the Ecuadorean Andes (Chimner and Karberg 2008).

7.4.5 Peat Age–Elevation Relationship

Peat currently forms and/or has formed in the past across a large elevation range from 650 m to 2,600 m a.s.l. High elevation does not limit peat formation as rainfall continues to increase with altitude toward summits at 2,800 m a.s.l., but there might be a lower elevation threshold for peat formation as annual rainfall decreases and average temperature increases. The lowest sites visited at around 650–700 m elevation show peat formation to take place in historical and even contemporaneous times, reflecting environmental conditions close to the present ones. For instance, profile P16 located at 650 m a.s.l. has an hemic horizon at 13–40 cm depth with a ^{14}C age of 185 BP, indicating peat formation in subrecent times (1650–1810 calAD). Similarly, profile P18 at 700 m a.s.l. has a hemic horizon at 35–45 cm depth that started forming a few decades ago (105% modern). For the purpose of comparison, there is no report on low-elevation peat localities in the Chimantá massif, eastern Guayana Highlands, where all recorded peat sites are in the range of 1,800–2,600 m a.s.l. (Huber 1992), concerning thus the higher tepui environment.

In Fig. 7.11, peat ^{14}C ages are plotted against the elevation of tepui and dome summits. The relationship is weak ($R^2 = 0.23$), but some interesting features can be highlighted. For instance, at elevations slightly above 2,500 m a.s.l., peat ages vary from recent (170 BP) to old (7490 BP). The largest number of dated samples lies in the range of 1,400–2,100 m a.s.l. At lower elevations, tepuis are infrequent, and only a few have been visited showing contemporaneous peat accumulation.

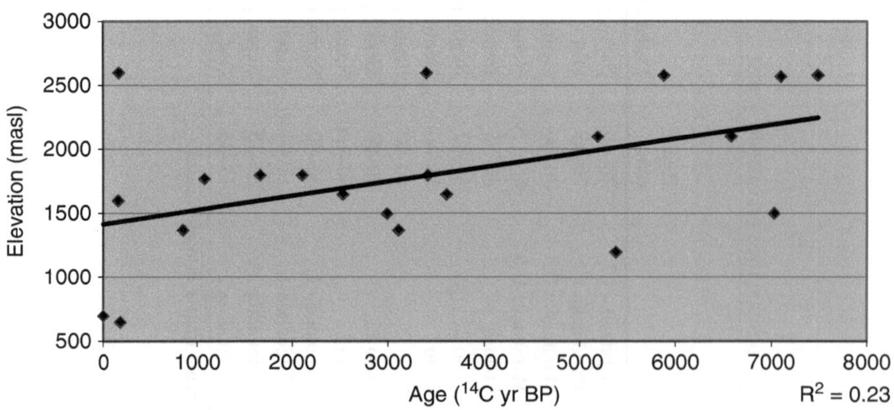

Fig. 7.11 Peat age–elevation relationship in the western Guayana Highlands (peat residue samples; $n = 21$)

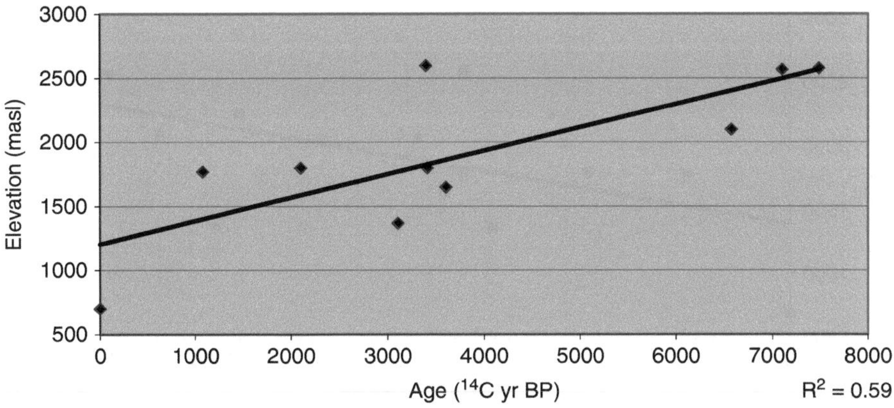

Fig. 7.12 Peat age–elevation relationship in the western Guayana Highlands (basal layers)

In Fig. 7.12, only basal layers were taken into account to describe the relationship between peat ^{14}C age and elevation ($R^2 = 0.59$). The age of the organic material in the bottom layers reflects the time of peat inception. The figure shows that peat started accumulating first at higher elevations when climate changed from dry to moist at the beginning of the Holocene. It seems to be that higher tepuis were more effective in those early times in promoting moist air masses advection and condensation. As moister conditions generalized in the regional context, peat formation propagated to tepuis of mid and lower elevations.

In Fig. 7.13, ^{14}C age-elevation data of the basal peat layers from western and eastern Guayana Highlands were pooled together. The trend that peat accumulation started first at higher elevations is confirmed, although the relationship between peat age and elevation loses strength ($R^2 = 0.28$).

7.4.6 Peat Decomposition and Age

The stages of decomposition of the organic materials found in tepui and dome peats range from raw folic to slightly decomposed fibric, to moderately decomposed hemic, and to strongly decomposed sapric. Only hemic and sapric materials were dated. Additionally, a few humus-containing mineral horizons were dated. The oldest ^{14}C ages determined in our area correspond to this kind of organic matter of advanced decomposition in two mineral horizons (7100 and 7490 ^{14}C BP, respectively) (Fig. 7.14). Mineral sediments are found in narrow floodplains where mineral debris washed away from surrounding rocky slopes have accumulated. The humus present in the basal layers of these mineral deposits has not formed in situ: it has been either transported together with the mineral material or washed in by groundwater flow from upslope. In either case, the humus of the basal mineral

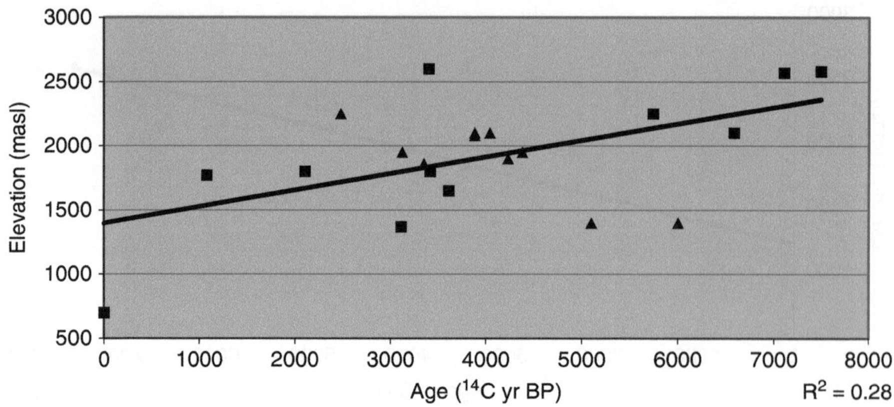

Fig. 7.13 Peat age–elevation relationship in the western (*squares*; $n = 11$) and eastern (*triangles*; $n = 10$) Guayana Highlands (basal layers). Data from Schubert and Fritz (1985) and Schubert et al. (1994) for eastern Guayana

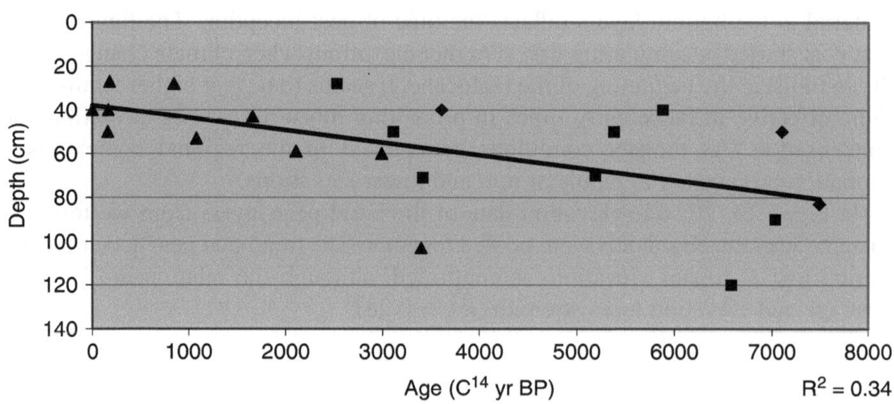

Fig. 7.14 Relationship between degree of peat decomposition and age in the western Guayana Highlands (*triangle* = hemic material; *square* = sapric material; *rhombus* = humus mixed with mineral sediments)

horizons in floodplains is an allochthonous evidence of the earliest peat accumulation and decomposition in the area.

Decomposition of the organic material into hemic and sapric is related to peat age. Hemic layers are in general younger than 3500 ^{14}C BP, while sapric layers are older than 2500 ^{14}C BP (Fig. 7.14). There is no strong relationship between degree of peat decomposition and depth. Hemic and sapric materials occur at variable depths, although hemic materials tend to be more surficial (mainly between 20 and 60-cm depth) than sapric materials.

7.4.7 Variations in Peat Nature

Peat kinds vary with the vegetation source. As peat occurs over a large altitudinal range (650–2,600 m a.s.l.), flora and vegetation types can be expected to vary according to the regional temperature gradient of −0.6°C per 100 m elevation increase (Galán 1992), while rainfall might be less differentiating. Climate changes during the Quaternary may have caused vertical vegetation displacements and interchange between lowland and highland flora (Steyermark and Dunsterville 1980; Huber 1988; Rull 2004; Gröger and Huber 2007). However, climate oscillations that occurred during the Holocene had certainly less impact on vegetation shifts than the Pleistocene glacial-interglacial cycles (Rull 2004). In the Holocene period, meso-relief and micro-relief have played a fundamental role in determining the vegetation composition. Vegetation on rock outcrops might have been more sensitive to climate change than the vegetation in peat sites more controlled by water stagnation (Rull 2005). The influence of vegetation on peat characteristics and stages of evolution varies between forb-herb cover and shrub-forest cover (e.g., low *Bonnetia* forest), the former being largely dominant in peat sites.

Peat material is a product of the vegetation cover that existed during a given time at a given site. This relationship can be analyzed using the $\delta^{13}C$ values that reflect the photosynthetic pathway of the plants contributing organic material to peat formation (Table 7.1). In Fig. 7.15, $\delta^{13}C$ values are plotted against elapsed time. With exception of a few outliers, the majority of $\delta^{13}C$ data lies between −25‰ and −28‰. A proportion of 65% of the $\delta^{13}C$ values fits in the range of −25‰ to −27‰, and 87% in the range of −25‰ to −27.4‰. This suggests that the vegetation cover was mainly composed of C3 plants, in a mixture of predominantly woody (from −26 to −28‰) and subordinate herbaceous species (from −24‰ to −26‰), with no significant variations all along the time period from 7500 BP to

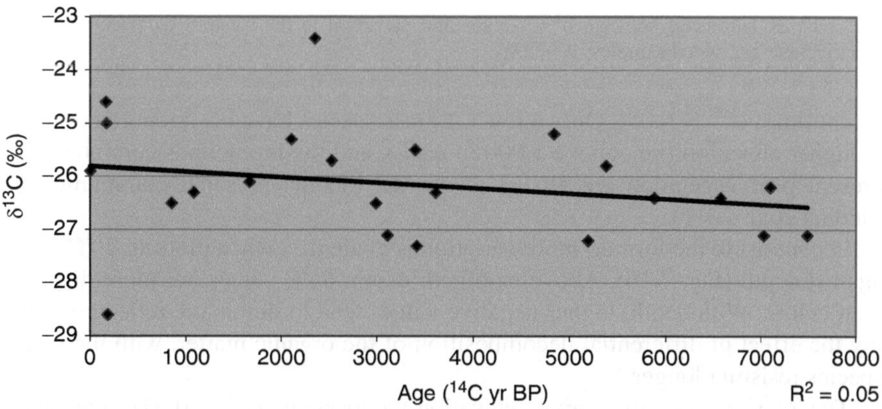

Fig. 7.15 Relationship between $\delta^{13}C$ values and peat age in the western Guayana Highlands (peat residue and wood samples; $n = 23$)

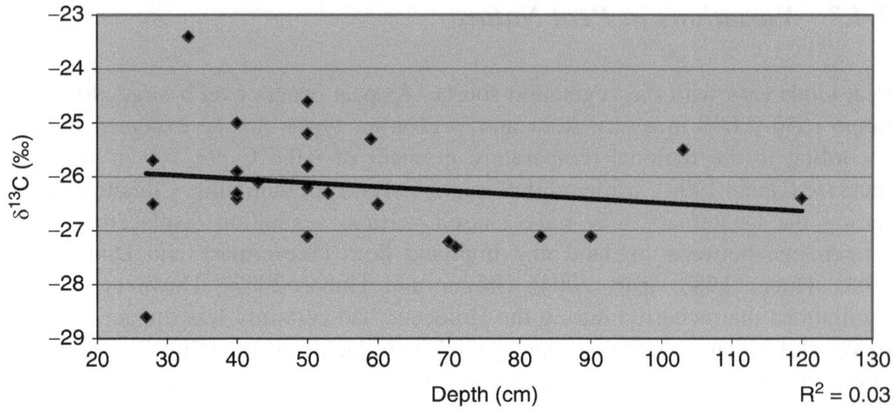

Fig. 7.16 Relationship between δ^{13}C values and peat depth in the western Guayana Highlands (peat residue and wood samples; $n = 23$)

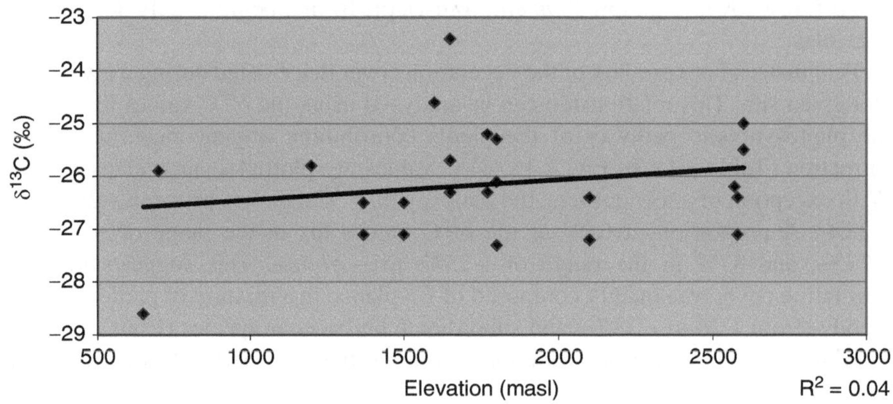

Fig. 7.17 Relationship between δ^{13}C values and elevation in the western Guayana Highlands (peat residue and wood samples; $n = 23$)

recent times ($R^2 = 0.05$). Only a few C4 grass species have been reported to occur at higher elevation (i.e., above 1,000 m a.s.l.), and the bromeliads common in the present peat vegetation are all C3 plants (E. Medina personal communication; Medina et al. 2011).

In contrast to the former, more variation is evidenced when plotting δ^{13}C values against depth (Fig. 7.16). Above ca 60 cm depth, δ^{13}C values are more dispersed than below. With depth, higher negative values tend to dominate, reflecting possibly the effect of differential decomposition of the organic matter, with the woody species resisting longer.

The δ^{13}C values are mainly elevation-independent ($R^2 = 0.04$) (Fig. 7.17). Except two outliers at −23.4‰ and −28.6‰, the δ^{13}C values lie between −24.6‰

and −27.3‰ all along the elevation range from 700 m to 2,600 m a.s.l. Values higher than −26‰ occur at a variety of elevations in all three study areas. This means that the temperature gradient has little influence on the composition of the vegetation cover at peat sites and that the main factor controlling it is peat hydrology. This holds for the whole period since the inception of peat accumulation at ca 7500 BP (8385–8195 calBP). Vertical displacement of vegetation belts resulting from climate variations in the Holocene, as evidenced by pollen analysis in the Chimantá tepuis (Rull 2004), could not be inferred from our $\delta^{13}C$ peat data. Altitudinal vegetation shift is probably more a response to climate-forcing at regional level, while vegetation at peat sites seems to be more controlled by the local-forcing of peatland hydrology. Peatland vegetation is relatively insensitive to regional climate changes that do not alter substantially the water regime at peat sites. Long-term water saturation sustained by abundant rainfall causes inertia buffering against temperature changes.

7.4.8 The Fire Effect

Inference of the effect of fire on the subsidence, degradation, or full destruction of peat deposits is not made here, as no morphological field evidence was found to substantiate this effect. However, dry periods or even shorter dry spells causing temporary lowering of the water table in peatlands might have favored the onset of occasional wildfires struck by lightening. For instance, Nogué et al. (2009) have identified a charcoal peak at 3.9 kyr calBP in the pollen record of the Eruoda-tepui summit (2,627 m a.s.l.) in the Chimantá massif, eastern Guayana Highlands. This coincides with the beginning of an extended dry phase in the neighboring plateau of the Gran Sabana (900–1,000 m a.s.l.) where forest fire events have been documented around 3.5 kyr calBP (Fölster 1992). The Gran Sabana is considered by Nogué et al. (2009), the most probable source area of the charcoal recorded at the Eruoda-tepui summit. The fact that charcoal at the Eruoda site is not authigenic allows assuming that herbaceous and shrubby peatlands at higher elevations were/are less exposed to burning than forest-covered peatlands at lower elevations.

7.4.9 History of Peat Formation and Climate Change

7.4.9.1 Peat Inception

The chronology of the ^{14}C-dated organic deposits ranges from 7490 BP to recent times. The two oldest ^{14}C ages determined in the study area (7490 BP at site Marahuaka I-2 and 7100 BP at site Marahuaka I-1) do not refer to peat material proper but to humus that impregnates basal mineral horizons found at the lower edge of footslopes covered with peat or in narrow floodplains. There is no evidence that this humus has formed in situ, because there are large time gaps between the age of the organic matter in the mineral horizons and that of the overlying peat layers. Therefore, we assume that it corresponds to old allochthonous organic substances that have moved within the groundwater flowing atop rocky slopes, from early upslope peats toward lower landscape positions where they have been trapped in mineral layers. This process can be observed working in current situations. As a consequence of the former, peat must have started accumulating upslope before 7490 BP. The oldest peat layer identified at the bottom of a closed bog position (P17) was dated 7030 BP. Thus, the three oldest ^{14}C dates in our area provide a range of ca 7500–7000 BP (ca 8400–7800 calBP), which can be considered the time span of peat inception in the area. This would mark the turn from late glacial aridity to the start of increasing moisture in the tepui summit environment. For northern South America, this major climate change at the Pleistocene-Holocene boundary has been documented on the basis of geomorphic, paleontological, and sedimentological evidence that is summarized in Schubert et al. (1994). Thus, in the Guayana Highlands, the oldest records of peat accumulation so far available suggest that the Holocene period started somewhat later than the usually accepted beginning at 10 kyr BP (12 kyr calBP). The peat starting range of 8400–7800 calBP might vaguely correlate with the end of the only Holocene Bond event (8.2 kyr cooling event or Bond event 5) that has a clear temperature signal in the Greenland ice cores (Bond et al. 1997). A recent study carried out on the Eruoda-tepui summit in the Chimantá massif, eastern Guayana Highlands, reports a ^{14}C age of 10630 ± 60 BP (12821–12573 calBP) determined on pollen residues at 175–177 cm depth, while the oldest ^{14}C age determined on macroscopic plant remains is 7690 ± 20 BP (8523–8420 calBP) at 172–176 cm depth (Nogué et al. 2009). The latter is an age quite similar to the oldest ^{14}C age of 7490 ± 50 BP (8385–8195 calBP) determined at 70–95 cm depth in our study area of the western Guayana Highlands.

Peat inception has not been uniform throughout the area. If the age of the basal layer at any given site is considered to reflect the starting of peat accumulation at that site, then peat initiation time in the area was random. In Fig. 7.10 (Sect. 7.4.3), the depth of basal layers is plotted against age showing that peat started accumulating at variable times throughout the Holocene, with an important data gap between ca 6600 and 3600 BP. This large variability of peat inception dates is related with the geodynamics operating at tepui summits. On slopes, sliding causes peat scarring and fragmentation or even full dismantlement, and this leads to new phases of peat initiation on freshly exposed rock surfaces. In flat areas, where peat forms mainly in closed depressions, rock weathering and karstic activity went on throughout the Holocene, favored by the perhumid tepui summit environment. Thus, geomorphic and topographic conditions never ceased to be favorable to continuous peat initiation since the earlier peat formation around 7500–7000 BP. The interference of this geodynamics makes it difficult to infer climate changes or minor variations during the Holocene.

At a few places, peat started forming in subrecent historical times. Three sites, geographically well distributed, have peat layers with remarkably concordant ^{14}C dates and calendar ages at relatively similar depths. At site Marahuaka II.5

(P5), a 40-cm layer (20–60 cm) was dated 170 BP (1650–1880 calAD) above a much older peat layer (3400 BP). At site Cuao 3 (P9), a 20-cm layer (40–60 cm) was dated 165 BP (1665–1880 calAD) above a floating fiber mat. At site Cuao 11 (P16), a 27-cm layer (13–40 cm) was dated 185 BP (1650–1810 calAD) immediately above the bedrock. As the three layers lie on unconformable substrata, they correspond to the inception of a new phase of organic material accumulation around the seventeenth to nineteenth centuries and might thus reflect a major environmental change. The peat layer closest in time is as old as 850 BP (1050–1260 calAD) at site Cuao 7 (P12). There is thus a time span of several centuries within which no peat inception was detected. This could reflect increasing aridity correlated with the Little Ice Age period that lasted from about 1350 to 1850 AD in temperate regions. The return to a moister environment at the beginning of the seventeenth century would have allowed peat formation to resume. Using pollen analysis of a ^{14}C-dated bog peat in the Venezuelan Andes, Rull and Schubert (1989) identified a general altitudinal depression of ecological belts, with prevailing dry conditions, between the fifteenth and eighteenth centuries, and correlated this event with the Little Ice Age.

7.4.9.2 Peat Development

Figure 7.7 (Sect. 7.4.3) that relates peat age and depth shows three segments: generally a linear relationship between the Present (i.e., 1950 AD) and ca 3600 BP, a data gap between ca 3600 and 5200 BP, and data dispersion between ca 5200 and 7400 BP. Additionally, there are several small dating hiatuses that could possibly be filled with more intensive sampling (now 23 radiocarbon determinations at 14 different sites). The gap between 3610 and 5190 BP splits the data set into two clusters. The lack of recorded dates for this time span could be a sampling artifact, because observation sites were not located using a formal random sampling scheme. The exploration of the area used satellite images at the scale of 1:250,000 and, in the case of the Cuao-Sipapo massif, a schematic geomorphic map at the scale of 1:100,000, mainly for navigation purpose via helicopter. The two basic criteria to locate observation points were as follows: (1) geographic distribution of the points as even as possible and (2) variation of the vegetation types. Thus, sampling was practically close to random, and the above data gap might not be totally accidental. As such, it suggests that peat areas had considerably shrunk in the period of ca 5200–3600 BP because of decreasing moisture or more frequent dry spells. With moister conditions restarting around 3600 BP, peat formation resumed and the peat cover expanded again. Thus, in the context of this hypothesis, two main peat formation phases would have taken place during the Holocene in the study area: 7500–5200 BP and 3600 BP-Present (1950 AD).

Minor dry spells might have occurred throughout the Holocene as suggested, for instance, by the presence of a colluvial glacis on the top of the Autana tepui (Cuao 6 site). There is no evidence that the mechanical erosion–deposition process that originated the glacis is working in the current moist environment of the region; it possibly reflects a former drier period. Subsequent peat deposition fossilizing the

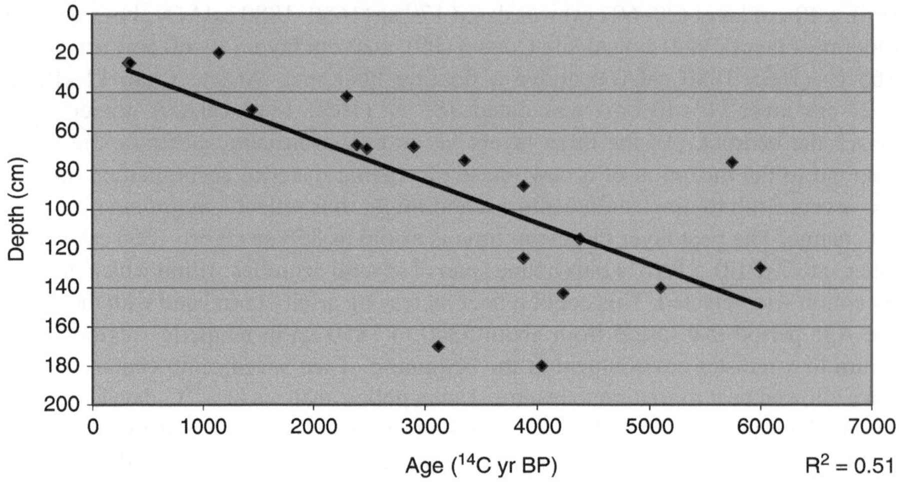

Fig. 7.18 Peat age–depth relationship in the eastern Guayana Highlands. Data from Schubert and Fritz (1985) and Schubert et al. (1994)

distal part of the sandy colluvium (P11) may indicate a return to moister environmental conditions at around 5400 BP.

For the purpose of comparison, we have plotted against depth the available ^{14}C data referring to peat deposits on tepui summits in the eastern Guayana Highlands (Chimantá and Guaiquinima massifs), as published by Schubert and Fritz (1985) and Schubert et al. (1994). Radiocarbon dates of peat and wood samples from this region range from 6000 BP to modern (Schubert et al. 1994). The resulting scatter diagram in Fig. 7.18 shows a slightly better age–depth relationship ($R^2 = 0.51$) than that obtained from our data in the western Guayana Highlands ($R^2 = 0.34$). It does not corroborate the above hypothesis of peat formation interruption in the period of ca 5200–3600 BP, but shows well the following two distinct segments: (1) linear relationship from ca 4000 BP to the Present (1950 AD), and (2) broad data dispersion for the period before 4000 BP. Thus, there is a change in pattern of the peat age–depth relationship around 4000 BP. To support this trend in a larger regional context, we have pooled together the data sets from eastern and western tepuis and plotted age against depth as shown in Fig. 7.19. The resulting graph corroborates the aforementioned two segments: the linear relationship after ca 4000 BP clearly contrasts with the data cloud before this time threshold.

The first segment is enhanced in Fig. 7.20, taking into account only peat layers younger than 4000 BP. It shows a clear linear trend in peat age–depth relationship: peat depth is mainly 20–60 cm in the lapse from Present to 2000 BP, and 40–80 cm in the lapse from 2000 to 3900 BP. Peat depth decreases from maxima of 80–100 cm at ca 4000 BP to minima of 20–40 cm in contemporaneous times. When the outlier at 170 cm depth is disregarded, the coefficient of determination R^2 improves from 0.4 to 0.5. There is no evidence of relevant data gaps over the whole period, suggesting

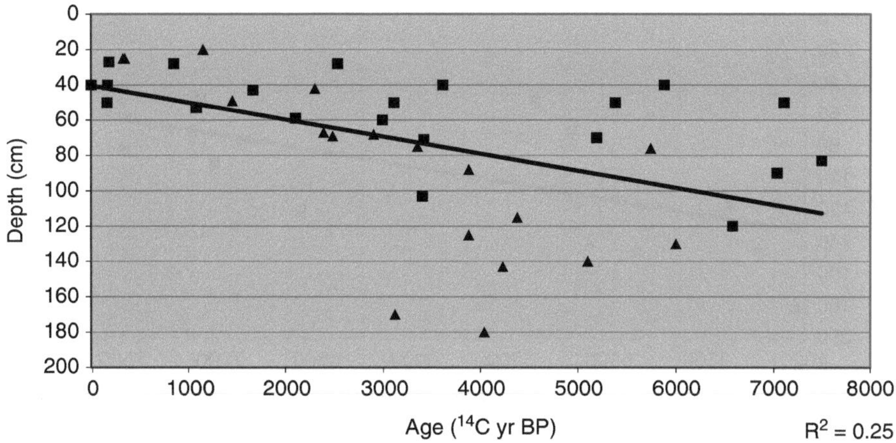

Fig. 7.19 Peat age–depth relationship in the western (*squares*; $n = 21$) and eastern (*triangles*; $n = 18$) Guayana Highlands. Data from Schubert and Fritz (1985) and Schubert et al. (1994) for eastern Guayana

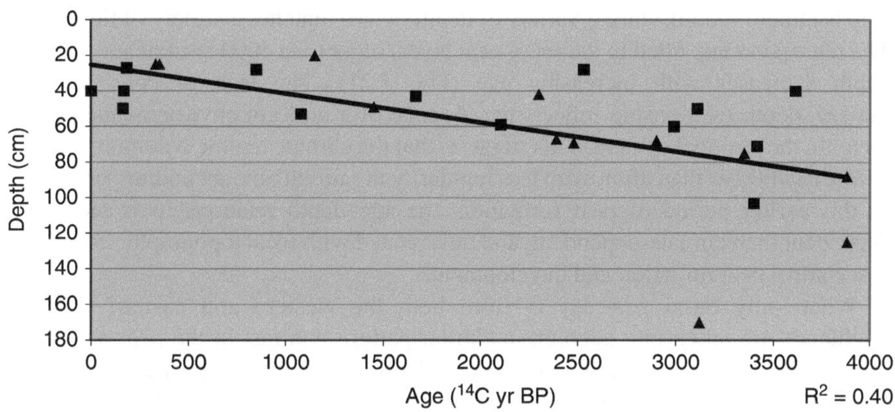

Fig. 7.20 Peat age–depth relationship in the western (*squares*; $n = 14$) and eastern (*triangles*; $n = 12$) Guayana Highlands. Only peat layers younger than 4000 BP

that organic material might have been accumulating almost continuously throughout time and space. The linear behavior of the peat age–depth relationship reflects a constant rate of organic material accumulation, and such a long-term peat development over the whole region requires stable formation conditions. The forcing factor securing such conditions in the regional context over a long period of time can only be climatic. Thus, the data do not allow detecting any significant climate variation in the upper Holocene.

In the second segment of the graph in Fig. 7.19, the linear age–depth relationship shown on the first segment is replaced by a loose cloud of age–depth points. There is no trend evidence, except that the data from western Guayana are above

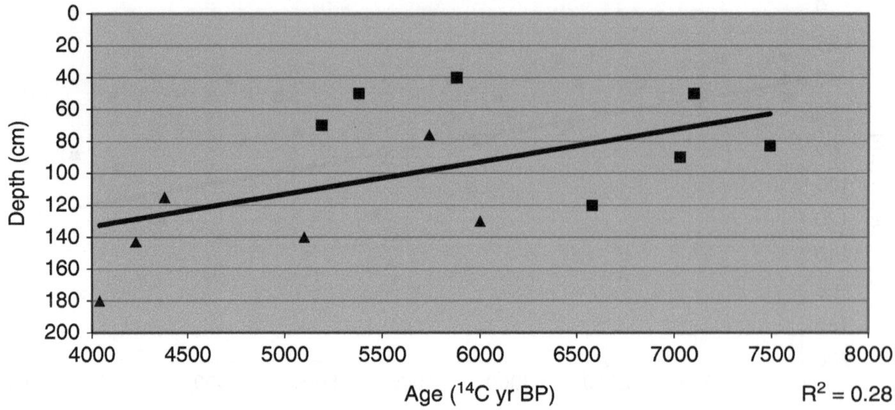

Fig. 7.21 Peat age–depth relationship in the western (*squares*; $n = 7$) and eastern (*triangles*; $n = 6$) Guayana Highlands. Only peat layers older than 4000 BP

the regression line, while those from eastern Guayana are below. Age–depth relationship is rather random, since a variety of depth ranges matches a variety of time spans. The regression line fitted to the set of peat layers older than 4000 BP indicates that peat depth decreases with increasing age (Fig. 7.21). The lack of consistency in the age–depth relationship reflects the absence of a uniform environmental forcing factor in the regional context. This suggests that the climate regime was more variable before ca 4000 BP than after, with less regularity in rainfall amount and/or periodicity. In this earlier period of peat formation, the age–depth relationship is more site-dependent than climate-dependent, and thus related with local topographic conditions that control peat initiation and development.

When only basal peat layers from both the western and eastern Guayana Highlands are taken into account, a similar pattern contrast in the peat age–depth relationship between younger and older deposits appears (Fig. 7.22). The data gap between ca 3600 and 6600 BP shown in Fig. 7.10 (Sect. 7.4.3) for the western Guayana Highlands is filled with data from eastern Guayana, where some deeper basal layers have been recorded.

In conclusion, there is a clear difference in peat age–depth pattern before and after ca 4000 BP. Before that time threshold, peat formation seems to be more related to site conditions (local forcing) and, afterwards, more related to climate conditions (regional forcing). The period around 4000 BP represents a time threshold of broader significance in the regional as well as in the continental context. Working with pollen data from peats of the Gran Sabana (eastern Guayana Highlands), Rull (1991) reached the conclusion that the interval 4500–4000 BP could represent the beginning of a humid phase. Similarly, climate change around 4000 BP has been reported from Africa and South America (Marchant and Hooghiemstra 2004). Using peat records from southern Africa, Meadows (1988) signals the beginning of a moister period between 5000 and 3500 BP conducive to peat formation in the mid-late Holocene.

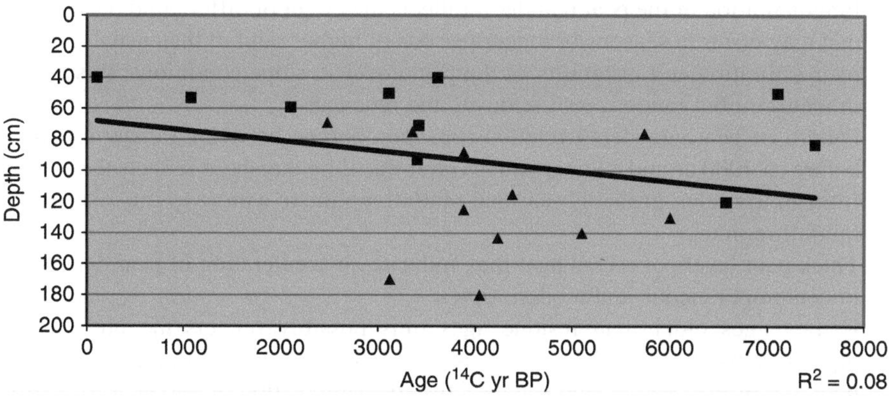

Fig. 7.22 Peat age–depth relationship in basal layers from the western (*squares*; $n = 10$) and eastern (*triangles*; $n = 11$) Guayana Highlands. Data from Schubert and Fritz (1985) and Schubert et al. (1994) for eastern Guayana

7.4.10 Indicators of Environmental Changes

The idea that organic material has been pilling up at a regular pace, steadily aggradating at constant rate from the onset of the Holocene until today, is not supported by peat morphology and age in the study area. It appears that peat formation has been essentially a discontinuous phenomenon in time and space. The peat mantle as a whole shows the integrated result of organic matter cumulization through time, but it is more difficult to disentangle the chronology and the formation context of individual component layers. Many peat profiles are composite, polygenetic profiles. Environmental changes seem to have been frequent in the local context (at specific sites) and/or in the regional context (in the Guayana Highlands), but probably not rhythmic, causing peat formation to proceed by diachronic pulsations rather than by gentle, sustained evolution or synchronic cycles.

Characteristics of the peat cover that indicate environmental changes lead to formulate the following relationships and inferences:

- Large time gaps between consecutive peat layers: they indicate important depositional hiatuses, especially around 4000 BP, which might have been induced by climate change.
- Sapric layers in the lower tiers of the peat cover: they indicate that, in the inception phase of peat accumulation, water-saturation in bogs and on slopes was much less permanent than in the present conditions, allowing exposure of the organic material to decomposition.
- Diachronic peat initiation dates in basal layers: they reflect the variable start of vegetation colonization at scattered sites that offered more favorable microtopographic conditions than others, in the form of (pseudo-)karstic alveoli and depressions promoting water concentration and organic matter accumulation from pioneer plants.

- Fragmentation of the peat mantle: it reflects the effect of differential peat sliding that may occur in seasons or longer periods of higher rainfall than usual, causing over-saturation and instability of the peat cover in slope conditions; this results in recurrent but random peat removal and renewal.
- Pattern of peat age–depth relationship: the contrast between a chaotic pattern before ca 4000 BP and a linear pattern after this time threshold reflects the change from an irregular, probably seasonal rainfall regime to a more evenly distributed rainfall regime.
- Thick peat layers of recent age: they indicate an acceleration of peat deposition in contemporaneous and modern times.
- Extract ages older than residue ages: this possibly reflects the presence of organic substances in the extract inherited from a peat forming phase older than that of the residue, and suggests the superimposition of two peat formation events within a single layer.
- Humus impregnating mineral horizons beneath the peat cover: this indicates the extraction of soluble organic substances from upslope peats, their mobilization in the groundwater flowing along the slopes, and their trapping in the mineral horizons in floodplains. There is no genetic relationship between the peat cover and the humus in the underlying basal mineral horizon.
- Inclusion, in peat layers, of wood fragments older than the wrapping peat material: this indicates long-term survival of wooden components in acid herbaceous peat, and possibly the removal of the fragments from their original position and incorporation in more recent layers.
- Sandy colluvial glacis in footslope position: this indicates mechanical erosion–deposition activity and suggests an interval of drier environmental conditions.
- Mineral horizons of water-transported sediments beneath the peat cover or inserted between peat layers: they indicate intervals of mechanical erosion of rock weathering debris from slopes and deposition of mineral sediments in narrow floodplains.

There is thus a multiple environmental causality context, including climatic, vegetational, geodynamic, and hydrodynamic factors that induce variations in the peat cover. This makes it difficult to infer strict climatic changes at regional level and distinguish them from local-forcing causes. The occurrence of Bond events based on 1500-year climate cycles is not evident, except for the peat starting range of 7500–7000 BP. This calls for a multicriteria interpretation model to disentangle the history of the peat deposits in the Guayana Highlands.

7.4.11 Fate of Tepuian Peats: A Vulnerable Ecosystem

Peats are fragile natural bodies and, as such, they are vulnerable to the impact of environmental changes, whether these are natural or human-induced. Potential changes in temperature and/or rainfall will affect the sustainability of peat formation in the near future. When studying glacier retreat in Ecuador, Perú, and Bolivia,

Fig. 7.23 Peatland vulnerability to trampling. Cerro Autana is an emblematic tower-like tepui (1,200 m a.s.l.) at the western edge of the Guayana Highlands (**a**). Easy to access by helicopter, this isolated tepui has been visited over the last decades by large tourist groups. Trampling by visitors has damaged the fragile *Brocchinia*-dominated tubiform meadow cover (**b**). (**c**) Shows a degraded patch of peat surface that remained bare after the passage of a group of visitors, with no sign of recovery of the original vegetation (*Brocchinia, Kunhardtia, Lagenocarpus*). The peatsoil is a Lithic Terric Haplosaprist (Profile 11) (**d**) (photos Zinck)

Vuille et al. (2008) found that the climate in the tropical Andes had significantly changed over the past 50–60 years, including rising temperatures and rainfall more seasonally concentrated. They predict that this change is going to continue in the present century. Extrapolating the Andean trend to the tepui environment, a warmer atmosphere and a more seasonal water regime would cause faster decomposition of the organic material and lead to the shrinkage of peat areas. This might affect the hydrological function of the peat cover as a water reservoir that retards the redistribution of rainwater to the tepui surroundings. Water storage capacity of the tepui summits will be diminished, and water will be released faster to the lowlands causing flooding in the whitesand plains and limiting water availability during the dry season. Shrinkage of the tepui peat cover will also contribute to the global depletion of the peatland carbon store, which represents some 30% of the carbon stored in world soils (Immirzi et al. 1992). Monitoring is needed to assess the potential retreat of the peat mantle and the drying up of the smaller peatland units. As the rims of peat bogs will dry up, wildfire hazard by lightening and the risk of extension of human-induced fires from lower to higher elevations will increase. Gorzula and Huber (1992) note that many plants leaving on tepui summits, from herbs to shrubs, are highly inflammable. Once set on, fire would easily propagate through the usually dense *Stegolepis* and *Chimantaea* communities.

Uncontrolled (eco-)tourism will also be a threat to the integrity of the peatland ecosystem. The effect of trampling by visitors on soil compaction at the summit of Cerro Autana (1,200 m a.s.l.) was assessed by comparing bulk density at a disturbed spot (Cuao 4) and a nearby undisturbed spot (Cuao 5), both with similar soil conditions (P10, Cuao 4 and Cuao 5 in Table A.6, in the Appendix). Bulk density values for the first two horizons (0–5 and 5–10 cm depth) were 0.49 and 0.95 Mg m^{-3} at the pristine spot compared with 0.78 and 1.34 Mg m^{-3} at the trampled spot, representing a density increase of 40–60%. Thus, visiting and camping activities on tepui summits can cause soil compaction, which leads to increased water runoff and favors rill and gully formation, as all kinds of tepui vegetation recover very slowly after trampling, or do not recover at all (Fig. 7.23).

7.5 Conclusions

7.5.1 Peat Formation and Dating

- Peat started forming in the lower Holocene at about 7500 BP, thus not immediately after the commonly recognized climate change from late Pleistocene to early Holocene (ca 10 kyr BP).
- Calibrated calendar ages stretch the initiation of peat accumulation back to ca 8400 calBP, instead of the conventional age of ca 7500 BP. Above ca 5000 BP, calibration generated calendar ages that are older than the corresponding ^{14}C ages, while after that threshold calibration resulted in calendar ages that are either older or younger than the related ^{14}C ages.
- The oldest peat layer is 7030 ± 50 BP old (=7955–7755 calBP). However, allochthonous humus impregnating basal mineral horizons in narrow floodplains goes back to 7490 ± 50 BP (=8385–8195 calBP). Thus, early peat materials in upslope positions that provided these organic substances must be somewhat older than 7500 BP.
- Peat has formed during the major part of the Holocene. However, peat deposition was not constant. In some places, peat accumulation was interrupted by depositional hiatuses, some of which might have been induced by climate forcing and others by truncation through sliding or cutting by creek migration.

- Organic material started accumulating at different times according to the sites visited. Peat initiation and development were strongly dependent on site conditions (local forcing), together with favorable climate conditions (regional forcing). Ongoing (pseudo-)karstic activity provided regularly new depressions for water to concentrate and organic material to accumulate.
- The age of the (pseudo-)karstic depressions influences the age of the peat deposits. It can be assumed that not all depressions started forming at the same time. Probably, the increase of moisture at the onset of the Holocene favored rock weathering, but depression formation is distributed over time, creating new sites for peat accumulation. This resulted in diachronic rather than cyclic peat inception, with starting dates of organic material accumulation spread over time.
- Sustained by abundant rainfall, water percolation through the peat mantle and underlying rock fissures causes the labile organic compounds to be extracted from the bottom of the peat reservoirs, especially from the sapric layers, and exported through joints, crevices, and fractures to karstic underground pipes, ultimately feeding the blackwater rivers in the lowlands.
- The leaching of younger organic substances from upper to lower peat layers may cause natural contamination of the basal layers.
- In a few geographically well-distributed places, a new phase of organic matter accumulation initiated around the seventeenth to nineteenth centuries. This might reflect a major environmental change that can be correlated with the end of the Little Ice Age period.
- In historical times, moderately thick peat deposits have been formed in relatively short time lapses: up to 60 cm in two centuries (P9 at the Cuao 3 site); up to 45 cm in contemporaneous times (P18 at the Cuao 13 site). Peat continues forming nowadays.
- There are often important time gaps between consecutive dated peat layers, which are as large as 2000–4000 ^{14}C years. These gaps reflect depositional hiatuses that can result either from the interruption of peat accumulation because of climate change or from the truncation/removal of the peat cover through sliding. As a consequence, peat profiles are often polygenetic, resulting from more than one single accumulation phase.
- There are no significant differences in terms of peat age and depth between the three study areas and between individual tepuis. The earliest organic material accumulation started at similar ^{14}C dates (ca 7500–7000 BP) that calibrate into ca 8400–7800 calBP in both the Marahuaka-Huachamakari and Cuao-Sipapo areas.

7.5.2 *Relationships*

- In general, peat age increases with depth. However, variable ages were determined at similar depth ranges, and similar dates occurred at variable depth ranges. This suggests that the visited sites have undergone different evolution and have thus different peat formation histories.

- Deposition rates expressed in calendar years are strongly dispersed, ranging from 0.065 mm yr^{-1} to 0.333 mm yr^{-1}. The average peat accumulation rate for the whole study area and time period is 0.164 mm yr^{-1}. When only the highest rates are taken into account, then the average depositional rate is 0.229 mm yr^{-1}. The highest rates refer to peats older than 6000 calBP, while the lower rates are characteristic of peats younger than 4000 calBP.
- Peat currently forms and has formed in the past over a large elevation range from 650 m to 2,600 m a.s.l., but the relationship between peat age and elevation is in general weak except for the basal layers. The age of the organic material in the basal layers, which reflects the time of peat inception, shows that peat started accumulating first at higher elevations when climate changed from dry to moist at the beginning of the Holocene.
- The vegetation cover was mainly composed of C3 plants, in a mixture of predominantly woody ($\delta^{13}C$ from −26‰ to −28‰) and subordinate herbaceous species ($\delta^{13}C$ from −24‰ to −26‰), with no significant variations all along the time period from 7500 BP to the Present (1950 AD).
- The scatter diagram relating ^{14}C age and depth of the peats shows three segments: a weak linear relationship between the Present and ca 3600 BP, a data gap between ca 3600 and 5200 BP, and data dispersion between ca 5200 and 7400 BP.
- In the hypothesis of random distribution of the sampling points, the above data gap suggests that peat areas had considerably shrunk in the period of ca 5200–3600 BP because of decreasing moisture or more frequent dry spells. Thus, two main peat formation phases would have taken place during the Holocene in the study area: 7500–5200 BP and 3600 BP-Present.
- When the ^{14}C data from western Guayana Highlands are merged with similar data from eastern Guayana Highlands, the graphic does not corroborate the hypothesis of peat interruption in the period of ca 5200–3600 BP, but shows a clear difference in the peat age–depth pattern before and after ca 4000 BP. Before this time threshold, peat formation seems to be more related to site conditions (local forcing) and, afterwards, more related to climate conditions (regional forcing).
- The period around 4000 BP seems to be a time threshold in peat formation of broader significance in the regional context (Gran Sabana) as well as in the continental context (South America and Africa).
- In the hypothesis of current climate change, a warmer atmosphere and a more seasonal rainfall regime would cause faster decomposition of the organic matter and lead to the shrinkage of peatland areas. This might affect the hydrological function of the peat cover as a water reservoir. Water will be released faster to the lowlands, causing flood in the whitesand plains and thus limiting the water availability during the dry season. Monitoring is needed to assess the potential retreat of the peat mantle.

References

Beven K, Lawson A, McDonald A (1978) A landslip/debris flow in Bilsdale, North York moors, September 1976. Earth Surf Process 3:407–419

Bond G, Showers W, Cheseby M, Lotti R, Almasi P, deMenocal P, Priore P, Cullen H, Hajdas I, Bonani G (1997) A pervasive millenial-scale cycle in North Atlantic Holocene and Glacial climates. Science 278:1257–1266

Chimner RA, Karberg JM (2008) Long-term carbon accumulation in two tropical mountain peatlands, Andes Mountains, Ecuador. Mires and Peat 3: Art. 4. http://www.mires-and-peat.net/

D'Amore DV, Lynn WC (2002) Classification of forested Histosols in Southeast Alaska. Soil Sci Soc Am J 66:554–562

Dehling H, Van der Plicht J (1993) Statistical problems in calibrating radiocarbon dates. Radiocarbon 35(1):239–244

Dykes AP, Kirk KJ (2006) Slope instability and mass movements in peat deposits. In: Martini IP, Martinez-Cortizas A, Chesworth W (eds) Peatlands: evolution and records of environmental and climate changes, vol 9, Developments in earth surface processes. Elsevier, Amsterdam, pp 377–406

Dykes AP, Warburton J (2007) Mass movements in peat: a formal classification scheme. Geomorphology 86(1–2):73–93

Dykes AP, Gunn J, Convery (née Kirk) KJ (2008) Lanslides in blanket peat at Cuilcagh Mountain, northwest Ireland. Geomorphology 102(3–4):325–340

Eijkelkamp (1989) User manual for "Wardenaar-peatsampler". Eijkelkamp Agrisearch Equipment, Giesbeek, The Netherlands

Fölster H (1992) Holocene autochthonous forest degradation in southeast Venezuela. In: Goldhammer JG (ed) Tropical forests in transition. Birkhäusser, Berlin, pp 25–44

Galán C (1992) El clima del macizo del Chimantá. In: Huber O (ed) El macizo del Chimantá, Escudo de Guayana, Venezuela. Un ensayo ecológico tepuyano. Oscar Todtmann Editores, Caracas, pp 37–52

Gorzula S, Huber O (1992) Consideraciones finales. In: Huber O (ed) El macizo del Chimantá, Escudo de Guayana, Venezuela. Un ensayo ecológico tepuyano. Oscar Todtmann Editores, Caracas, pp 325–330

Gröger A, Huber O (2007) Rock outcrop habitats in the Venezuelan Guayana lowlands: their main vegetation types and floristic components. Rev Bras Bot 30(4):599–609

Holden J (2006) Peatland hydrology. In: Martini IP, Martinez-Cortizas A, Chesworth W (eds) Peatlands: evolution and records of environmental and climate changes, vol 9, Developments in earth surface processes. Elsevier, Amsterdam, pp 319–346

Huber O (1988) Guayana Highlands versus Guayana Lowlands, a reappraisal. Taxon 37:595–614

Huber O (1992) El macizo del Chimantá: el área de estudio y su exploración. In: Huber O (ed) El macizo del Chimantá, Escudo de Guayana, Venezuela. Un ensayo ecológico tepuyano. Oscar Todtmann Editores, Caracas, pp 23–36

Huber O, García P (2011) The Venezuelan Guayana region and the study areas: geo-ecological characteristics. In: Zinck JA, Huber O (eds) Peatlands of the Western Guayana highlands, Venezuela, Chap. 3. Springer, Heidelberg

Immirzi CP, Maltby E, Clymo RS (1992) The global status of peatlands and their role in carbon cycling. Report 11, Friends of the Earth. London

Marchant R, Hooghiemstra H (2004) Rapid environmental change in African and South American tropics around 4000 years before present: a review. Earth Sci Rev 66:217–260

Meadows ME (1988) Late Quaternary peat accumulation in southern Africa. CATENA 15:459–472

Medina E, Cuevas E, Huber O (2011) Origin of organic matter leading to peat formation in the southeastern Guayana uplands and highlands. In: Zinck JA, Huber O (eds) Peatlands of the Western Guayana Highlands, Venezuela, Chap. 8. Springer, Heidelberg

Mook WG, Streurman HJ (1983) Physical and chemical aspects of radiocarbon dating. In: Mook WG, Waterbolk HT (eds) Proceedings of the Groningen Conference on ^{14}C and Archaeology, PACT Publication 8, pp 31–55

Mook WG, Waterbolk HT (1985) Radiocarbon dating, vol 3, Handbooks for archaeologists. European Science Foundation, Strasbourg

Nogué S, Rull V, Montoya E, Huber O, Vegas-Vilarrúbia T (2009) Paleoecology of the Guayana Highlands (northern South America): Holocene pollen record from the Eruoda-tepui, in the Chimantá massif. Palaeogeogr Palaeoclimatol Palaeoecol 281:165–173

Reimer PJ, Baillie MGL, Bard E, Bayliss A, Beck JW, Bertrand CJH, Blackwell PG, Buck CE, Burr GS, Cutler KB, Damon PE, Edwards RL, Fairbanks RG, Friedrich M, Guilderson TP, Hogg AG, Hughen KA, Kromer B, McCormac G, Manning S, Bronk Ramsey C, Reimer RW, Remmele S, Southon JR, Stuiver M, Talamo S, Taylor FW, Van der Plicht J, Weyhenmeyer CE (2004) IntCal04 terrestrial radiocarbon age calibration, 0–26 cal kyr BP. Radiocarbon 46(3):1029–1058

Rull V (1991) Contribución a la paleoecología de Pantepui y la Gran Sabana (Guayana Venezolana): clima, biogeografía y ecología. Scientia Guaianae 2, CVG-EDELCA, Caracas

Rull V (2004) An evaluation of the Lost World and the vertical displacement hypotheses in the Chimantá Massif, Venezuelan Guayana. Glob Ecol Biogeogr 13:141–148

Rull V (2005) Vegetation and environmental constancy in the Neotropical Guayana Highlands during the last 6000 years? Rev Palaeobot Palynol 135:205–222

Rull V, Schubert C (1989) The Little Ice Age in the tropical Venezuelan Andes. Acta Cient Venez 40(1):71–73

Schargel R, García P, Jiménez D (2011) Laboratory methods for characterization of peat materials. In: Zinck JA, Huber O (eds) Peatlands of the Western Guayana Highlands, Venezuela, Chap. 5. Springer, Heidelberg

Schubert C, Fritz P (1985) Radiocarbon ages of peat, Guayana Highlands (Venezuela). Some paleoclimatic implications. Naturwissenschaften 72:427–429

Schubert C, Fritz P, Aravena R (1994) Late Quaternary paleoenvironmental studies in the Gran Sabana (Venezuelan Guayana Shield). Quatern Int 21:81–90

Steyermark JA, Dunsterville GCK (1980) The lowland floral element of the summit of Cerro Guaiquinima and other cerros of the Guayana Highlands of Venezuela. J Biogeogr 7:285–303

Van der Plicht J (1993) The Groningen radiocarbon calibration program. Radiocarbon 35(1): 231–237

Van der Plicht J (2004) Radiocarbon calibration – past, present and future. Nucl Instrum Methods Phys Res B 223–224:353–358

Van der Plicht J (2007) WinCal25. The Groningen radiocarbon calibration program. Center for Isotope Research (CIO), University of Groningen, The Netherlands

Van der Plicht J, Hogg A (2006) A note on reporting radiocarbon. Quat Geochronol 1:237–240

Van der Plicht J, Mook WG (1987) Automatic radiocarbon calibration: illustrative examples. Palaeohistoria 29:173–182

Van der Plicht J, Mook WG (1989) Calibration of radiocarbon ages by computer. Radiocarbon 31 (3):805–816

Vuille M, Francou B, Wagnon P, Juen I, Kaser G, Mark B, Bradley R (2008) Climate change and tropical Andean glaciers: past, present and future. Earth Sci Rev 89:79–96

Chapter 8
Origin of Organic Matter Leading to Peat Formation in the Southeastern Guayana Uplands and Highlands

E. Medina, E. Cuevas, and O. Huber

8.1 Introduction

High rainfall in tropical mountains above ca 1,000 m a.s.l. frequently promotes the formation of wetlands with permanent water bodies of variable depth and seasonally flooded swamps. The vegetation growing within and around these sites contributes to the accumulation of organic matter that decomposes slowly because of the hypoxic water environment and the relatively low average annual temperature (ca 15°C). The organic matter contributing to peat formation in tropical mountains is mainly derived from higher plants, in contrast to the oligotrophic peats of the temperate regions in which *Sphagnum* provides most of the organic matter (Clymo 1983).

In the Guayana Highlands exposed to abundant rainfall, peat bogs occur in areas of impeded drainage atop the Roraima sandstone table-mountains called tepuis by local people. A particular herbaceous (savannoid peat bog) or woody vegetation (shrubby peat bog) grows on these high-elevation peatlands (above 1,500 m a.s.l.). Peat bogs also occur at lower elevation (800–1,500 m a.s.l.) as for instance on the Gran Sabana plateau (Schubert and Huber 1990), but here vegetation is mainly grasses with a few woody components (Rinaldi et al. 1990; Rull 1991). The content of ^{13}C in plants varies with the type of photosynthetic pathway. During the processes of CO_2 uptake, C3 plants discriminate against $^{13}CO_2$ more than C4 plants do (Farquhar et al. 1989). Therefore, the assessment of the natural abundance of carbon-13 (expressed as $\delta^{13}C$ values) may help determine the photosynthetic pathway of the plants contributing organic matter to peat bogs (Hillaire-Marcel et al. 1989; Aucour et al. 1993). Herbaceous vegetation above 1,500 m elevation in our study area is dominated by C3 species of the Bromeliaceae, Rapateaceae, Xyridaceae, and Cyperaceae families (Huber 1992; Medina 1990, 1996). In the Gran Sabana plateau, Poaceae dominate both the wet and dry herbaceous cover with C4 species of the genera *Trachypogon, Axonopus, Thrasya*, and others (Hattersley 1987; Berry et al. 1995; Knapp and Medina 1999; Medina et al. 1999). The only

J.A. Zinck and O. Huber (eds.), *Peatlands of the Western Guayana Highlands, Venezuela*, Ecological Studies 217, DOI 10.1007/978-3-642-20138-7_8,

exception is *Echinolaena inflexa*, a codominant C3 grass species that occurs throughout the central Cerrado region of Brazil (Klink and Joly 1989; Medina et al. 1999). As these herbaceous species contribute organic matter accumulated in peat bogs, a differentiation in the isotopic signatures of the upland peats (i.e., below 1,500 m elevation) may be expected. None of the woody species in the Guayana Highlands has been reported as a C4 or a CAM plant.

More than 95% of plant organic matter is constituted by polysaccharides, proteins, lipids, and lignins. During decomposition, this organic matter undergoes significant chemical changes due to differences in biochemical breakdown rates of the various substances. Soluble sugars and aminoacids are quickly consumed by the soil microflora, whereas polysaccharides (cellulose, hemicellulose, starch) and proteins have to be hydrolyzed before they are consumed by microorganisms. Lignins are very resistant to decomposition. Therefore, as the process of decomposition proceeds, the proportion of lignins compared to that of polysaccharides and proteins markedly increases (Swift et al. 1979). After a few years, remaining organic matter from plants is mostly lignin and its derivatives. In peat layers of different ages, within a particular bog, $\delta^{13}C$ values could provide information on the relative composition of the organic matter, particularly the ratio of cellulose to lignin (Benner et al. 1987). Lignin-rich organic matter is relatively impoverished in ^{13}C compared to the cellulose content in the same plant (Benner et al. 1987). Additionally, $\delta^{13}C$ values of peat materials differing in age by hundreds or thousands of years may help detect increase in atmospheric CO_2 concentration over the last 10,000 years (Pearman et al. 1986; Stauffer et al. 2002).

Schubert and Fritz (1985) and Schubert et al. (1994) used carbon isotope data to date peat deposits in several localities of the southeastern Guayana Highlands. Peat chronology together with pollen analysis allowed outlining the paleoclimatic and paleoecological history of the region (Schubert et al. 1986; Rull 1991). Reworking these earlier data, we evaluate in the present chapter the relationship between peat $\delta^{13}C$ and peat age as estimated by ^{14}C dating. The aim is to relate the variability of $\delta^{13}C$ values to vegetation composition, changes in atmospheric CO_2 concentration, and chemical changes occurring during peat aging.

8.2 Site Description

All peat deposits sampled are located in the Guayana highland region of southeastern Venezuela, including (1) table-mountain sites (tepui mesetas) above 1,200 m a.s.l. described in Schubert and Fritz (1985) and Rull (1991, 2004), and (2) plateau sites (Gran Sabana) at 800–950 m a.s.l. described in Rinaldi et al. (1990) and Rull (1991). A brief description of the sites is provided hereafter and their location is shown in Fig. 8.1.

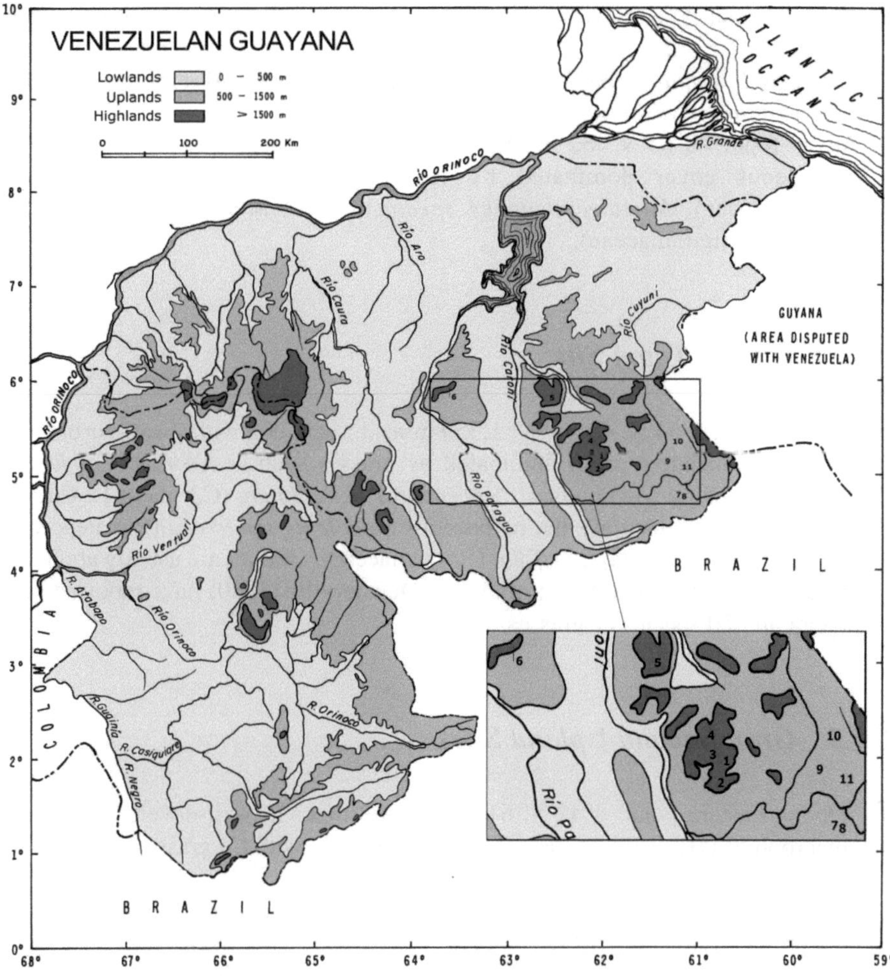

Fig. 8.1 Venezuelan Guayana region with the location of the sampling sites

8.2.1 Highland Tepui Sites

1. Churí-tepui: 5°15′N, 62°01′W; 2,250 m a.s.l; shrubby peat bog (Cuevas 1987) characterized by *Chimantaea humilis* (Asteraceae) and dominated by an herbaceous layer of *Stegolepis ligulata* (Rapateaceae) and *Brocchinia hechtioides* (Bromeliaceae).
2. Acopán-tepui: 5°11′N, 62°05′W; 1,950 m a.s.l.; peat bog with herbaceous cover dominated by *S. ligulata* and *Panicum* cf. *eligulatum* (Poaceae).
3. Toronó-tepui: 5°16′N, 62°09′W; 2,100 m a.s.l.; peat bog with herbaceous (savannoid) cover (Cuevas 1987) dominated by *S. ligulata*, *P. eligulatum*, and *Lagenocarpus rigidus* (Cyperaceae).

4. Apacará-tepui: 5°19′N, 62°12′W; 2,150 m a.s.l.; peat bog with dense, shrubby clusters of *Chimantaea mirabilis* (Asteraceae), and an herbaceous layer dominated by *S. ligulata, Myriocladus steyermarkii* (Poaceae-Bambusoideae), and *Syngonanthus obtusifolius* (Eriocaulaceae).
5. Auyán-tepui: 5°55′N, 62°38′W; 1,860 m a.s.l.; peat bog with shrubby-herbaceous cover dominated by the woody species *Bonnetia sessilis* (Bonnetiaceae), *Maguireothamnus speciosus* (Rubiaceae), and *Digomphia laurifolia* (Bignoniaceae).

8.2.2 Upland Tepui Site

6. Guaiquinima: 5°54′N, 63°42′W; 1,350 m a.s.l.; peat bog with dwarf shrubland-herbaceous meadow cover dominated by grasses such as *Axonopus* cf *flabellifolius* (C4 plant; Medina et al. 1999) and *P. eligulatum* (C3 plant; E. Cuevas personal communication), the herbaceous *Stegolepis squarrosa* (Rapateaceae), and the woody *Bonnetia lancifolia* (Bonnetiaceae). Grasses are usually abundant in this type of peatland. Steyermark and Dunsterville (1980) have reported nine genera and 20 species of grasses.

8.2.3 Gran Sabana Upland Sites

7. Divina Pastora pond: 4°42′N, 61°04′W; 800 m a.s.l.; well-developed aquatic macrophytes (Rull 1991) surrounded by grassland with C4 grasses and clusters of *Mauritia minor* (Arecaceae).
8. Santa Teresa pond: 4°43′N, 61°05′W; 880 m a.s.l.; vegetation similar to that of the previous site (Rull 1991).
9. Urué: 5°02′N, 61°10′W; 940 m a.s.l.; headwater area with a small *M. flexuosa* grove within a general landscape of savanna grassland (Rull 1991).
10. Quebrada Arapán: 5°10′N, 61°06′W; 900 m a.s.l; small peat bog with vegetation transitional between typical savanna grassland and tepui meadow (Rull 1991), located within a large savanna grassland.
11. Santa Cruz de Mapaurí: 4°56′N, 61°06′W; 940 m a.s.l.; savanna grassland dominated by C4 grasses (Rinaldi et al. 1990; Medina et al. 1999).

8.3 Results: Relationship Between $\delta^{13}C$ Values and Peat Age

The plotting of the isotope data from Schubert and Fritz (1985) and Schubert et al. (1994) in Fig. 8.2 shows four groups: two of them cluster upper mountain sites and two others pool the Gran Sabana upland sites. The first group including

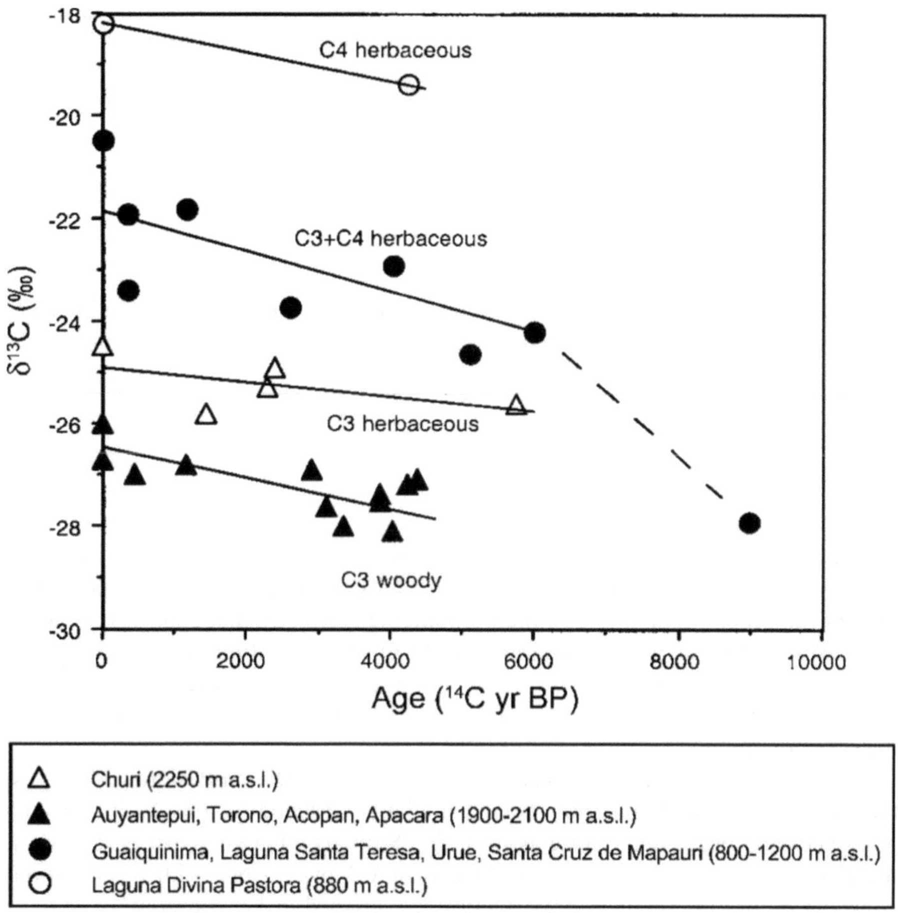

Fig. 8.2 Distribution of $\delta^{13}C$ values of peat samples according to ^{14}C age. Original data from Schubert and Fritz (1985) and Schubert et al. (1994)

the Auyán-tepui, Toronó, Acopán, and Apacará sites (Auyán-tepui group) has the most negative current $\delta^{13}C$ values (−26.6‰ on average) (Table 8.1 and Fig. 8.2). The second group representing the Churí samples (Churí group) has current $\delta^{13}C$ values that are on average 1.7‰ more positive than the values of the first group. The third group (Guaiquinima group), including the Guaiquinima and Gran Sabana sites except the Divina Pastora pond, shows current $\delta^{13}C$ values that are about 3‰ more positive than those of the Churí group. The fourth group corresponding to the Divina Pastora pond (Divina Pastora group) has the least negative current $\delta^{13}C$ values, being 3.6‰ more positive than those of the third group. In all groups there is a tendency for $\delta^{13}C$ values to become more negative with peat age (i.e., increasing sampling depth). The rate of reduction of $\delta^{13}C$ values ranges from 0.1 to 0.4‰ per 1,000 years (Table 8.1). One outlier corresponding to the oldest peat sample collected in the Gran Sabana area (Rinaldi et al. 1990) was disregarded because its

Table 8.1 Current average $\delta^{13}C$ values and average decrease per unit time (i.e., increasing sampling depth) in surface peat layers from the southeastern Venezuelan Guayana uplands and highlands (calculated from data in Fig. 8.2)

Peat sample group	Current $\delta^{13}C$ ‰	Decrease ‰/10^3 years
Auyán-tepui	−26.6	0.24
Churí	−24.9	0.14
Guaiquinima	−21.8	0.45
Divina Pastora	−18.2	0.28

$\delta^{13}C$ value is more similar to those measured at the Auyán-tepui site. To interpret the significance of this sharp reduction in $\delta^{13}C$, more data covering the early Holocene would be needed.

8.4 Discussion

The analysis of the carbon isotope values reported by Schubert and coworkers leads to the hypothesis that the natural abundance of ^{13}C in peat deposits reflects a set of factors and processes including (a) environmental variables that control vegetation composition, such as water availability and CO_2 concentration in the atmosphere, (b) photosynthetic properties of the vegetation producing the organic matter, and (c) changes in biochemical composition during the process of peat formation under hypoxic conditions.

The $\delta^{13}C$ values of the youngest peat layers from the Divina Pastora site suggest that they derive predominantly from C4 plants, essentially grasses in savanna grassland that discriminate little against ^{13}C. Similarly, the youngest peat layer at the Santa Teresa site shows the influence of C4 plants, but with a substantial contribution of organic matter relatively impoverished in ^{13}C, possibly from trees and palms. The similarity between the Guaiquinima and Gran Sabana sites may be attributed to flora convergence resulting from the lower elevation of the Guaiquinima uplands (Steyermark and Dunsterville 1980). The difference between the Churí and Auyán-tepui groups is probably related to the proportion of herbaceous and woody C3 plants. Herbaceous plants may have more positive $\delta^{13}C$ values because they are more sensitive to short-term water deficit. Lower stomatal conductance in C3 plants leads to reduction in ^{13}C discrimination (Farquhar et al. 1989). Comparatively, the trees with more extensive root systems are less sensitive to short-term droughts. In addition, the ratio of lignin to cellulose is higher in woody roots, contributing also to more negative $\delta^{13}C$ values (Benner et al. 1987; Hobbie and Werner 2004).

During the time span measured by Schubert and coworkers (around 9,000 ^{14}C years) (Rinaldi et al. 1990) there has been a steady increase of CO_2 concentration in the atmosphere. Measurements in ice cores from Antarctica indicate that atmospheric CO_2 increased slowly from about 260 to 280 ppmv in the period comprised between 9,000 and 1,000 years BP and remained at that level until the 1700s (Stauffer et al. 2002). Increases of the CO_2 concentration in the atmosphere were accompanied by a reduction of the $\delta^{13}C$ values of the air (Friedli et al. 1986; Siegenthaler and Oeschger 1987) because additional CO_2 originated from organic matter decomposition. Therefore, $\delta^{13}C$ values can be expected to increase with peat age. They should also increase when assuming that the lighter carbon is preferentially mineralized during the decomposition process. In contrast to this, we actually observed a slow decrease of the $\delta^{13}C$ values with peat age. It seems as if the effect of past changes in atmospheric CO_2 concentration and preferential mineralization of organic matter had been obscured by other processes that influence carbon isotopic composition.

The predominance of C3 plants in the history of a peat deposit would generate more negative $\delta^{13}C$ values, while the opposite should occur with increasing abundance of C4 plants (Bender 1971; Schwartz et al. 1986; Ambrose and Sikes 1991; Aucour et al. 1993). If the species composition remains unchanged throughout peat formation, then differential decomposition rates of the plant material may result in changes in $\delta^{13}C$ values. Cellulose and proteins decompose faster than compounds such as lignin and waxes (Akagi et al. 2004). As the latter are impoverished in ^{13}C (Benner et al. 1987), the $\delta^{13}C$ of the peat would decrease with age. Vegetation appears to have been stable over the last 6,000 years, maintaining the relative difference in proportion between woody and herbaceous plants, and between C3 and C4 plants (Rull 2004; Nogué et al. 2009).

8.5 Conclusion

We suggest that the reduction in $\delta^{13}C$ values with peat age is due to the differential decomposition of organic matter constituents. Complex carbohydrates (cellulose and hemicellulose) dissapear at a faster rate than lignins relatively impoverished in ^{13}C. The outlier peat sample from Santa Cruz de Mapaurí may reflect recent changes in the relative proportion of C3 and C4 plants.

References

Akagi T, Minomo K, Kasuya N, Nakamura T (2004) Variation in carbon isotopes of bog peat in the Ozegahara peatland, Japan. Geochem J 38:299–306

Ambrose SH, Sikes NF (1991) Soil carbon isotope evidence for Holocene habitat change in the Kenya rift valley. Science 253:1402–1404

Aucour AM, Hillaire-Marcel C, Bonnefille R (1993) A 30,000 year record of ^{13}C and ^{18}O changes in organic matter from an equatorial peatbog. In: Swart PK, Lohman KC, McKenzie J, Savin S (eds) Climate change in continental isotopic records. Geophys Monogr 78:343–351

Bender MM (1971) Variations in the $^{13}C/^{12}C$ ratios of plants in relation to the pathway of photosynthetic carbon dioxide fixation. Phytochemistry 10:1239–1244

Benner R, Fogel ML, Sprague EK, Hodson RE (1987) Depletion of ^{13}C in lignin and its implication for stable carbon isotope studies. Nature 329:708–710

Berry P, Huber O, Holst B (1995) Floristic analysis and phytogeography. In: Steyermark J, Berry P, Holst B (eds) Flora of the Venezuelan Guayana, vol 1. The Missouri Botanical Garden, Timber Press, Portland, OR, pp 161–191

Clymo RS (1983) Peat. In: Gore AJP (ed) Mires: swamp, bog, fen and moor, vol 4A, Ecosystems of the World. Elsevier, Amsterdam, pp 159–224

Cuevas E (1987) Perfil nutricional de la vegetación de turberas en el Macizo de Chimantá, Edo. Bolívar, Venezuela. Resultados preliminares. Acta Cient Venez 38:366–375

Farquhar GD, Hubick KT, Condon AG, Richards RA (1989) Carbon isotope fractionation and plant water use efficiency. In: Rundel PW, Ehleringer JR, Nagy KA (eds) Stable isotopes in ecological research, vol 68, Ecological Studies. Springer, New York, NY, pp 21–40

Friedli H, Lötscher H, Oeschger H, Siegenthaler U, Stauffer B (1986) Ice core record of $^{13}C/^{12}C$ ratio of atmospheric CO_2 in the past two centuries. Nature 324:237–238

Hattersley PW (1987) Variations in photosynthetic pathway. In: Soderstrom TR, Hilu KW, Campbell CS, Barthworth ME (eds) Grass systematic and evolution. Smithsonian Institution Press, Washington, DC, pp 49–64

Hillaire-Marcel C, Aucour A-M, Bonnefille R, Riollet G, Vincens A, Williamson D (1989) $^{13}C/^{12}C$ palynological evidence of differential residence times of organic carbon prior to its sedimentation in East African rift lakes and peat bogs. Quatern Sci Rev 8:207–212

Hobbie EA, Werner RA (2004) Intramolecular, compound-specific, and bulk carbon isotope patterns in C3 and C4 plants: a review and synthesis. New Phytol 161:371–385

Huber O (1992) La flora: helechos y plantas fanerógamas. In: Huber O (ed) El macizo del Chimantá, Escudo de Guayana, Venezuela. Un ensayo ecológico tepuyano. Oscar Todtmann Editores, Caracas, pp 139–160

Klink CA, Joly CA (1989) Identification and distribution of C3 and C4 grasses in open and shaded habitats in São Paulo State, Brazil. Biotropica 21:30–34

Knapp AK, Medina E (1999) Success of C4 photosynthesis in the field: lessons from communities dominated by C4 plants, chapter 8. In: Sage RF, Monson RK (eds) C4 plant biology. Academic Press, San Diego, CA, pp 251–284

Medina E (1990) Eco-fisiología y evolución de las Bromeliaceae. Boletín de la Academia Nacional de Ciencias de Córdoba, Argentina 59:71–114

Medina E (1996) CAM and C4 plants in the humid tropics, chapter 2. In: Mulkey SS, Chazdon RL, Smith AP (eds) Tropical forest plant eco-physiology. Chapman & Hall, New York, NY, pp 56–88

Medina E, Martinelli LA, Barbosa E, Victoria RL (1999) Natural abundance of ^{13}C in tropical grasses from the INPA, Instituto Nacional de Pesquisas da Amazônia, herbarium. Revista Brasileira de Botanica, São Paulo 22:43–51

Nogué S, Rull V, Montoya E, Huber O, Vegas-Vilarrúbia T (2009) Paleoecology of the Guayana Highlands (northern South America): Holocene pollen record from the Eruoda-tepui, in the Chimantá massif. Palaeogeogr Palaeoclimatol Palaeoecol 281:165–173

Pearman GI, Etheridge DM, de Silva F, Fraser PJ (1986) Evidence of changing concentrations of atmospheric CO_2, N_2O and CH_2 from air bubbles in Antarctic ice. Nature 320:248–250

Rinaldi M, Rull V, Schubert C (1990) Análisis paleoecológico de una turbera en la Gran Sabana (Santa Cruz de Mapaurí), Venezuela: resultados preliminares. Acta Cient Venez 41:66–68

Rull V (1991) Contribución a la paleoecología de Pantepui y la Gran Sabana (Guayana Venezolana): clima, biogeografía y ecología. Scientia Guaianae 2. Caracas

Rull V (2004) An evaluation of the Lost World and vertical displacement hypotheses in the Chimantá Massif, Venezuelan Guayana. Glob Ecol Biogeogr 13:141–148

Schubert C, Fritz P (1985) Radiocarbon ages of peat, Guayana Highlands (Venezuela). Naturwissenschaften 72:427–429

Schubert C, Huber O (1990) The Gran Sabana. Panorama of a region. Lagoven Booklets, Caracas

Schubert C, Briceño HO, Fritz P (1986) Paleoenvironmental aspects of the Caroní-Paragua river basin (southeastern Venezuela). Interciencia 11:278–289
Schubert C, Fritz P, Aravena R (1994) Late Quaternary paleoenvironmental studies in the Gran Sabana (Venezuelan Guayana Schield). Quatern Int 21:81–90
Schwartz D, Mariotti R, Guillet B (1986) $^{13}C/^{12}C$ ratios of soil organic matter as indicators of vegetation changes in the Congo. Geoderma 39:97–103
Siegenthaler U, Oeschger H (1987) Biospheric carbon dioxide emissions during the past 200 years reconstructed by deconvolution of ice core data. Tellus 39B:140–154
Stauffer B, Flückiger J, Monnin M, Schwander J, Barnola J-M, Chappellaz J (2002) Atmospheric CO_2, CH_4 and N_2O records over the past 60,000 years based on the comparison of different polar ice cores. Ann Glaciol 35:202–208
Steyermark JA, Dunsterville GCK (1980) The lowland floral elements on the summit of Cerro Guaiquinima and other cerros of the Guayana highlands of Venezuela. J Biogeogr 7:285–303
Swift MJ, Heal OW, Anderson JM (1979) Decomposition in terrestrial ecosystems. University of California Press, Berkeley, CA

Chapter 9
Synthesis: The Peatscape of the Guayana Highlands

J.A. Zinck

9.1 Introduction

The present study is concerned with the peatlands and peats that have developed on sandstone–quartzite mesetas (tepuis) and igneous–metamorphic massifs in the western Guayana Highlands of Venezuela. Peatland refers to the physical and biotic landscape where peats have formed, while peat refers to the organic material deposits that constitute the main component of the peatland landscape, in short the *peatscape*. Visited sites are distributed over three study areas, all located in the western portion of the Amazonas state: the Marahuaka-Huachamakari massif, the Cuao-Sipapo massif, and the western border of the Maigualida massif. The volume starts with an overview on tropical and subtropical peats, followed by a large introduction to the geo-ecological features of the Guayana region as a whole, with emphasis on the diversity of the vegetation cover from lowlands to uplands to highlands. The core subject of the book focuses on the properties and dating of the peat deposits and the interpretation of the chronological record in terms of past environmental changes. A synthesis of the main findings of the research carried out and the specific features of the study area is hereafter provided. This synthesis is built upon the relevant conclusions drawn from each chapter of the volume.

9.2 Why Do Peatlands and Peats Matter?

Compared with the areal importance of peats in temperate and especially in boreal regions, peatlands are much less extensive in tropical and subtropical environments. Only 0.36–0.5 M km^2 peatland, representing 8.5–11.7% of the global 4.26 M km^2, occurs in the warm and moist regions of the world, especially in Southeast Asia in the areas surrounding the South China Sea and areas in Papua-New Guinea that together concentrate 68% of the known tropical peats.

Worldwide, peats and peatlands are increasingly used as agricultural resource, source of energy, water regulation body, biodiversity reservoir, carbon pool, and

J.A. Zinck and O. Huber (eds.), *Peatlands of the Western Guayana Highlands, Venezuela*, Ecological Studies 217, DOI 10.1007/978-3-642-20138-7_9,

provider of other environmental services. However, knowledge on tropical peats is still lagging behind when compared to the development of peat and peatland studies in temperate and boreal areas, although considerable progress has been made in mapping tropical peats, identifying their specific characteristics, assessing their use potentials, and calling the attention on their vulnerability.

Indigenous people developed early experience in managing peatlands in a variety of tropical and subtropical areas (e.g., chinampas in Xochimilco, central Mexico; raised fields in poorly drained tropical lowlands such as in the Orinoco llanos). However, large-scale reclamation and use of peats were constrained for long time by factors such as poor drainage, low fertility, risk of disease, and limited accessibility. Nowadays, tropical peatlands are becoming new agricultural frontier areas because of the need to meet increasing food demands. This leads to severe farming challenges as colonizers are often not familiar with the management of peatsoils. Surface fires, underground peat combustion, subsidence, and loss of biodiversity are some of the main factors threatening the sustainability of peatland use.

Tropical lowland peats are used in agricultural production and as a source of energy. The vegetation cover of peatlands provides raw materials used by local people, such as raffia palm and papyrus from African swamps and timber from coastal forest swamps in Southeast Asia. From an environmental point of view, peats and peatlands perform regulating functions that play a role in the carbon cycle as carbon stocks, in hydrology as water reservoirs and catchment areas for flood mitigation, in the adsorption of heavy metals and organic pollutants, and in the buffering between salt- and freshwater systems in coastal marshes. Comparatively, peats in tropical highlands are poorly documented and so far little used by local populations. In the Guayana Highlands, peatlands are still mostly pristine landscapes. However, fire in drier conditions and uncontrolled (eco-)tourism may represent potential threats to this unique ecosystem.

Since the inception of peat formation after the last glacial period, about 10–12 kyr ago, considerable amounts of organic material have slowly built up in peatlands. With the current relevance of climate change on the global agenda, the role of peats as carbon stores and the impact of the potential release of greenhouse gases from peatlands are given increasing importance. Climate warming may cause changes in the balance and annual distribution of rainfall and evapotranspiration that, in turn, will affect peatland productivity and peat decomposition. However, peatlands are not yet formally included in global climate models and predictions of future climate change.

9.3 Setting and Formation of the Peatlands: The Peatscape

Peatland is a type of landscape where peat has formed under the interaction of a complex of factors including geologic setting, geomorphic processes, climatic conditions, water regime, and biota. Derived from the contraction of peat and landscape, the *peatscape* is the assemblage of the peat bodies that have formed on a terrain surface in a particular landscape.

Tropical lowland peats occur mainly in coastal areas, including tidal flats and river deltas, and to a lesser extent in depressional inland river basins. Tropical highland peats are much less extensive and correspond frequently to organic fillings of earlier glacial lakes. Within this context, the peatlands and peats found in the Guayana Highlands are very special as they are intimately related to karstic and pseudokarstic landscapes. So far there are no known analogs of this kind elsewhere in the tropics.

Peatlands cover about 30% of the tepui and dome summits in the Guayana Highlands, with the remaining surface area corresponding to forest and shrub cover (40%), bare rock terrain (25%), and water bodies (5%). The peatscape shows essentially a discontinuous, mosaic pattern. Peat occurs in discrete landscape units surrounded and fragmented by rock outcrops. Organic materials have deposited mainly in depressions from small to large in two types of landscape. The most characteristic peatlands occur at the summit of sandstone–quartzite mesetas called tepuis, distributed over a large elevation range from about 600 m to 2,800 m a.s.l. Small peat areas occur also atop dome-like hills developed from igneous–metamorphic rocks at elevations of 600–2,400 m a.s.l. In both cases, there is a unique relationship between peat formation and the formation of depressions through karstic and pseudokarstic rock weathering.

Karst-like landscape features and relief forms, similar in appearance to real karst morphology, have been reported to occur on noncalcareous rocks. There is evidence that dissolution affects not only the amorphous silica cement but also the skeletal quartz grains of quartzose sandstones and quartzites. On the basis of the strong resemblance of the surface and underground features resulting from silica dissolution with those caused by carbonate dissolution, we have opted to call the former karstic instead of pseudokarstic, although the rock weathering processes and dissolution rates are quite different. However, we call pseudokarstic the karst-like surface features developed on igneous–metamorphic rocks, as these features are more related to the hydrolysis of primary minerals than to dissolutional weathering and are not connected to any underground drainage system.

Microrelief is the initial controlling factor. Small alveoli at the beginning and larger depressions in later times allow water to concentrate, vegetation to colonize, and organic residues to be preserved in situ starting the process of peat formation. Topographic irregularities result from rock physical disintegration and chemical dissolution along joints and other fissures causing the disaggregation of the sandstone–quartzite cement. Small ponds show always sand-size debris at the bottom serving as foothold to pioneer vegetation that takes up the few nutrients resulting from rock weathering and eolian particles trapped in the water.

Because rain and runoff waters tend to concentrate in the central parts of the individual spoon-shaped mesetas, peat deposits are more frequent in these areas, while the meseta rims are generally bare rock surfaces. There is a variety of landscape positions where peat occurs: in depressions, on slopes, in small valleys, in narrow tectonic crevices and fissures, and on narrow floodplains, but peat occurs typically in depressions (i.e., dolines and polje-like flats) encircled by rock outcrops. Peat deposits rest usually directly on the rock substratum. The process of organic matter accumulation starts in small pans and ponds colonized by pioneer vegetation (algae, lichens, cyanobacteria). As the depressions enlarge and deepen through rock weathering along joints and fractures, higher plants including herbs, forbs, shrubs, and small trees, develop. Peat formation at tepui summits responds to the peat-forming-upwards mechanism (i.e., paludification) that better explains rapid peat accumulation than the commonly assumed peat-forming-downwards mechanism (i.e., terrestrialization).

In general, individual peatland units have an elongated configuration. The average length-to-width ratio is 3.5:1 but can be as large as 14:1. The length of peat areas varies from 140 to 325 m, while the width ranges from 10 to 50 m. The extent of the bogs varies from a few square meters in the case of small alveoli to several hectares in the case of polje-like depressions. The surface area of the larger bogs is about 0.5–3 ha. Thickness of the organic deposits varies among and within peatland areas. A maximum thickness of 170–200 cm was observed, but the most common range is 45–150 cm. Landscape position and the nature of the bedrock mainly control the spatial variations of peats and peatlands. The peat mantle is thicker in bogs than on slopes. Slope peats are unstable and exposed to fragmentation through differential sliding.

Organic matter under permanent water table seems to accelerate rock weathering and arenization, resulting in the occurrence of small strata of coarse sand and fine gravel along the interface between the base of the peat mantle and the underlying rock surface. Microtopography under the peat mantle is very irregular, with sharp highs and narrow troughs forming a crypto-karren. The peat-buried sandstone profile is more indented than the granite profile. Although the pre-existence of small alveoli on the terrain surface is an initial condition for rainwater to concentrate and organic matter to accumulate, there is a retro-action of peat on rock weathering leading to the enlargement and deepening of the karstic and pseudo-karstic depressions.

Peat currently forms and has formed in the past over a large elevation range from 600 to 2,800 m a.s.l., but the relationship between peat age and elevation is in general weak except for the basal layers. The age of the organic material in the basal layers, which reflects the time of peat inception, shows that peat started accumulating first at higher elevations when climate changed from dry to moist at the beginning of the Holocene.

9.4 Morphological, Physical, Chemical, and Taxonomic Characteristics of the Peats

Organic soils have characteristics which distinguish them markedly from mineral soils. The analytical procedures used for their characterization and classification are relatively recent. Not many organic soils in the world have been analyzed by the methods of the USDA Soil Taxonomy, and those that have, are mostly in the United

States of America. These methods were originally developed on soils occurring in geomorphic environments very different from that of the Guayana Highlands. Consequently, adjustments of the analytical procedures were necessary.

Peatlands are generally small because of strong geostructural control. Organic and organo-mineral soils are associated in space, both resting on a lithic substratum of mainly sandstone and granitic rocks that commonly occur at less than 150 cm depth. Surrounding stony and rocky terrains are locally covered by sandy colluvial mantles.

Peats are soft and unconsolidated deposits, water-saturated the largest part of the year. The water table is usually near or above the ground surface. Organic layers, including folic Oo, fibric Oi, hemic Oe, and sapric Oa, were recognized using field and laboratory criteria. The most common peat material is slightly-to-moderately decomposed fibric and hemic organic material. Strongly decomposed sapric material is less frequent. In tepui environment, the upper peat layers are less decomposed than the deeper ones, in contrast to what happens usually in most tropical and subtropical lowland peats where stages of decomposition decrease with depth. Variations in layer arrangement are largely controlled by profile position on the landscape. The most common layer sequence is Oo–Oi–Oe. Locally, Oi–Oe and Oi–Oe–Oa sequences occur. In general, root content decreases with soil depth, especially in the deeper Oa layers. The presence of thick surface root mats in the Oo and upper portion of the Oi and Oe layers is an important feature of the Guayana peats, indicating that nutrient cycling basically occurs in the upper leaf and fibrous litter layers rather than in the deeper permanently waterlogged peat.

Dry matter content increases with depth from 35 to 47% in Oi layers and from 43 to 87% in Oe layers. In general, unrubbed and rubbed fibers decrease with soil depth, indicating that physical fiber decomposition is weak in the surface peat layers but increases with depth close to the mineral C horizon or the lithic contact. Rubbed fiber amounts to about 16% in both Oi and Oe layers but only 7% in Oa layers. Mineral matter content in organic layers is highly variable, ranging from nothing to about 80%. Variations in mineral content are mostly related to the geomorphic processes that favored the accumulation of mineral matter in low-lying landscape positions, together with the accumulation and preservation of the organic residues.

Wet bulk density is the physical parameter with lowest variability, ranging from 0.54 to 0.78 Mg m^{-3} without significant differences between organic layer types. Dry bulk density varies with depth from 0.11 to 0.16 to 0.20 Mg m^{-3} in Oi, Oe, and Oa layers, respectively. Wet and dry bulk density values increase with depth in contrast to tropical lowland peats which have higher bulk density in the surface than subsurface layers because of more advanced organic matter decomposition. Average values of field water content are 1,026%, 1,170%, and 1,532% for Oi, Oe, and Oa layers, respectively.

Average organic matter content is 74% by loss on ignition and 28% using the Walkley–Black method. In general, organic matter contents decrease with peat depth. Changes in organic matter content are often more related to the mixture of organic and mineral materials along the profiles than to the degree of decomposition of the organic materials. The highest amounts of organic matter occur at elevations of 650–1,800 m a.s.l.

Soil pH is the least variable chemical property of the organic layers. The pH values vary from ultra acid (pH 3.1–3.5) to strongly acid (pH 5.2–5.3). Total acidity is about 3–43 times higher than salt-replaceable acidity, which indicates the high pH-dependent surface charge of the organic compounds. Although organic materials have high total exchange capacity, as high as 187 cmol(+) kg^{-1}, the effective cation exchange capacity is low (5 cmol(+) kg^{-1}) at the very acid pH of these peat soils. Exchangeable bases are mainly stored in the surface layers through nutrient cycling, while often reaching only trace values in deeper layers. The relative abundance of exchangeable bases is $Ca^{+2} > Na^{+1} > K^{+1} > Mg^{+2}$, with average levels of 0.71, 0.52, 0.42, and 0.38 cmol(+) kg^{-1}, respectively.

All organic soil profiles support the requirements of the Histosol order in both the USDA Soil Taxonomy and the World Reference Base for Soil Resources (WRB). At the suborder level of the Soil Taxonomy, most soils classify as Saprists (50%), followed by Hemists (25%) and Fibrists (25%), on the basis of the degree of decomposition of the organic soil material. According to the WRB framework, the organic soils classify as Sapric, Hemic, and Fibric Histosols in equal proportions.

9.5 Vegetation of the Peatlands: Present and Past

The Guayana region has a varied vegetation cover, including forests, shrublands, meadows and savannas, and pioneer plant communities colonizing essentially open rock surfaces. Forests are by far the largest formations of the Guayana region as a whole, estimated at about 70% of the surface area. In Huber and García (2011), these formations are extensively described from lowlands to uplands to highlands, together with the physiographic, geomorphic, and climatic conditions specific to each elevation belt. Hereafter, reference is made mainly to shrublands and meadows that are the most common vegetation types on peatlands. The influence of vegetation on peat characteristics and stages of evolution varies between forb–herb cover and shrub–low forest cover, the first one being largely dominant in peat sites.

Depression peats (i.e., peat bogs) on sandstone–quartzite as well as on igneous–metamorphic rocks are usually covered by broadleaved meadows or tubiform meadows. According to the vegetation identified at the visited sites, broadleaved meadows were more frequent above 1,600–1,800 m a.s.l. and tubiform meadows below. Broadleaved meadows consist of a dense, continuous herbaceous layer, dominated by one or more members of the family Rapateaceae. In most cases, the plants belong to the genus *Stegolepis* (Rapateaceae) present on almost all tepuis, together with a variety of forbs and small herbs from the families Xyridaceae, Eriocaulaceae, Cyperaceae, Sarraceniaceae, Liliaceae, and Droseraceae, and scattered shrubs belonging mainly to Bonnetiaceae, Asteraceae, Rubiaceae, Malpighiaceae, Ericaceae, Melastomatacaeae, and Ochnaceae. Tubiform meadows are amongst the most impressive herbaceous plant communities of the Guayana Highlands. They are characterized by a cylindrical growth habit well developed in the genera *Brocchinia* (Bromeliaceae) and *Heliamphora* (Sarraceniaceae), both endemic to the Guayana region. Small peatlands interrupted by rock outcrops are often covered by shrubby meadows or even open shrubland with the characteristic *Bonnetia* shrubs and trees. Tepui grassland occurs in places where organic layers are intercalated with mineral horizons, such as in narrow floodplains. Slightly larger floodplains may have low riparian forest. Many of the peatland plants are endemic to the Guayana Highlands. Some monospecific communities living in broadleaved meadows occur only on specific tepui massifs. Among the most emblematic peatland plants, small pitcher plant species of *Heliamphora* are frequently site-specific, while *Brocchinia* is widely distributed over all high-tepui environments.

From the few sampled sites only loose relationships can be established between vegetation cover, peat depth, and the nature of the substratum. Deep peats (80–100 cm or deeper) in karstic depressions on sandstone–quartzite mesetas are usually covered by broadleaved or tubiform meadows. Common plants of broadleaved meadows include species of *Everardia*, *Xyris*, *Orectanthe*, mixed with *Drosera*, *Cortaderia*, *Utricularia*, *Cephalocarpus*, *Lycopodium*, among others. Characteristic species of tubiform meadows belong to *Brocchinia*, *Kunhardtia*, *Stegolepis*, *Lagenocarpus*, *Abolboda*, sometimes mixed with isolated shrubs of *Cyrilla*, *Maguireothamnus*, *Phyllanthus*, *Heteropterys*, among others. On moderately deep peat deposits (50–60 cm), meadows tend to be shrubby with a lower story of *Brocchinia*, *Heliamphora*, *Stegolepis*, *Utricularia*, *Xyris*, *Orectanthe*, *Kunhardtia*, *Myriocladus*, *Abolboda*, mixed wit shrubs of *Cyrilla*, *Bonnetia*, *Acanthella*, *Phyllanthus*, *Clusia*, *Spathelia*, *Ilex*, among others. Shallow peats (less than 40 cm deep) are often covered by open shrubland including characteristic species of *Cyrilla*, *Terminalia*, *Stenopadus*, *Bonnetia*, and *Tepuianthus*. On granitic domes, deep peat deposits are unusual; they are covered by short grasses, forbs, and isolated shrubs. In the granitic Maigualida mountains, the tepui species of broadleaved herbs of *Stegolepis*, so abundant elsewhere, are replaced by *Kunhardtia*, another genus of the same family Rapateaceae, and the shrubs of the genus *Bonnetia* are entirely missing. Rocky islands within peatlands and rock outcrops along peatland rims show species of *Schefflera*, *Ilex*, *Podocarpus*, *Geonoma*, and *Daphnopsis*.

As peat occurs over a large altitudinal range (650–2,600 m a.s.l. in our study areas), flora and vegetation types can be expected to vary according to the regional temperature gradient of −0.6°C per 100 m elevation increase, while rainfall might be less differentiating. Climate changes during the Quaternary may have caused vertical vegetation displacements and interchange between lowland and highland flora. However, climate oscillations that occurred during the Holocene had certainly less impact on vegetation shifts than the Pleistocene glacial–interglacial cycles. The $\delta^{13}C$ values are mainly elevation-independent and lie between −24.6‰ and −27.3‰ all along the elevation range. Values higher than −26‰ occur at a variety of elevations in all three study areas. This means that the temperature gradient has little influence on the composition of the vegetation cover at peat sites and that the main factor controlling it is peat hydrology. This holds for the whole period since the inception of peat accumulation at the beginning of the Holocene. Vertical displacement of vegetation belts resulting from climate variations during the Holocene, as evidenced by pollen data in the eastern Guayana Highlands, could not be inferred from our $\delta^{13}C$ peat data. Altitudinal vegetation shift is a response to climate forcing at regional level, while vegetation at peat sites is rather controlled by the local forcing of peatland hydrology. Peatland vegetation is relatively insensitive to regional climate changes that do not alter substantially the water regime at peat sites. Long-term water saturation sustained by abundant rainfall causes inertia buffering against temperature changes.

The earliest organic material found in the western Guayana Highlands dates back to about 8,400 calBP. Peatlands before the beginning of the Holocene must have been very scarce or even just missing. In the absence of pollen data, the $\delta^{13}C$ ratio can be used as a proxy indicator of past vegetation leading to peat formation. Past vegetation cover of the peatlands was mainly composed of C3 plants, in a mixture of predominantly woody ($\delta^{13}C$ from −26‰ to −28‰) and subordinate herbaceous species ($\delta^{13}C$ from −24‰ to −26‰), with no significant variations all along the time period from lower Holocene to nowadays. Vegetation on rock outcrops might have been more sensitive to climate change than the vegetation in peat sites more controlled by the water regime. Thus, based on $\delta^{13}C$ data, the current peat vegetation cover in the western highlands might not be fundamentally different from past vegetation. Also in the southeastern uplands and highlands, $\delta^{13}C$ values indicate that vegetation was stable over the last 6,000 years, maintaining the relative difference in proportion between woody and herbaceous plants, and between C3 and C4 plants. Similarly, earlier palynological research conducted in the eastern Guayana Highlands and in the Gran Sabana plateau concluded that the vegetation cover in those areas did not change substantially during the mid- and upper Holocene.

9.6 Environmental Changes: Diachronic Inception and Polygenetic Evolution of Peats

Peat inception in the western Guayana Highlands roughly coincides with peat starting dates obtained by earlier studies in the eastern Guayana Highlands. Calibrated calendar ages stretch the initiation of peat accumulation back to about 8,400 calBP (= ca 7,500 BP). Peat initiation and development were strongly dependent on site conditions (local forcing), together with favorable climate conditions (regional forcing). Ongoing karst activity, especially with increased moisture at the onset of the Holocene favoring rock weathering, provided regularly new depressions for water to concentrate and organic material to accumulate. This resulted in diachronic rather than cyclic peat inception, with starting dates of organic material accumulation spread over time. Indeed, variable ages were determined at similar depth ranges, while similar dates occurred at variable depth ranges, suggesting that the visited sites have undergone different evolution and have thus different peat formation histories.

Most peats are composite soil bodies including unconformable organic layers and, sometimes, interstratified mineral horizons that reveal discontinuity in peat formation. Although peat has formed during the major part of the Holocene, peat deposition was not constant. In some places, peat accumulation was interrupted by depositional hiatuses, some of which might have been induced by climate change and others by truncation through sliding or cutting by creek migration. Time gaps between consecutive dated peat layers are as large as 2,000–4,000 ^{14}C years at some sites. Another indicator of complex evolution is the fact that extract ages are older than residue ages, in contrast to what usually occurs. This possibly reflects the presence of organic substances in the extract inherited from a peat forming phase older than that of the residue, and suggests the superimposition of two peat formation events within a single layer. Furthermore, soil formation in organic deposits departs substantially from soil formation in mineral deposits. In the latter, pedogenesis usually starts when deposition stops, whereas peatsoil formation is an ongoing process of deposition and decomposition. While organic residues continue accumulating in the top folic and fibric layers, the basal sapric layers are depleted in polysaccharides and proteins, leading to the concentration of resistant lignins impoverished in ^{13}C. As a consequence, peat profiles are often polygenetic, resulting from more than one single accumulation and/or transformation phase.

The correlation between ^{14}C age and depth of the peats reveals a significant data gap between ca 5,200 and 3,600 BP that suggests that peat areas could have considerably shrunk in this period because of decreasing moisture or more frequent dry spells. Thus, two main peat formation phases would have taken place during the Holocene in the study area: 7,500–5,200 BP and 3,600 BP–Present. The deposition rates expressed in calendar years were higher in the first period (0.271–0.333 mm yr^{-1}) than in the second (as low as 0.065 mm yr^{-1}), while the average peat accumulation rate for the whole study area and time period was 0.164 mm yr^{-1}.

When the ^{14}C data from western Guayana Highlands are merged with similar data from eastern Guayana Highlands, the hypothesis of peat interruption in the period of ca 5,200–3,600 BP is not corroborated. There is, however, a clear difference in the peat age-depth pattern before and after ca 4,000 BP. Before this time threshold, peat formation seems to be more related to site conditions (local forcing) and, afterwards, more related to climate conditions (regional forcing). The period around 4,000 BP seems to be a time threshold in peat formation of broader significance in the regional context (Gran Sabana) as well as in the continental context (South America and Africa).

In a few places geographically well distributed, a new phase of organic matter accumulation initiated around the seventeenth to nineteenth centuries, with moderately thick peat deposits forming in relatively short time lapses. This might reflect a major environmental change that can be correlated with the end of the Little Ice Age period.

Peat formation reflects a multiple environmental causality context, including climatic, vegetational, geodynamic, and hydrodynamic factors that induce variations in the peat cover. This makes it difficult to infer strict climatic changes at regional level and distinguish them from local-forcing causes, and calls for a multicriteria approach and interpretation model to disentangle the history of the peat deposits in the Guayana Highlands.

9.7 The Highland–Lowland Connection: A System Approach

From a geochemical point of view, the most reactive sites at tepui summits are the peatlands. These are the places where physical rock disintegration takes place leaving a residue of bleached sand upon leaching of the iron oxides, where chemical silica dissolution develops generating karstic features, and where organic matter decomposes producing labile organic substances. Sustained by abundant rainfall, water percolation through the peat mantle and underlying rock fissures causes detrital sand grains, dissolved silica, and dissolved/suspended organic compounds to be extracted from the bottom of the peat reservoirs and exported to endokarst conduits. Ultimately, these migrating mineral and organic debris contribute to feeding the blackwater rivers and whitesand deposits in the lowlands (Fig. 9.1).

One remarkable feature is the infrequency of oxido-reduction mottles in the peat profiles in spite of water saturation. Color variegation from brown to gray to black is essentially due to increasing stages of root decomposition. This suggests that the peat environment at tepui summits is not fully anoxic because water is regularly renewed and does not stagnate long enough. A single peatland unit functions like an open system where water enters at the top, from runoff during the rheic phase and from rain during the ombric phase, and water exits at the bottom via deep percolation through joints and fractures to the underground karstic drainage network. During this filtering process, the most decomposed organic substances are translocated downwards through the peat mantle, with a lateral component in the case of slope peats. The enlargement of the rock fissures through mechanical and chemical weathering and the high hydraulic gradient resulting from the elevation of the tepui summits above the base level of the lowlands secure the system's sustainability. This peculiar water regime of the tepui peatlands may also explain why the upper peat layers are less decomposed than the deeper ones. Usually, in tropical lowland peats, decompositional stages decrease with increasing depth due to decreasing oxidative conditions at the bottom of the peats. In tepui environment, however, water does not stagnate in the lower peat tier for longer periods and, therefore, decomposition is less controlled by the drainage conditions than by the age of the organic deposits. Hemic layers are in general younger than 3,500 ^{14}C BP, while sapric layers are older than 2,500 ^{14}C BP. Thus, the oldest peat layers, which are the deepest ones, are the most decomposed, and this is another distinctive feature of tepui peats. This model applies well to the sandstone–quartzite substratum of the tepui mesetas. It does not apply to the impermeable rock substratum of

Fig. 9.1 Whitesand deposits and blackwater rivers in the lowlands. The weathering of sandstones and quartzites at tepui summits generates clastic materials that are transported stepwise from highlands to lowlands where they accumulate as whitesand deposits (**a**). Whitesands are often reworked by blackwater rivers (**b**) and spread over large floodplains where they form Quartzipsamments (**c**) and "giant podzols" (Spodosols) (photos Zinck)

the granitic domes in spite of the presence of curved joints caused by pressure release.

Peatlands are source of clastic debris that is stepwise transferred from the highlands to the lowlands. Whitesands in the lowlands have multiple origin. They are partly generated in the lowlands themselves from weathering of the igneous–metamorphic basement rocks, in particular gneiss, in the large peneplains of the Orinoco-Río Negro basins. But a substantial part comes from the uplands and highlands (see Quartzipsamment profiles 4 and 10 in Appendix). In the highlands, the weathering of the Roraima sandstones and quartzites produces sand that is washed away from the rock outcrops and flushed out of the tepui summits through surface runoff, followed by transport through creeks and underground pipes. In the uplands as well as in the lowlands, large amounts of sandy materials are stored in the piedmonts that circumscribe the base of the tepui walls. Thus, the lowland whitesand deposits in which the well-known Amazonian giant podzols (Spodosols) have formed are related to the weathering and natural erosion processes taking place in the highlands (see Quartzipsamment profiles 14 and 15 in Appendix). Uncontrolled and overloaded (eco-)tourism activities at tepui summits can trigger anthropogenic erosion upon destruction of the fragile peatland vegetation trough trampling (see Fig. 7.23).

Peatsoils have formed under waterlogged and nutrient-deficient conditions with limited organic matter decomposition. During the minerotrophic stage, the main sources of nutrients are rainfall and a tiny provision of minerals from bedrocks. During the ombrotrophic stage, some of these nutrients are fixed by the vegetation successions, then released and incorporated into the peat material during biomass decomposition, and finally recycled again by the vegetation. Most of the nutrients are washed out by runoff and deep percolation. However, the concentration of exchangeable bases in the topsoil (0–30 cm) indicates that nutrient transfer and recycling occur within the upper peat layers, where organic matter decomposition increases temporarily, releasing nutrients during the short dry season. In general, available phosphorus is low, often no more than traces.

Peatlands play an important role as water reservoir and regulator. Rainwater is temporarily stored in the organic deposits of the peatlands, having high water retention capacity, and progressively released to the lowlands via surface runoff and underground conduction. This contributes in particular to sustain the water regime of the lowland rivers during the short dry season from January to March. In the hypothesis of current climate change, a warmer atmosphere and a more seasonal rainfall regime would cause faster decomposition of the organic material and lead to the shrinkage of peat areas. This might affect the hydrological function of the peat cover as a water reservoir. Water will be released faster to the lowlands, causing flood in the whitesand plains and limiting water availability during the dry season. Monitoring is needed to assess the potential retreat of the peat mantle.

9.8 Concluding Remark

Tepui peatlands constitute a unique ecosystem, where peat has formed from specialized, largely endemic plant communities on (pseudo-)karstic landforms that have developed on noncalcareous rocks in a high-altitude tropical humid environment. Karst and peat formation are intimately related by retro-action effects. The formation of karstic depressions allows water and organic matter to accumulate, while the accumulation of organic matter and water leads to rock weathering and the formation of depressions. Joints and fractures in rocks, especially in sandstones, favor the exportation of clastic products and humus to the lowlands.

Peat formation was discontinuous in time and space, often suspended for sometime and then resuming again, and proceeded at variable deposition rates. Deconstructing and deciphering the complex paleogeographic fabric archived in peats are a challenging endeavor.

Full insight and understanding of the formation and functioning of the tepui peatlands, especially in terms of environmental services and vulnerability to climate change and human-induced hazards, need more inventory and research in a region of limited accessibility.

Reference

Huber O, García P (2011) The Venezuelan Guayana region and the study areas: geo-ecological characteristics. In: Zinck JA, Huber O (eds) Peatlands of the Western Guayana Highlands, Venezuela, Chap. 3. Springer, Heidelberg, doi: 10.1007/978-3-642-20138-7_3

Appendix: Site and Profile Characteristics

A.1 Introduction

The Appendix provides an overview of the primary data collected in the field as site and profile descriptions and the full set of data obtained from laboratory determinations. The cartographic location of the visited sites is shown in Chap. 3 (Figs. 3.8, 3.9, 3.13, and 3.16).

Peatsoils including organic and organomineral soils are estimated to cover about 30% of the tepui summits, while about 25% is bare rock terrain (O. Huber personal communication). Mineral soils, mainly sandy glacis deposits on quartzite–sandstone outcrops and medium-textured lixiviated soils on igneous–metamorphic rocks, are among the other main components of the soil cover. Erosion products from the highlands are widespread in the lowlands as whitesand deposits. For this reason, some representative whitesand profiles are included here. Out of 12 full peat profiles, eight lie on sandstone–quartzite mesetas and four lie on granitic substratum on domes and in small valleys. Additional four profiles have organic layers interstratified with mineral material. The remaining profiles are mineral soils, two from the highlands and two from the lowlands.

Organic and mineral soils were described following the FAO guidelines (FAO 1990) and classified according to the USDA Soil Taxonomy (Soil Survey Staff 2006) and the World Reference Base for Soil Resources (FAO 2006). Soil samples were analyzed at the Regional Laboratory for Soil and Water, Ministry of Environment and Renewable Natural Resources (MARNR), Guanare, Venezuela, using the methods described in Chap. 5. Carbon-14 determination was carried out at the Center for Isotope Research (CIO) of the University of Groningen, the Netherlands (see Chap. 7).

A.1.1 Organic and Organomineral Soils

The visited peatland sites are spread over three mountain massifs in the western Amazonas state of Venezuela: the Marahuaka-Huachamakari massif (including the

J.A. Zinck and O. Huber (eds.), *Peatlands of the Western Guayana Highlands, Venezuela,* Ecological Studies 217, DOI 10.1007/978-3-642-20138-7, northern edge of the adjoining Duida massif), the Cuao-Sipapo massif, and the western border of the Maigualida massif. Organic soils are mainly found on tepui mesetas formed from siliceous sedimentary rocks (sandstone and quartzite of the Roraima Group), but also on dissected hills formed from igneous and metamorphic rocks of the Guayana Shield basement (granite and gneiss).

From the 18 soil profiles described in the highlands, 12 are Histosols (Table A.1). The identification of and distinction between folic, fibric, hemic, and sapric materials were primarily done in the field by visual appraisal of the amount and degree of decomposition of the fibers, and adjusted in the case of borderline profiles using the technique of fiber rubbing as recommended in the Soil Taxonomy.

According to the Soil Taxonomy, most soils classify as Saprists (50%), followed by Hemists (25%) and Fibrists (25%), on the basis of the degree of decomposition of the organic soil material. All soils belong to the Haplo-great groups and most soils are Lithic extragrades because they have a lithic contact within the control section (0–130 cm). According to the WRB framework, the organic soils classify as Sapric, Hemic, and Fibric Histosols in equal proportions. Rheic Histosols are more frequent (60%) than Ombric Histosols. All soils are hyperdystric.

Some peatland soils did not comply with the requirements of the Histosols in terms of organic layer thickness and/or organic matter content. Soils formed from local alluvium on tepui summits often include humus-rich mineral horizons or peat layers at depth.These organomineral soils classify as Endoaquents and Humaquepts according to the Soil Taxonomy, and as Gleysols and Fluvisols according to the WRB framework (Table A.2).

Table A.1 Classification of the organic soils

Profile	USDA classification (2006)	WRB classification (2006)
P2	Lithic Haplosaprist	Endoleptic Rheic Sapric Histosol (Hyperdystric)
P5	Lithic *Terric* Haplofibrist	Rheic Fibric Histosol (Hyperdystric)
P6	Lithic Haplofibrist	Endoleptic Rheic Fibric Histosol (Hyperdystric)
P8	Lithic Haplohemist	Endoleptic Rheic Hemic Histosol (Hyperdystric)
P9	Lithic *Terric* Haplofibrist	Rheic Fibric Histosol (Hyperdystric)
P11	Lithic *Terric* Haplosaprist	Ombric Sapric Histosol (Hyperdystric)
P12	Lithic Haplosaprist	Endoleptic Ombric Hemic Histosol (Hyperdystric)
P16	Lithic Haplohemist	Epileptic Ombric Hemic Histosol (Hyperdystric)
P17	Typic Haplosaprist	Ombric Hemic Histosol (Hyperdystric, *Thaptosapric*)
P18	Lithic Haplohemist	Epileptic Rheic Fibric Histosol (Hyperdystric)
P19	Typic Haplosaprist	Ombric Sapric Histosol (Hyperdystric)
P20	Lithic Haplosaprist	Epileptic Rheic Sapric Histosol (Hyperdystric)

Proposed: *Terric, Thaptosapric*

Table A.2 Classification of the organomineral soils

Profile	USDA classification (2006)	WRB classification (2006)
P1	Typic Endoaquent	*Endoleptic* Gleysol (Hyperhumic, Hyperdystric, Episiltic)
P7	Cumulic Humaquept	Histic *Endoleptic* Fluvisol (Hyperdystric)
P13	Typic Humaquept	Histic *Epileptic* Fluvisol (Hyperdystric, Epiarenic)

Proposed: *Endoleptic, Epileptic*

Table A.3 Classification of the mineral soils

Profile	USDA classification (2006)	WRB classification (2006)
P3	Typic Kanhapludult	Cutanic Acrisol (Hyperdistryc, Profondic, Chromic)
P4	Aquic Quartzipsamment	Protic Hyperalbic Arenosol (Hyperdystric)
P10	Udoxic Quartzipsamment	Albic Arenosol (Hyperdystric)
P14	Aquic Quartzipsamment	Protic Hyperalbic Arenosol (Hyperdystric)
P15	Aquic Quartzipsamment	Protic Hyperalbic Arenosol (Hyperdystric)

A.1.2 Mineral Soils

Some mineral profiles were described in the highlands and lowlands to provide a complementary view of the regional soil diversity (Table A.3). In the highlands, Entisols have formed from short-transport sandy materials, including sandy colluvium or sandy depression filling (P4 and P10). Ultisols have developed from weathering of igneous–metamorphic rocks in summit areas less exposed to erosion (P3).

The lowlands surrounding the mountain massifs are sinks of the rock weathering debris and organic substances removed from the highlands by surface runoff and underground karstic drainage. Resulting from the former, whitesand deposits with black groundwater are widespread in the lowland plains (P14 and P15). Whitesands are nutrient-poor Quartzipsamments with open herbaceous or shrub cover including a variety of endemic plants. Some have deep to very deep spodic horizons and, for this reason, have been called giant podzols. Whitesands and blackwaters are related to each other and occur together on the landscape, while both are also intimately related to karst and peat formation in the highlands. There is thus a strong landscape relationship between highlands and lowlands.

References

FAO (1990) Guidelines for soil profile description. Food and Agriculture Organization of the United Nations, Rome

FAO (2006) World reference base for soil resources. A framework for international classification, correlation and communication. World Soil Resources Reports 103. FAO, ISRIC and IUSS, Rome

Soil Survey Staff (2006) Keys to soil taxonomy. Natural Resources Conservation Service, US Department of Agriculture, Washington, DC

A.2 Site and Profile Descriptions

Profile 1

Area: Venezuelan Guayana, Marahuaka tepui, Duida-Marahuaka National Park
Site: Marahuaka I-1; N 03°40′, W 65°26′; 2,570 m a.s.l.
USDA classification (2006): Typic Endoaquent
WRB classification (2006): *Endoleptic* Gleysol (Hyperhumic, Hyperdystric, Episiltic)
Landscape: table-shaped plateau highlands (2,000–2,800 m a.s.l.), developed on horizontally lying to slightly folded or tilted sandstone and quartzite formations of the Roraima Group (Precambrian)
Relief type: rocky tepui meseta with karstic landforms
Lithology/facies: water-transported sediments of local origin on top of sandstone substratum
Landform: narrow floodplain along a small creek running on bare rock
Parent material: colluvio-alluvium
General topography gently undulating with rugged ruiniform rock outcrops and deep crevices
Local topography almost flat (<1% slope); microtopography flat
Poorly drained, periodically flooded; water-saturated from 10 cm downward; slow permeability
Erosion and human influence: none
Surface characteristics: patches of sandy colluvial material (10–15 cm thick) on the terrain surface and between rock outcrops
Vegetation: high-tepui grassland, 0.3–0.5 m high, 100% coverage (*Cortaderia roraimensis*, *Aulonemia* sp., and *Hypericum marahuacanum*); occasional 2–3 m high shrub islands (*Cyrilla racemiflora*, *Mycerinus sclerophyllus*, *Geonoma appuniana*, and *Myriocladus* sp.)
Described on 24.03.1992 by A. Zinck, P. Garcia, O. Huber

Profile description (depth in cm; colors are for moist soil unless otherwise stated; soil moist at 0–60 cm depth and saturated below, at description date):

Oi	0–13	Brown (7.5YR 4/3); root mat with strata of sandy loam mineral material in the lower part
2C1	13–40	Dark gray (10YR 4/1) with frequent light olive brown (2.5Y 5/6) mottles; silty clay loam; massive; plastic
2C2	40–60	Dark gray (2.5Y 4/1); silty clay loam; massive; plastic; ^{14}C age 7,100 ± 60 BP
3Cr	60–61	White (10YR 8/1); loose bleached fine sand; free water
3R	61+	Sandstone

Profile 2

Area: Venezuelan Guayana, Marahuaka tepui, Duida-Marahuaka National Park
Site: Marahuaka I-2, 200 m north of site Marahuaka I-1; N 03°40′, W 65°26′; 2,580 m a.s.l.
USDA classification (2006): Lithic Haplosaprist
WRB classification (2006): Endoleptic Rheic Sapric Histosol (Hyperdystric)
Landscape: table-shaped plateau highlands (2,000–2,800 m a.s.l.), developed on horizontally lying to slightly folded or tilted sandstone and quartzite formations of the Roraima Group (Precambrian)
Relief type: rocky tepui meseta
Lithology/facies: sandstone
Landform: karstic depression, 30 m × 20 m in size, surrounded by rock outcrops that dominate the depression by 1–5 m elevation
Parent material: peat over sandy loam alluvial sediments
General topography gently undulating with rugged ruiniform rock outcrops and deep crevices
Local topography flat to slightly concave (0.3–0.5% slope); microtopography slightly hummocky
Very poorly drained; water-saturated during the largest part of the year, with patches of stagnating water at the soil surface; water table at 10 cm depth; slow permeability
Erosion and human influence: none
Surface characteristics: low supporting capacity
Vegetation: high-tepui meadow dominated by *Everardia* sp., mixed with *Drosera* sp., *Xyris* sp., *Orectanthe sceptrum*, *Nietneria paniculata*, *Isidrogalvia schomburgkiana*, *Sauvagesia imthurniana*, and Eriocaulaceae
Described on 24.03.1992 by A. Zinck, P. Garcia, O. Huber

Profile description (depth in cm; colors are for moist soil unless otherwise stated; soil saturated at description date):

Oi	0–5	Very dark gray (10YR 3/1); root mat
Oe	5–20	Very dark brown (10YR 2/2); moderately decomposed organic material, with 10–20% fibers
Oa1	20–60	Very dark brown (10YR 2/2); decomposed organic material, with 5–10% fibers; ^{14}C age 5,880 ± 50 BP
Oa2	60–70	Black (10YR 2/1); decomposed organic material, with 5–10% fibers
2C	70–95	Dark gray (10YR 4/1); sandy loam mixed with decomposed organic material; few decomposed fine roots; ^{14}C age 7,490 ± 50 BP
3R	95+	Sandstone

Profile 3

Area: Venezuelan Guayana, Duida massif (eastern edge), Duida-Marahuaka National Park
Site: Duida I-3; N 03°31′, W 65°32′; 1,200 m a.s.l.
USDA classification (2006): Typic Kanhapludult
WRB classification (2006): Cutanic Acrisol (Hyperdistryc, Profondic, Chromic)
Landscape: table-shaped plateau highlands (1,200–2,000 m a.s.l.), developed on horizontally lying to slightly folded or tilted sandstone and quartzite formations of the Roraima Group (Precambrian) with areas of hilly dome-like relief corresponding to igneous–metamorphic basement rocks (granite) of the Guayana Shield
Relief type: horseshoe-shaped, steep-slope bare rocky dome
Lithology/facies: granodiorite
Landform: dome shoulder
Parent material: weathering material from granodiorite
General topography steeply dissected
Local topography rolling (10–15% slope); microtopography smooth
Well drained; moderate permeability
Erosion: slight; human influence: none
Surface characteristics: generalized litter cover, partially decomposed
Vegetation: moderately tall submontane forest, with 20–30 m high emergent trees; between 1,160 and 1,200 m elevation, succession of several vegetation types including tepui meadow, tepui shrub with *Bonnetia*, 5–6 m low forest, and tall submontane forest
Described on 24.03.1992 by A. Zinck, P. Garcia

Profile description (depth in cm; colors are for moist soil unless otherwise stated; soil moist at description date):

Oi	0–10	Litter of dead and partially decomposed leaves and live roots with mycorrhizae
A	10–20	Yellowish brown (10YR 5/6); sandy loam; weak blocky structure; few fine roots
Bt1	20–50	Strong brown (7.5YR 5/6); sandy clay loam; few thin clay cutans; weak to moderate blocky structure; few fine roots
Bt2	50–120	Strong brown (7.5YR 5/8); sandy clay loam; few dark red mottles; few thin clay cutans; moderate blocky structure; rests of burnt vertical roots at 65–70 cm depth
R	120+	Granodiorite

Profile 4

Area: Venezuelan Guayana, Duida massif (eastern edge), Duida-Marahuaka National Park
Site: Duida I-4; N 03°31′, W 65°32′; 1,160 m a.s.l.
USDA classification (2006): Aquic Quartzipsamment
WRB classification (2006): Protic Hyperalbic Arenosol (Hyperdystric)
Landscape: table-shaped plateau highlands (1,200–2,000 m a.s.l.), developed on horizontally lying to slightly folded or tilted sandstone and quartzite formations of the Roraima Group (Precambrian), with areas of hilly dome-like relief corresponding to igneous–metamorphic basement rocks (granite) of the Guayana Shield
Relief type: small valley (vale), deeply incised between a horseshoe-shaped, steep-slope bare rocky dome, on one side, and a sandstone meseta on the other side
Lithology/facies: colluvial material
Landform: vale bottom
Parent material: mixed colluvium coming from the weathering of surrounding granite and sandstone exposures
General topography slightly concave
Local topography almost flat (<1% slope); microtopography smooth
Poorly drained; water-saturated from the surface; water table at 40 cm depth; rapid permeability
Erosion: none; human influence: site visited by occasional indigenous hunters
Surface characteristics: loose whitesand
Vegetation: small glade of tepui meadow (*Amphiphyllum rigidum*, *Brocchinia* sp., Eriocaulaceae, Tepuianthaceae, Orchidaceae, and Rapateaceae, among others), surrounded by dense tepui scrub
Described on 24.03.1992 by A. Zinck, P. Garcia

Profile description (depth in cm; colors are for moist soil unless otherwise stated; soil saturated at description date):

A	0–10	Grayish brown (10YR 5/2); fine sand; frequent fine roots
C1	10–20	Light brownish gray (10YR 6/2); fine sand; few medium and fine roots; rests of charcoal
C2	20–50	White (7.5YR 8/1); medium sand; few fine and very fine roots; blackwater table at 40 cm depth
C3	50–100+	White (7.5YR 8/1); medium sand

Profile 5

Area: Venezuelan Guayana, Marahuaka tepui (southern edge), Duida-Marahuaka National Park
Site: Marahuaka II-5; N 03°37′, W 65°24′; 2,600 m a.s.l.
USDA classification (2006): Lithic *Terric* Haplofibrist
WRB classification (2006): Rheic Fibric Histosol (Hyperdystric)
Landscape: table-shaped plateau highlands (2,000–2,800 m a.s.l.), developed on horizontally lying to slightly folded or tilted sandstone and quartzite formations of the Roraima Group (Precambrian); ripple marks and flute casts on rock surface
Relief type: rocky tepui meseta, with 50–100 m high scarps between internal meseta subdivisions
Lithology/facies: sandstone
Landform: large karstic amphitheater-like depression (500 m diameter), dominated by a rock outcrop crest; subdivided into smaller alveoli (60 m × 30 m) surrounded by rock outcrops carved with microchannels by chemical erosion (karren field)
Parent material: peat over sandstone
General topography gently undulating (3–5% slope), with rugged ruiniform rock outcrops and deep crevices
Local topography flat to slightly concave (<0.5% slope); microtopography hummocky
Very poorly drained; water-saturated during the largest part of the year, with patches of stagnating water at the soil surface; slow permeability
Erosion and human influence: none
Surface characteristics: patches of sandy colluvial cover, 10–15 cm thick
Vegetation: high-tepui meadow, very dense, 0.5–0.7 m high (dominant: *Xyris* sp., *Everardia montana*, and *Orectanthe sceptrum*; frequent: *Cortaderia roraimensis*, *Drosera* sp., *Xyris* sp., *Utricularia* sp., *Cephalocarpus rigidus*, *Burmannia foliosa*, *Lycopodium* sp., *Hypericum* sp., *Cyrilla racemiflora*, and *Baccharis* sp.); shrub vegetation on rocky islands close to the peat site with *Marahuacaea schomburgkiana*, *Myriocladus* sp., *Schefflera* sp., *Ilex retusa*, *Podocarpus* sp., and *Geonoma appuniana*
Described on 25.03.1992 by A. Zinck, P. Garcia, O. Huber

Profile description (depth in cm; colors are for moist soil unless otherwise stated; soil saturated at description date):

Oo	0–5	Very dark gray (10YR 3/1); root mat from herbs and shrub plants
Oi1	5–20	Very dark brown (10YR 2/2); dense root mat and moderately decomposed organic material (40–50%)
Oi2	20–60	Very dark brown (10YR 2/2); dense root mat and moderately decomposed organic material (40–50%); coarse tubular roots (*Brocchinia*); ^{14}C age 170 ± 40 BP
Oe	60–85	Black (10YR 2/1); moderately decomposed organic material, with 20–25% root fibers
2C	85–110/120	Very dark gray (10YR 3/1); moderately decomposed organic material, with 5–10% root fibers; ^{14}C age 3,400 ± 50 BP
3R	110/120+	Sandstone

Profile 6

Area: Venezuelan Guayana, Huachamakari tepui, Duida-Marahuaka National Park
Site: Huachamakari I-6; N 03°51′, W 65°44′; 1,770 m a.s.l.
USDA classification (2006): Lithic Haplofibrist
WRB classification (2006): Endoleptic Rheic Fibric Histosol (Hyperdystric)
Landscape: table-shaped plateau highlands (1,500–2,800 m a.s.l.), developed on horizontally lying to slightly folded or tilted sandstone and quartzite formations of the Roraima Group (Precambrian)
Relief type: dissected rocky tepui meseta
Lithology/facies: sandstone
Landform: sequence of shelf-like sandstone slabs 20–30 m wide, flat to slightly sloping (2–3%), arranged in consecutive steps separated by small scarps 1–2 m high.
Parent material: peat over sandstone
General topography gently undulating (3–5% slope), with rugged ruiniform rock outcrops and deep crevices
Local topography of flat steps (<1% slope); microtopography slightly hummocky
Very poorly drained; water-saturated during the largest part of the year, with patches of stagnating water at the soil surface; slow permeability
Erosion and human influence: none
Surface characteristics: frequent small, flat rock outcrops (<10 m^2)
Vegetation: shrubby tepui meadow (*Brocchinia tatei*, *Brocchinia acuminata*, *Maguireothamnus* sp., *Bonnetia* sp., *Heliamphora tatei*, *Utricularia humboldtii*, *Duidania montana*, *Xyris* sp., *Orectanthe sceptrum*, and *Cyrilla racemiflora*)
Described on 26.03.1992 by A. Zinck, P. Garcia, O. Huber

Profile description (depth in cm; colors are for moist soil unless otherwise stated; soil saturated at description date):

Oi1	0–12	Brown (7.5YR 4/3); root mat with little moderately decomposed organic material
Oi2	12–45	Very dark brown (10YR 2/2); root mat with little moderately decomposed organic material
Oe	45–60	Black (10YR 2/1); moderately decomposed organic material and some roots with, at 50 cm depth, a fragment of horizontally lying *Bonnetia* stem with ^{14}C age 4,840 ± 40 BP
2R	60+	Sandstone

Profile 7

Area: Venezuelan Guayana, center-west sector of the Cuao-Sipapo massif
Site: Cuao 1; N 05°03′, W 67°22′; 1,800 m a.s.l., close to a peak of the massif (1,900 m a.s.l.)
USDA classification (2006): Cumulic Humaquept
WRB classification (2006): Histic *Endoleptic* Fluvisol (Hyperdystric)
Landscape: table-shaped plateau highlands (600–2,000 m a.s.l.), developed on horizontally lying to slightly folded or tilted sandstone and quartzite formations of the Roraima Group (Precambrian) with areas of hilly dome-like relief corresponding to igneous–metamorphic basement rocks (granite) of the Guayana Shield
Relief type: rocky tepui meseta
Lithology/facies: water-transported sediments of local origin over peat and sandstone
Landform: karstic depression, exposed to flooding by a small creek running on bare rock
Parent material: alluvium over peat or mixed sequence of alluvium and peat
General topography table-shaped
Local topography almost flat (1–2% slope), surrounded by steeper slopes (30%); microtopography irregular
Very poorly drained; water-saturated during a large part of the year, with patches of stagnating water at the soil surface; slow permeability
Erosion and human influence: none
Surface characteristics: no particular evidence
Vegetation: low dense riparian forest with *Clusia* sp., *Graffenrieda fantastica*, *Phyllanthus* sp., and *Neurolepis* sp.
Described on 10.02.1993 by A. Zinck, P. Garcia, O. Huber

Profile description (depth in cm; colors are for moist soil unless otherwise stated; soil saturated from 11 cm downward at description date):

Oe	0–11	Black (10YR 2/1); loam mixed with organic material; very friable, slightly plastic, and nonsticky; many fine and medium roots (1–3 mm); moist
2A1	11–27	Very dark brown (10YR 2/2); loam to sandy loam mixed with organic material; friable, slightly plastic, and slightly sticky; common fine and very fine roots (1–2 mm)
2A2	27–48	Black to very dark brown (10YR 2.5/1); sandy loam (medium sand); nonplastic and nonsticky; few fine roots
3C	48–70	Black (10YR 2/1); fairly well-decomposed organic material with loamy touch; slightly plastic and slightly sticky; common fine and medium roots (1–3 mm) and few coarse roots (10 mm); ^{14}C age 2,105 ± 45 BP (residue) and 2,730 ± 50 BP (extract)
4R	70+	Sandstone

Profile 8

Area: Venezuelan Guayana, center-west sector of the Cuao-Sipapo massif
Site: Cuao 2 (close to Cuao 1); N 05°03′, W 67°22′; 1,800 m a.s.l., close to a peak of the massif (1,900 m a.s.l.)
USDA classification (2006): Lithic Haplohemist
WRB classification (2006): Endoleptic Rheic Hemic Histosol (Hyperdystric)
Landscape: table-shaped plateau highlands (600–2,000 m a.s.l.), developed on horizontally lying to slightly folded or tilted sandstone and quartzite formations of the Roraima Group (Precambrian) with areas of hilly dome-like relief corresponding to igneous–metamorphic basement rocks (granite) of the Guayana Shield
Relief type: rocky tepui meseta, with 10–100 m high scarps between internal meseta subdivisions
Lithology/facies: sandstone
Landform: narrow interfluve with many small karstic alveoli (30–50 m^2), between two small creeks running on bare rock
Parent material: peat
General topography table-shaped
Local topography almost flat (1–2% slope); microtopography slightly hummocky
Poorly drained; slow permeability
Erosion and human influence: none
Surface characteristics: no particular evidence
Vegetation: high-tepui meadow with *Kunhardtia rhodantha*, *Brocchinia* div. sp., and *Stegolepis choripetala; Bonnetia* bushes close to rock outcrops
Described on 10.02.1993 by A. Zinck, P. Garcia, O. Huber

Profile description (depth in cm; colors are for moist soil unless otherwise stated; soil moist with water leaking at 30 cm depth but not saturated, at description date):

Oo	0–5	Loose folic material, including dead leaves, twigs, mosses, and lichens
Oe1	5–30	Black (5YR 2.5/1); moderately decomposed organic material; abundant fine and common medium roots (1–3 mm), and some very coarse roots (up to 3 cm); massive structure; slightly plastic and slightly sticky
Oe2	30–55	Black (10YR 2/1); moderately decomposed organic material; strong concentration of roots, mainly coarse (up to 5 cm), with reddish surface, often horizontally lying; massive structure; slightly plastic and slightly sticky; ^{14}C age 1,665 ± 35 BP
Oa	55–87	Black (10YR 2/1); decomposed organic material with loamy touch; massive structure; slightly plastic and slightly sticky; few roots; ^{14}C age 3,415 ± 30 BP
2R	87+	Sandstone

Profile 9

Area: Venezuelan Guayana, northwest sector of the Cuao-Sipapo massif
Site: Cuao 3; N 05°04′, W 67°24′; 1,600 m a.s.l.
USDA classification (2006): Lithic *Terric* Haplofibrist
WRB classification (2006): Rheic Fibric Histosol (Hyperdystric)
Landscape: table-shaped plateau highlands (600–1,800 m a.s.l.), developed on horizontally lying to slightly folded or tilted sandstone and quartzite formations of the Roraima Group (Precambrian) with areas of hilly dome-like relief corresponding to igneous–metamorphic basement rocks (granite) of the Guayana Shield
Relief type: rocky tepui meseta
Lithology/facies: sandstone
Landform: karstic depression, at close distance to a small creek running on bare rock
Parent material: peat
General topography gently undulating (3–5% slope), with rugged ruiniform rock outcrops and deep crevices
Local topography almost flat to slightly concave (1–2% slope); microtopography smooth
Very poorly drained; water-saturated during the largest part of the year, with patches of stagnating water at the soil surface; free water table at 40 cm depth; slow permeability
Erosion and human influence: none
Surface characteristics: no particular evidence
Vegetation: tubiform high-tepui meadow (*Brocchinia hechtioides*, *Kunhardtia rhodantha*, *Brocchinia acuminata*, *Stegolepis* sp., and *Neurolepis* sp.) with shrubs of *Maguireothamnus tatei* and *Graffenrieda* sp.
Described on 11.02.1993 by A. Zinck, P. Garcia, O. Huber

Profile description (depth in cm; colors are for moist soil unless otherwise stated; soil saturated at description date):

Oi	0–10	Black (5YR 2.5/1); loose folic material at the soil surface, including dead leaves, twigs, mosses, and lichens; fibric root mat with many fine to coarse roots (1–6 mm); massive structure; slightly plastic and slightly sticky
Oe	10–20	Black to very dark gray (10YR 2.5/1); root mat with many fine to coarse roots (1–6 mm); massive structure; slightly plastic and slightly sticky
2A1	20–40	Black (10YR 2/1); slightly to moderately decomposed organic material; frequent coarse roots (up to 1 cm); massive structure; plastic and sticky
2A2	40–60	Black (10YR 2/1); slightly to moderately decomposed organic material; many medium roots (2–4 mm), essentially from *Brocchinia;* massive structure; plastic and sticky; ^{14}C age 165 ± 30 BP
3Oi	60–115	Black (10YR 2/1); very loose mix of slightly decomposed organic material and roots, floating in water
4R	115+	Sandstone

Profile 10

Area: Venezuelan Guayana, Cuao-Sipapo massif, Cerro Autana
Site: Cuao 4; N 04°53′, W 67°26′; 1,210 m a.s.l.
USDA classification (2006): Udoxic Quartzipsamment
WRB classification (2006): Albic Arenosol (Hyperdystric)
Landscape: low-lying piedmont and peneplain area with a few outstanding isolated tepuis, located south of the Cuao table-shaped sandstone highlands (600–1,800 m a. s.l.) and north of the Autana river valley
Relief type: rocky tepui meseta; summit
Lithology/facies: sandstone–quartzite of the Roraima Group (Precambrian)
Landform: glacis at the footslope of a sandstone–quartzite hill (1,240 m a.s.l.)
Parent material: colluvium from the erosion of sandstone–quartzite outcrops
General topography strongly inclined (10–15% slope)
Local topography slightly inclined (3–5% slope); microtopography irregular, with erosion rills and anthropogenic disturbance
Somewhat excessively drained; rapid permeability
Erosion: slight sheet and rill erosion; human influence: vegetation disturbance and soil trampling caused by tourist activity
Surface characteristics: locally severely disturbed by trampling
Vegetation: tubiform tepui meadow (*Brocchinia hechtioides*, *Kunhardtia rhodantha*, and *Lagenocarpus* sp.) with isolated shrubs of *Heteropterys steyermarkii* and *Cyrilla racemiflora*
Described on 12.02.1993 by A. Zinck, P. Garcia, O. Huber

Profile description (depth in cm; colors are for moist soil unless otherwise stated; soil moist at description date; water leaking at 90 cm depth):

A1	0–5	Black (2.5Y 2/1); sandy loam; compact; common fine roots; blocky structure; slightly plastic and slightly sticky
A2	5–10/13	Very dark gray (10YR 3/1); loamy sand; weak angular blocky; few fine roots
A3	10/13–28	Dark grayish brown (10YR 4/2); loamy sand; massive; common medium roots
C1	28–70	Brown (7.5YR 4/2); fine sand; single grain; many dead medium and coarse roots
C2	70–90	Brown (7.5YR 5/3); fine sand; single grain; few fine roots
C3	90–135	Light brown (7.5YR 6/4); fine sand; single grain
2R	135+	Sandstone–quartzite

NOTE: profile Cuao 5 is similar to profile Cuao 4, but without surface trampling and vegetation disturbance. Cover is tubiform meadow with *Brocchinia hechtioides* and *Kunhardtia rhodantha*. The upper two horizons (0–5 and 5–10 cm) were sampled for bulk density analysis.

Profile 11

Area: Venezuelan Guayana, Cuao-Sipapo massif, Cerro Autana
Site: Cuao 6; N 04°53′, W 67°26′; 1,200 m a.s.l.
USDA classification (2006): Lithic *Terric* Haplosaprist
WRB classification (2006): Ombric Sapric Histosol (Hyperdystric)
Landscape: low-lying piedmont and peneplain area with a few outstanding isolated tepuis, located south of the Cuao table-shaped sandstone highlands (600–1,800 m a. s.l.) and north of the Autana river valley
Relief type: tepui meseta summit
Lithology/facies: sandstone–quarzite of the Roraima Group (Precambrian)
Landform: distal part of glacis fossilized by peat formation
Parent material: peat over glacis colluvium
General topography almost flat
Local topography flat to slightly concave; microtopography slightly hummocky
Poorly drained; slow permeability
Erosion and human influence: none
Surface characteristics: no particular evidence
Vegetation: tubiform tepui meadow (*Brocchinia hechtioides*, *Kunhardtia rhodantha*, and *Lagenocarpus* sp.) with isolated shrubs of *Heteropterys steyermarkii* and *Cyrilla racemiflora*
Described on 13.02.1993 by A. Zinck, P. Garcia, O. Huber

Profile description (depth in cm; colors are for moist soil unless otherwise stated; soil moist at 0–60 cm depth and saturated below 60 cm, at description date):

Oo	0–3	Loose folic material, including dead leaves, twigs, mosses, and lichens
Oi	3–15	Very dark brown (10YR 2/2); root mat with little moderately decomposed organic material
Oa1	15–40	Black (2.5Y 2/1); decomposed organic material; abundant fine and medium roots; massive structure; slightly plastic and slightly sticky
Oa2	40–60	Black (2.5Y 2/1); fairly well-decomposed organic material; frequent fine and very fine roots; massive structure; slightly plastic and slightly sticky; ^{14}C age 5,380 ± 50 BP (residue) and 6,745 ± 45 BP (extract)
2C	60–70	Black (2.5Y 2/1); fairly well-decomposed organic material; frequent fine and very fine roots; massive structure; slightly plastic and slightly sticky
3C	70–100	Brown (7.5YR 5/3); fine and medium sand; single grain
4R	100+	Sandstone–quartzite

Profile 12

Area: Venezuelan Guayana, Cuao-Sipapo massif, northwest of Autana river
Site: Cuao 7; N 04°55′, W 67°11′; 1,370 m a.s.l.
USDA classification (2006): Lithic Haplosaprist
WRB classification (2006): Endoleptic Ombric Hemic Histosol (Hyperdystric)
Landscape: deeply incised hilland surrounded by mountain ranges
Relief type: rocky domes, with grooves incising the backslopes
Lithology/facies: granodiorite with feldspar phenocrysts (rapakivi texture)
Landform: pseudokarstic depression at the summit of a bare dome
Parent material: peat
General topography steeply dissected (>100% slope)
Local topography slightly convex (3–5% slope); microtopography smooth
Poorly drained; slow permeability
Erosion and human influence: none
Surface characteristics: no particular evidence
Vegetation: shrubby broadleaved meadow (*Kunhardtia rhodantha*, *Myriocladus* sp., *Neurolepis* sp., and *Brocchinia tatei*) with isolated shrubs of *Cyrilla racemiflora* and *Bonnetia* sp.
Described on 13.02.1993 by A. Zinck, P. Garcia, O. Huber

Profile description (depth in cm; colors are for moist soil unless otherwise stated; soil moist at 0–40 cm depth and saturated below 40 cm, at description date):

Oo	0–3	Loose folic material, including dead and live roots, dead leaves, twigs, mosses, and lichens
Oe1	3–15	Brown (7.5YR 4/2); moderately decomposed organic material; abundant medium and fine dead and live roots; massive structure; slightly plastic and slightly sticky
Oe2	15–40	Very dark gray (5YR 3/1); moderately decomposed organic material; abundant medium and few coarse dead and live roots; massive structure; slightly plastic and slightly sticky; ^{14}C age 850 ± 35 BP
Oa	40–60	Black (5YR 2/1); fairly well-decomposed organic material; abundant medium and few coarse dead and live roots; massive structure; slightly plastic and slightly sticky; ^{14}C age 3,110 ± 35 BP
2R	60+	Granodiorite

Profile 13

Area: Venezuelan Guayana, northern edge of the Cuao-Sipapo massif
Site: Cuao 8; N 05°06′, W 67°18′; 1,240 m a.s.l.
USDA classification (2006): Typic Humaquept
WRB classification (2006): Histic *Epileptic* Fluvisol (Hyperdystric, Epiarenic)
Landscape: plateau highlands with inclusion of rocky domes
Relief type: horseshoe-shaped, steeply dissected rocky domes
Lithology/facies: granodiorite
Landform: pseudokarstic depression at the top of a bare dome
Parent material: peat over weathered granodiorite
General topography steeply dissected (>100% slope)
Local topography slightly convex (3–5% slope); microtopography irregular
Poorly drained; low to moderate permeability
Erosion and human influence: none
Surface characteristics: no particular evidence
Vegetation: tubiform tepui meadow (*Brocchinia hechtioides*, *Brocchinia melanacra*, *Abolboda macrostachya*, and *Kunhardtia rhodantha*) with shrubs (*Acanthella sprucei*, *A. pulchra*, *Phyllanthus* sp., and *Decagonocarpus cornutus*)
Described on 13.02.1993 by A. Zinck, P. Garcia, O. Huber

Profile description (depth in cm; colors are for moist soil unless otherwise stated; soil moist at description date):

Oe	0–16	Black (10YR 2/1); moderately decomposed organic material; abundant medium and fine dead and live roots; few coarse roots (2–4 mm); massive structure; slightly plastic and slightly sticky
2A	16–45	Very dark gray (10YR 3/1); coarse and medium sand; single grain; few fine roots
2Cr	45–47	Dark reddish gray (5YR 4/2); coarse sand from weathered granodiorite
2R	47+	Granodiorite

Profile 14

Area: Venezuelan Guayana, middle Orinoco lowlands, western border of the Cuao-Sipapo massif
Site: Cuao 9, approximately 1.5 km south of the Sipapo river; ca N 04°30′, W 67°30′; 90–95 m a.s.l.
USDA classification (2006): Aquic Quartzipsamment (possible spodic horizon in depth)
WRB classification (2006): Protic Hyperalbic Arenosol (Hyperdystric)
Landscape: river valley
Relief type: floodplain
Lithology/facies: alluvium
Landform: whitesand patch
Parent material: whitesands
General topography almost flat (1–3% slope)
Local topography flat (<1% slope); microtopography smooth
Moderately well drained; water table as from 90 to 100 cm depth; rapid permeability
Erosion and human influence: none
Surface characteristics: loose sand cover
Vegetation: open, medium-to-high shrubland with *Humiria balsamifera*, *Ilex divaricata*, *Heteropterys oblongifolia*, and *Pradosia schomburgkiana*, among others
Described on 14.02.1993 by A. Zinck, P. Garcia

Profile description (depth in cm; colors are for moist soil unless otherwise stated; soil moist at 100 cm depth):

Cover	0–3	Light gray (10YR 7/1); fine to medium sand; single grain; dry
A	3–17	Light brownish gray (10YR 6/2); fine to medium sand; single grain; frequent medium and fine roots
C1	17–43	Light gray (10YR 7/1); fine to medium sand; single grain; few yellowish brown mottles along roots; few fine roots
C2	43–80	Light gray (10YR 7/1); loamy sand; single grain; common yellowish brown mottles along roots; very few fine roots
C3	80–105	White (10YR 8/1); medium to coarse sand; single grain
C4	105–135+	White (10YR 8/1); medium to coarse sand; single grain; free water

Profile 15

Area: Venezuelan Guayana, middle Orinoco lowlands, western border of the Cuao-Sipapo massif
Site: Cuao 10, close to Raudal Ceguera, on the left bank of the Autana river; ca N 04°48′, W 67°30′; 95–100 m a.s.l.
USDA classification (2006): Aquic Quartzipsamment (possible spodic horizon in depth)
WRB classification (2006): Protic Hyperalbic Arenosol (Hyperdystric)
Landscape: river valley
Relief type: floodplain
Lithology/facies: alluvium
Landform: whitesand patch
Parent material: whitesands
General topography almost flat (1–2% slope)
Local topography flat (<1% slope); microtopography smooth
Moderately well drained; water table as from 100 cm depth; rapid permeability
Erosion: none; human influence: fire
Surface characteristics: loose sand cover
Vegetation: open, medium-to-high (1–3 m) burnt shrubland; grassland islands with Cyperaceae, *Abolboda* sp. (10–15% coverage)
Described on 14.02.1993 by A. Zinck, P. Garcia

Profile description (depth in cm; colors are for moist soil unless otherwise stated; soil moist at description date):

Cover	0–3	Light gray (10YR 7/1); coarse to medium sand; single grain; dry
A1	3–12	Gray (10YR 5/1); medium to coarse sand; single grain; common fine and medium roots
A2	12–32	Gray (10YR 6/1); medium to coarse sand; single grain; common fine horizontal roots
C1	32–55	Light gray (10YR 7/1); medium to fine sand; single grain; very few fine roots
C2	55–100	White (10YR 8/1); medium to fine sand; single grain
C3	100–113+	White (10YR 8/1); medium to fine sand; single grain; free water

Profile 16

Area: Venezuelan Guayana, center-north of the Cuao-Sipapo massif
Site: Cuao 11; N 04°59′, W 67°18′; 650 m a.s.l.
USDA classification (2006): Lithic Haplohemist
WRB classification (2006): Epileptic Ombric Hemic Histosol (Hyperdystric)
Landscape: narrow, elongated valley, ca. 1 km wide, deeply incised between alignments of granite domes with nearly vertical rocky walls and convex summits at 1,200–1,300 m a.s.l.
Relief type: concave valley bottom between domes
Lithology/facies: granodiorite
Landform: bottom of the valley
Parent material: peat over granodiorite
General topography almost flat (1% slope toward valley axis)
Local topography flat; microtopography smooth
Very poorly drained; water-saturated during the largest part of the year, with patches of stagnating water at the soil surface; free water table at 30 cm depth; slow permeability
Erosion and human influence: none
Surface characteristics: presence of bare granite slabs
Vegetation: glades with dense cover of *Brocchinia* (2 m high) and *Kunhardtia* (1.5 m high), together with 10% shrub, interspersed within open woodland 6–8 m high, including *Bonnetia* trees with straight, 5–10 cm thick stems
Described on 16.02.1993 by A. Zinck, P. Garcia, A. Gröger

Profile description (depth in cm; colors are for moist soil unless otherwise stated; soil saturated at description date):

Oo	0–2	Litter of moss-grown dead leaves; very few coarse (5–6 cm) woody roots at the soil surface
Oi	2–13	Dark reddish brown (5YR 2.5/2); mat of weaved live fine to coarse (0.5–5 mm) roots, tubular, air-conducting, compressible, mainly horizontally lying, forming 65–70% of the horizon; together with moderately decomposed organic material with loamy, slightly sticky touch
Oe	13–40	Dark reddish brown (5YR 2.5/2); moderately decomposed organic material with loamy, slightly sticky touch; fine to coarse (0.5–5 mm) live tubular roots, mainly horizontally lying (40–50%); ^{14}C age 185 ± 30 BP
2Cr	40–43	Very coarse angular sand and fine angular gravel (2–4 mm) mixed with some decomposed organic material
2R	43+	Granite–granodiorite

Profile 17

Area: Venezuelan Guayana, center-south of the Cuao-Sipapo massif
Site: Cuao 12; N 04°54′, W 67°21′; 1,500 m a.s.l.
USDA classification (2006): Typic Haplosaprist
WRB classification (2006): Ombric Hemic Histosol (Hyperdystric, *Thaptosapric*)
Landscape: table-shaped plateau highlands (700–2,000 m a.s.l.), developed on horizontally lying to slightly folded or tilted sandstone and quartzite formations of the Roraima Group (Precambrian)
Relief type: tepui meseta
Lithology/facies: sandstone
Landform: karstic depression
Parent material: peat over sandstone
General topography gently undulating, with rugged ruiniform rock outcrops and deep crevices
Local topography flat to slightly concave (2–3% slope); microtopography slightly hummocky
Very poorly drained; water-saturated during the largest part of the year, with patches of stagnating water at the soil surface; free groundwater from 30 cm downward; slow permeability
Erosion and human influence: none
Surface characteristics: low supporting capacity (high "*n*" value)
Vegetation: tubiform tepui meadow (*Brocchinia hechtioides*, *Kunhardtia rodantha*, and *Abolboda macrostachya*) with shrubs of *Cyrilla racemiflora*, *Maguireothamnus tatei*, and *Phyllanthus* sp., among others
Described on 18.02.1993 by A. Zinck, C. Acosta, O. Huber

Profile description (depth in cm; colors are for moist soil unless otherwise stated; soil saturated at description date):

Oo	0–3	Mat of white and black algae
Oi1	3–20	Black to dark reddish brown (5YR 2.5/1.5); mat of fine and medium (1–3 mm) horizontally lying roots; little moderately decomposed organic material with loamy, slightly sticky touch
Oi2	20–40	Similar to above horizon, but with coarser roots (up to 5–6 mm)
Oe	40–80	Black (5YR 2.5/1); moderately decomposed organic material with loamy, slightly sticky touch; common roots; ^{14}C age 2,990 ± 40 BP
Oa1	80–100	Black (10YR 2/1); fairly well-decomposed organic material with loamy, slightly sticky touch; very few roots; ^{14}C age 7,030 ± 50 BP
Oa2	100–150	Black (10YR 2/1) sapric material
2R	150+	Sandstone

Profile 18

Area: Venezuelan Guayana, center-west of the Cuao-Sipapo massif
Site: Cuao 13; N 05°00′, W 67°26′; 720 m a.s.l.
USDA classification (2006): Lithic Haplohemist
WRB classification (2006): Epileptic Rheic Fibric Histosol (Hyperdystric)
Landscape: table-shaped plateau highlands (700–2,000 m a.s.l.), developed on horizontally lying to slightly folded or tilted sandstone and quartzite formations of the Roraima Group (Precambrian)
Relief type: tepui meseta
Lithology/facies: sandstone
Landform: small karstic depression on a rocky slab slightly inclined; profile located at the edge of the depression; peat cover is being cut by a small creek running on the sandstone; buried karstic microtopography is being exhumed by stream cutting into the peat mantle
Parent material: peat over sandstone
General topography gently undulating (3–5% slope) with rugged ruiniform rock outcrops and deep crevices
Local topography flat to slightly concave (1–3% slope); microtopography slightly hummocky
Poorly drained; water-saturated during the largest part of the year, with patches of stagnating water at the soil surface; free groundwater from 35 cm downward; slow permeability
Erosion and human influence: none
Surface characteristics: migrating creek that cuts into the peat cover; low supporting capacity (high "*n*" value)
Vegetation: open shrubland (*Cyrilla racemiflora*, *Terminalia* sp., *Stenopadus campestris*, *Bonnetia* sp., and *Tepuianthus savannensis*) with herbaceous layer (*Abolboda* sp., *Brocchinia hechtoides*, *Lagenocarpus* sp., and *Panicum* sp.)
Described on 18.02.1993 by A. Zinck, O. Huber

Profile description (depth in cm; colors are for moist soil unless otherwise stated; soil saturated at description date):

Oi1	0–15	Dark reddish brown (5YR 2.5/2); mat formed of elongated leaves (from *Abolboda*), roots, fibers, and algae; little moderately decomposed organic material
Oi2	15–35	Dark reddish brown (5YR 2.5/2); many fine and medium (1–3 mm), white and red air-conducting roots (75%); little moderately decomposed organic material with loamy, slightly sticky touch
Oe	35–45	Black (5YR 2.5/1); moderately decomposed organic material with loamy, slightly sticky touch (50–60% of the layer); common roots; ^{14}C age 105 ± 0.4%
2R	45+	Sandstone

Profile 19

Area: Venezuelan Guayana, Maigualida massif, Cerro Yudi
Site: Maigualida 1; N 05°34′, W 65°13′; 2,150 m a.s.l.
USDA classification (2006): Typic Haplosaprist
WRB classification (2006): Ombric Sapric Histosol (Hyperdystric)
Landscape: mountain of hilly dome-like relief (1,000–2,400 m a.s.l.), consisting of igneous–metamorphic basement rocks of the Guayana Shield, with inclusions of table-shaped plateau highlands developed on horizontally lying to slightly folded sandstone and quartzite formations of the Roraima Group (Precambrian)
Relief type: narrow elongated vale (325 m × 60 m), receiving smaller lateral vales, flanked by granite outcrops with 20–30% slope
Lithology/facies: coarse-grained rapakivi granite rich in feldspars (about 50%)
Landform: hummock area of 10 m × 5 m and 20–30 cm elevation in relation to surrounding water pools
Parent material: peat over granite
General topography gently undulating (1–3% slope)
Local topography flat to slightly concave; microtopography with small hummocks 20–30 cm high, alternating with small water-filled alveoli colonized by *Drosera rotundifolia*
Very poorly drained; water-saturated during the largest part of the year, with patches of stagnating water at the soil surface; free water at 0–60 cm depth and moist but no free water at 60–130 cm depth; slow permeability
Erosion and human influence: none
Surface characteristics: low supporting capacity
Vegetation: grassland with short grasses and forbs (*Axonopus* sp., *Orectanthe sceptrum, Xyris* sp., and *Cephalocarpus rigidus*) and isolated shrubs of *Spathelia* sp. nov., *Ilex retusa*, *Cyrilla racemiflora*, and *Phyllanthus* sp., among others; pioneer vegetation with Rapateaceae, *Kunhardtia*, and *Tillandsia* along joints and cracks in granite
Described on 08.03.1996 by A. Zinck, P. Garcia, O. Huber

Profile description (depth in cm; colors are for moist soil unless otherwise stated; soil saturated at description date):

Oo	0–10	Mat of live and dead folic material
Oe1	10–20	Black to very dark brown (10YR 2/1.5); mixture of many fine and medium roots (1–3 mm) and moderately decomposed organic material (40%)
Oe2	20–40	Black to very dark brown (10YR 2/1.5); mixture of many fine and medium roots (1–3 mm) and moderately decomposed organic material (30–40%)
Oa1	40–60	Black to very dark brown (10YR 2/1.5); mixture of many fine and medium roots (1–3 mm) and decomposed organic material (30%)
Oa2	60–80	Black (10YR 2/1); decomposed organic material with common roots; some mineral material; ^{14}C age 5,190 ± 40 BP
Oa3	80–110	Similar to the above horizon; fewer roots
Oa4	110–130	Similar to the above horizon; no roots; ^{14}C age 6,580 ± 50 BP
2R	130+	Granite

Profile 20

Area: Venezuelan Guayana, Maigualida massif, headwaters of the Asita river
Site: Maigualida 2; N 05°32′, W 65°09′; 1,700 m a.s.l.
USDA classification (2006): Lithic Haplosaprist
WRB classification (2006): Epileptic Rheic Sapric Histosol (Hyperdystric)
Landscape: mountain of hilly dome-like relief (1,000–2,400 m a.s.l.), consisting of igneous–metamorphic basement rocks of the Guayana shield, with inclusions of table-shaped plateau highlands developed on horizontally lying to slightly folded sandstone and quartzite formations of the Roraima Group (Precambrian)
Relief type: 300 m long peat glacis (slope peat) at the footslope of a granite hill
Lithology/facies: coarse-grained rapakivi granite rich in feldspars
Landform: contact area between the distal sector of a glacis and the left bank of a blackwater creek
Parent material: peat over granite
General topography undulating in the upper (proximal) part of the glacis (5–8% slope), gently undulating in the central part (3–5% slope) and almost flat in the lower (distal) part (1–3% slope)
Local topography almost flat (1–3% slope); microtopography slightly hummocky
Very poorly drained; water-saturated during the largest part of the year, with patches of stagnating water at the soil surface; free water table at 35 cm depth; slow permeability
Erosion and human influence: none
Surface characteristics: no particular evidence
Vegetation: broadleaved meadow, very dense (*Stegolepis albiflora*, *Orectanthe sceptrum*, and *Panicum chnoodes*) mixed with shrubs (*Clusia* sp., *Spathelia* sp. nov., *Cyrilla racemiflora*, and *Ilex retusa*)
Described on 09.03.1996 by A. Zinck, P. Garcia, O. Huber

Profile description (depth in cm; colors are for moist soil unless otherwise stated; soil saturated at description date):

Oo	0–10	Mat of live and dead folic material
Oe	10–20	Very dark brown (10YR 2/2); mixture of many fine and medium roots (1–3 mm) and decomposed organic material (40%)
Oa	20–35	Very dark brown (10YR 2/2); similar to the above horizon but with more roots; decomposed organic material with ^{14}C age 2,530 ± 40 BP; a flat, 100 cm long, 10 cm large, and 1–3 cm thick red woody root fragment, horizontally lying on top of the following horizon (30–35 cm depth), with ^{14}C age 2,350 ± 30 BP
2Cr	35–40/45	Coarse quartz-sand and fine gravel (1–5 mm), mixed with decomposed black (10YR 2/1) organic material (40%); ^{14}C age 3,610 ± 40 BP
2R	40/45+	Granite starting with a 1–2 cm thick weathered top layer

A.3 Laboratory Data

Table A.4 Physical properties of organic and organomineral soil materials

Profile	Site	Layer	Depth	Texture	Root1	Root2	Root3	Dry M	NRF1	NRF2	NRF3	RF	Bd1	Bd2	Min	Org	FWC	H_2O
P1	Marahuaka I-1	Oi	0–13	p/sl				30.2							77	23	231	4.9
		2C1	13–40	sicl				48.3							88.1	11.9	107	1.9
		2C2	40–60	sicl				71.9							97.8	2.2	39	0.6
P2	Marahuaka I-2	Oe	5–20	p				8.7							43.2	56.8	1,052	10.4
		Oa1	20–60	p				10.1							36.4	63.6	890	10.1
		Oa2	60–70	p				11.1							47.6	52.4	802	8.5
		2C	70–95	sl				20.3							83.6	16.4	392	4.0
P5	Marahuaka II-5	Oi1	5–20	p				7.4							30.2	69.8	1,250	8.6
		Oi2	20–60	p				13.1							62	38	662	7.4
		Oe	60–85	p				22.2							79.8	20.2	350	3.8
		2C	85–120	p				24.1							83.6	16.4	316	4.3
P6	Huachamakari I-6	Oi1	0–12	p													1,231	
		Oi2	12–45	p													1,400	
		Oe	45–60	p				12.0							70.4	29.6	733	8.5
P7	Cuao 1	Oe	0–11	p	3.1	0.44	3.5	12.0	40	30	30	20	0.70		5.8	94.2	732	
		2A1	11–27	p	1.1	0.18	0.3	59.0							92.9	7.1	70	
		2A2	27–48	sl				73.6							97.1	2.9	36	
		3C	48–70	l-sl				60.2							84	16	66	
P8	Cuao 2	Oe1	5–30	p	1.0	0.24	2.2	10.5	30	30	20	10	0.54		7.3	92.7	851	
		Oe2	30–55	p				11.2	30	30	20	5	0.60		5.9	94.1	791	
		Oa	55–87	p	1.2	0.16	1.1	14.1	20	10	10	5	0.56		7.3	92.7	609	
P9	Cuao 3	Oi	0–10	p	3.7	0.49	2.0	23.8	30	20	20	5	0.70		64.9	35.1	321	
		Oe	10–20	p	2.0	0.26	0.7	35.3	20	20	20	10	0.74		78.5	21.5	183	
		2A1	20–40	p	2.7	0.27	0.6	42.6	50	40	40	30	0.76		80.7	19.3	135	
		2A2	40–60	p	1.6	0.12	0.3	35.3	60	50	40	30	0.72		86.4	13.6	184	
P10	Cuao 4	A1	0–5	p				55.2	50	30	30	5	0.66	0.78	71.3	28.7	81	
		A2	5–10/13	ls				76.2						1.34	95.2	4.8	31	
P11	Cuao 6	Oi	3–15	p	14.2	1.23	8.7	12.9	40	30		5	0.60	0.11	16.5	83.5	677	
		Oa1	15–40	p	6.4	0.61	1.6	36.8	10	10	5	5	0.70	0.24	74.1	25.9	171	
		Oa2	40–60	p	1.4	0.1	0.2	48.7	50	50	40	20	0.76	0.23	84.8	15.2	106	
		2C	60–70	p	0.9	0.17	0.2	68.4							92.2	7.8	46	

P12	Cuao 7	Oe1	3–15	p	14.3	1.34	16.7	6.7	60	40	40	30	0.60	0.10	2.9	97.1	1,389	
		Oe2	15–40	p				8.8	10	10	5	5	0.78	0.23	2.3	97.7	1,042	
		Oa	40–60	p				15.6	10	10	5	5	0.58	0.14	7.6	92.4	542	
P13	Cuao 8	Oe	0–16	p	2.9	0.34	2.6	13.9	20	20	20	5	0.48		9.8	90.2	618	
		2A	16–45	s				85.4							95	5	17	
P16	Cuao 11	Oi	2–13	p	9.3	0.76	9.2	7.5	60	50	40	10	0.56		0.6	99.4	1,228	
		Oe	13–40	p	5.6	0.44	4.9	8.9	50	40	30	20	0.60		2	98	1,065	
		2Cr	40–43	s	0.9	0.13	0.3	46.1							42.7	57.3	117	
P17	Cuao 12	Oi1	3–20	p	7.0	0.36	4.6	7.5	30	10	10	5	0.58		0	100	1,237	
		Oi2	20–40	p				7.4	20	10	5	5	0.56		0	100	1,254	
		Oe	40–80	p				7.6	15	5	5	5	0.55		0.86	99.1	1,209	
		Oa1	80–100	p				9.0	5	5		5	0.60		12.6	87.4	1,008	
P18	Cuao 13	Oi1	0–15	p	13.1	0.89	13.9	5.3	60	50		50	0.62		36.9	63.1	1,782	
		Oi2	15–35	p	6.9	0.9	9.3	8.8	40	25	25	10	0.65		2	98	1,037	
		Oe	35–45	p	13.8	1.66	5.6	27.9	60	60	40	40	0.72		1.8	98.2	258	
P19	Maigualida 1	Oe1	10–20	p	9.4	0.72	11.0	5.8	60	30	20	20			9.2	90.8	1,628	14.9
		Oe2	20–40	p	16.8	1.17	31.0	2.6	40	40	30	30			8	92	3,710	13.6
		Oa1	40–60	p	1.6	0.13	6.1	2.0	20	10	5	<5			8.9	91.1	4,936	11.1
		Oa2	60–110	p				1.9	<5	<5	<5	<5			12.2	87.8	5,229	11.1
P20	Maigualida 2	Oe	10–20	p	15.1	2.0	34.5	3.8	50	20	10	5			9.3	90.7	2,562	16.3
		Oa	20–35	p	7.2	7.2	10.2	8.9	15	10	5	<5			26	74	1,025	8.7

Depth in cm; texture: *p* peaty, *s* sand, *ls* loamy sand, *sl* sandy loam, *l* loam, *sicl* silty clay loam; Root1: % moist roots in field moist soil material; Root2: % dry roots in field moist soil material; Root3: % dry roots in dry soil matter (dry matter + dry roots = 100%); Dry M: % oven-dry matter at 110°C in moist soil; NRF1 to NRF3: % unrubbed fiber content at three steps of increasing treatment strength; RF: % rubbed fiber content; Bd1: bulk density of field moist soil in Mg m^{-3}; Bd2: bulk density of dry soil in Mg m^{-3} (in both Bd cases, packing the field moist sample into a known volume); Min and Org: % mineral matter and organic matter by loss on ignition at 400°C; FWC: % field water content; H_2O: % moisture in air-dry soil upon oven drying

Table A.5 Chemical properties of organic and organomineral soil materials

Profile	Site	Layer	Depth	Texture	OM1	OM2	pH1	pH2	Ca	Mg	Na	K	TB	Acd1	Acd2	Al	CEC1	CEC2	BS1	BS2	P
P1	Marahuaka I-1	Oi	0–13	p/sl	6.7	23.0			0.2	0.3	0.2	0.3	1.0	40			41		2		
		2C1	13–40	sicl	6.7	11.9			0.1	T	0.1	T	0.2	26			26		1		
		2C2	40–60	sicl		2.2			T	T	0.1	T	0.1	8			8		1		
P2	Marahuaka I-2	Oe	5–20	p	14.1	56.8			0.4	0.6	0.8	0.4	2.2	54			57		4		
		Oa1	20–60	p	11.4	63.6			0.1	0.3	0.2	T	0.6	24			24		2		
		Oa2	60–70	p	10.7	52.4			0.2	0.2	0.2	T	0.6	24			24		2		
		2C	70–95	sl	4.0	16.4			0.2	0.1	0.2	T	0.5	13			13		4		
P5	Marahuaka II-5	Oi1	5–20	p	26.2	69.8			0.4	0.4	0.2	0.1	1.1	44			45		2		
		Oi2	20–60	p	9.5	38.0			0.2	0.3	0.2	0.1	0.8	53			54		1		
		Oe	60–85	p	12.1	20.2			0.1	0.2	0.2	0.1	0.6	43			44		1		
		2C	85–120	p	13.4	16.4			0.1	T	0.2	T	0.3	46			46		1		
P6	Huachamakari I-6	Oi1	0–12	p					0.2	0.8	0.3	0.3	1.6								
		Oi2	12–45	p					0.1	0.3	0.2	0.1	0.7								
		Oe	45–60	p		29.6								53							
P7	Cuao 1	Oe	0–11	p	35.0	94.2	3.4	3.2	0.7	0.8	0.5	1.1	3.1	61	3.8	T	64	6.9	5	45	T
		2A1	11–27	p	10.2	7.1	3.4	3.4	0.2	0.1	0.5	0.2	1.0	21	1.2	T	22	2.2	5	45	T
		2A2	27–48	sl	4.7	2.9	3.5	3.1	0.1	T	0.2	0.4	0.7	16	2.0	T	17	2.7	4	26	T
		3C	48–70	l-sl	8.1	16.0	3.4	3.1	0.3	T	0.6	0.1	1.0	35	3.4	T	36	4.4	3	23	T
P8	Cuao 2	Oe1	5–30	p	42.4	92.7	3.2	3.5	1.5	0.5	0.7	1.1	3.8	108	4.4	T	112	8.2	3	46	T
		Oe2	30–55	p	39.0	94.1	3.2	3.2						57							5.8
		Oa	55–87	p	30.9	92.7	3.2	3.3	0.6	0.1	0.6	0.3	1.6	45	3.8	T	47	5.4	3	29	1.8
P9	Cuao 3	Oi	0–10	p	18.1	35.1	4.0	3.6	2.1	0.5	0.4	0.3	3.3	45	1.6	T	48	4.9	7	67	T
		Oe	10–20	p	19.0	21.5	3.8	3.4	1.1	0.2	0.6	0.3	2.2	19	0.6	T	21	2.8	10	79	T
		2A1	20–40	p	16.0	19.3	3.5	3.3	0.7	0.1	0.6	0.1	1.5	21	1.4	T	23	2.9	7	54	T
		2A2	40–60	P	14.0	13.6	3.6	3.2	0.5	0.1	0.7	0.1	1.4	100	1.2	T	101	2.6	1	54	T
P10	Cuao 4	A1	0–5	p	30.9	28.7	4.0	3.6	0.2	T	0.6	0.2	1.0	110	9.8	4.8	111	10.8	1	9	T
		A2	5–10/13	ls	10.2	4.8	3.6	3.5	0.1	T	0.3	0.2	0.6	41	4.8	2.8	42	5.4	1	11	T
		A3	10/13–28	ls	2.1		4.5	4.0	T	T	0.1	T	0.1	8		7.7		7.8		1.3	
		C1	28–70	s	1.2		5.1	4.2	T	T	0.12	T	0.12	4		4.4		4.5		3	
		C2	70–90	s	0.34		5.5	4.3	T	T	0.12	T	0.12	0.6		0.6		0.7		17	
		C3	90–135	s	0.34		5.4	4.3	T	T	0.07	T	0.07	0.2		0.2		0.3		25	
P11	Cuao 6	Oi	3–15	p	53.1	83.5	4.2	3.6	0.2	0.1	0.8	0.6	1.7	185	9.0	5.2	187	10.7	1	16	0.5

		Oa1	15–40	p	29.0	25.9	4.2	3.6	T	T	1.1	0.2	1.3	117	10.2	7.0	118	11.5	1	11	T
		Oa2	40–60	p	22.2	15.2	4.3	3.0	0.2	T	0.3	0.1	0.6	81	6.4	4.0	82	7.0	1	9	T
		2C	60–70	p	3.4	7.8	4.2	3.5	0.2	T	0.3	0.1	0.6	31	1.4	T	32	2.0	2	30	T
		3C	70–100	s	1.0		5.0	4.2	T	T	0.07	T	0.07					3.0		2.0	
P12	Cuao 7	Oe1	3–15	p	50.5	97.1	4.6	3.6	1.7	0.9	1.1	1.6	5.3	136			141		4		
		Oe2	15–40	p	47.0	97.7	4.0	3.4	0.1	0.1	1.0	0.3	1.5	96	3.2	1.0	98	4.7	2	32	2.3
		Oa	40–60	p	30.9	92.4	4.2	3.4	0.1	T	0.8	0.2	1.1	66	5.2	3.4	67	6.3	2	17	5.8
P13	Cuao 8	Oe	0–16	p	45.7	90.2	4.0	3.6	0.5	0.1	0.5	0.5	1.6	58	5.2	1.6	60	6.8	3	24	3.8
		2A	16–45	s	10.7	5.0	5.0	3.7	0.1	T	0.7	0.1	0.9	58	6.0	4.0	59	6.9	1	13	
P16	Cuao 11	Oi	2–13	p	46.4	99.4	4.6	3.5	0.4	0.4	0.4	1.3	2.5	101	3.4	T	104	5.9	2	42	T
		Oe	13–40	p	27.6	98.0	4.3	3.2	T	T	0.2	0.5	0.7	98	3.8	T	99	4.5	1	16	T
		2Cr	40–43	s	8.6	57.3	4.5	3.8	T	T	0.3	0.1	0.4	12	4.2	3.4	12	4.6	3	9	T
P17	Cuao 12	Oi1	3–20	p	50.5	100	4.6	3.8	0.4	0.4	0.4	0.6	1.8	96	2.4	T	98	4.2	2	43	5.8
		Oi2	20–40	p	35.7	100	4.3	3.6	T	0.1	0.6	0.3	1.0	96	2.2	T	97	3.2	1	31	7.0
		Oe	40–80	p	27.7	99.1	4.3	3.6	T	0.1	1.0	0.2	1.3	96	2.8	T	97	4.1	1	32	8.2
		Oa1	80–100	p	29.7	87.4	4.2	3.6	T	0.1	0.3	0.1	0.5	60	2.6	T	61	3.1	1	16	6.8
P18	Cuao 13	Oi1	0–15	p	47.8	63.1	5.2	3.8	0.7	0.6	0.7	0.6	2.6	46	2.5	T	51	4.8	5	54	T
		Oi2	15–35	p	28.9	98.1	4.7	3.5	0.3	0.1	0.9	0.3	1.6	36	2.2	T	37	3.8	5	42	T
		Oe	35–45	p	17.2	98.2	4.5	3.5	0.1	0.1	0.3	0.1	0.6	29	1.8	T	30	2.4	2	25	T
P19	Maigualida 1	Oe1	10–20	p		90.8	5.0	5.2	2.7	0.9	0.5	1.4	5.5	41			47		12		4.0
		Oe2	20–40	p		92.0	4.6	5.0	2.2	1.2	0.6	0.2	4.2	49			54		8		24.0
		Oa1	40–60	p		91.1	4.4	4.7	1.6	0.5	0.3	0.1	2.5	47			50		5		6.0
		Oa2	60–110	p		87.8	4.5	4.7	1.4	0.1	0.2	T	1.7	49			51		3		1.0
P20	Maigualida 2	Oe	10–20	p		90.7	5.3	5.0	3.4	2.1	0.6	1.3	7.4	41			48		15		24.0
		Oa	20–35	p		74.0	5.0	4.6	2.1	0.7	0.4	0.2	3.4	16			20		17		18.0
		2Cr	35–40/45	s			5.2	4.6	0.3	T	0.2	0.1	0.6	8			9		7		4.0

Depth in cm; Texture: *p* peaty, *s* sand, *ls* loamy sand, *sl* sandy loam, *l* loam, *sicl* silty clay loam; OM1: % organic matter by Walkley–Black (= organic carbon × 1.724); OM2: % organic matter by loss on ignition at 400°C; pH1: measured at field moisture; pH2: measured in $CaCl_2$; Ca, Mg, Na, K, Al: exchangeable calcium, magnesium, sodium, potassium, aluminum; TB: total bases (all cations in cmol(+) kg^{-1}); Acd1: extractable (total) acidity determined with $BaCl_2$-trietanolamine; Acd2: exchangeable (salt-replaceable) acidity determined with KCl (cmol(+) kg^{-1}); CEC: cation exchange capacity; CEC1: sum of bases plus total acidity; CEC2: sum of bases plus salt-replaceable acidity (cmol(+) kg^{-1}); BS: % base saturation; BS1: from CEC1; BS2: from CEC2; P: available phosphorus (mg kg^{-1})

Table A.6 Physical properties of mineral soil materials

Profile	Site	Horizon	Depth	VCoS	CoS	MS	FS	VFS	Sand	Silt	Clay	Texture	Bd
P4	Duida I-4	A	0–10	0.2	0.7	10.5	59.6	17.5	88.5	7.4	4.1	s	
		C1	10–20	0	0.1	17.0	66.7	12.1	95.8	0.1	4.1	s	
		C2	20–50	6.7	23.9	56.0	6.0	3.2	95.8	0.1	4.1	s	
P10	Cuao 4	A1	0–5									sl	0.78
		A2	5–10/13									ls	1.34
		A3	10/13–28	0	3.3	27.0	45.8	9.6	85.7	6.3	8.0	sl	
		C1	28–70	0.1	3.2	24.7	52.3	8.8	89.1	5.5	5.4	s	
		C2	70–90	0.1	2.6	23.1	56.2	10.2	92.2	4.9	2.9	s	
		C3	90–135	0.1	2.5	22.0	55.3	11.6	91.5	5.6	2.9	s	
(P10)	Cuao 5	A1	0–5									s	0.49
		A2	5–10									s	0.95
P11	Cuao 6	3C	70–100	0.1	2.7	26.6	53.0	8.9	91.3	5.4	3.3	s	
P14	Cuao 9	Cover	0–3	1.8	7.4	25.6	44.6	13.2	92.6	4.1	3.3	s	
		A	3–17	3.9	13.6	28.5	37.9	10.5	94.4	2.3	3.3	s	
		C1	17–43	10.5	14.2	23.7	36.4	9.7	94.5	2.2	3.3	s	
		C2	43–80	13.0	15.8	17.7	26.5	6.6	79.6	19.6	0.8	ls	
		C3	80–105	14.7	25.9	27.6	23.0	4.6	95.8	0.9	3.3	s	
		C4	105–135	11.1	31.7	33.0	17.6	3.3	96.7	0	3.3	s	
P15	Cuao 10	Cover	0–3	14.6	47.7	29.1	3.9	0.8	96.1	0.6	3.3	s	
		A1	3–12	7.1	30.9	31.7	15.0	7.7	92.4	4.7	2.9	s	
		A2	12–32	9.0	28.5	35.1	17.9	6.0	96.7	0	3.3	s	
		C1	32–55	1.9	21.8	32.3	26.7	6.9	89.6	10.0	0.4	s	
		C2	55–100	15.2	18.4	30.7	25.6	5.6	95.5	1.2	3.3	s	
		C3	100–113	14.7	20.9	29.8	26.7	5.1	97.2	2.4	0.4	S	

Depth in cm; particle size distribution in %; VCoS: very coarse sand; CoS: coarse sand; MS: medium sand; FS: fine sand; VFS: very fine sand; Bd: bulk density of dry soil in Mg m^{-3}

Table A.7 Chemical properties of mineral soil materials

Profile	Site	Horizon	Depth	OC	N	C/N	pH1	pH2	EC	Ca	Mg	K	Na	TB	EA	CEC	BS
P4	Duida I-4	A	0–10	0.7						0.1	T	T	0.2	0.3	0.8	1.1	27
		C1	10–20	0.4						0.1	T	T	0.1	0.2	0.2	0.4	50
		C2	20–50	0.04						0.6	1.7	0.3	0.6	3.2			
P10	Cuao 4	A1	0–5				4.0										
		A2	5–10/13				3.6										
		A3	10/13–28	1.2	0.11	11	4.5	4.0	<0.1	T	T	T	0.1	0.1	7.7	7.8	1
		C1	28–70	0.7	0.06	12	5.1	4.2	<0.1	T	T	T	0.12	0.12	4.4	4.5	3
		C2	70–90	0.2	0.02	10	5.5	4.3	<0.1	T	T	T	0.12	0.12	0.6	0.72	17
		C3	90–135	0.2	0.02	10	5.4	4.3	<0.1	T	T	T	0.07	0.07	0.2	0.27	25
P11	Cuao 6	3C	70–100	0.6	0.05	12	5.0	4.2	<0.1	T	T	T	0.07	0.07	2.9	2.97	2
P14	Cuao 9	Cover	0–3	0.4	0.04	10	5.0	3.7	<0.1	T	T	T	0.07	0.07	0.6	0.67	10
		A	3–17	0.2	0.01	20	5.3	3.9	<0.1	T	T	T	0.07	0.07	1.0	1.07	6
		C1	17–43	0.1	0.01	10	5.7	4.3	<0.1	T	T	T	0.07	0.07	0.8	0.87	8
		C2	43–80	0.1	0.01	10	5.8	4.7	<0.1	T	T	T	0.07	0.07			
		C3	80–105	0.03	0.01	3	6.0	4.7	<0.1	T	T	T	0.07	0.07			
		C4	105–135	0.03	0.01	3	6.3	5.2	<0.1	T	T	T	0.07	0.07	0.8	0.87	8
P15	Cuao 10	Cover	0–3	0.5	0.05	10	5.2	4.0	<0.1	T	T	T	0.07	0.07	0.2	0.27	26
		A1	3–12	0.4	0.04	10	5.4	3.9	<0.1	T	T	T	0.07	0.07	1.5	1.57	4
		A2	12–32	0.4	0.04	10	5.4	4.2	<0.1	T	T	T	0.07	0.07	0.2	0.27	26
		C1	32–55	0.1	0.01	10	5.8	4.8	<0.1	T	T	T	0.07	0.07			
		C2	55–100	0.1	0.01	10	6.2	5.1	<0.1	T	T	T	0.07	0.07	0.2	0.27	26
		C3	100–113	0.03	0.01	3	6.4	5.4	<0.1	T	T	T	0.07	0.07	0.4	0.47	15

Depth in cm; OC: % organic carbon by Walkley–Black; N: % nitrogen; C/N: carbon/nitrogen ratio; pH1: measured in 1:2 water; pH2: measured in 1:2 $CaCl_2$; EC: electric conductivity in dS m^{-1}; Ca, Mg, K, Na: exchangeable bases; TB: total bases; EA: exchangeable acidity (H+Al); CEC: cation exchange capacity by sum of cations (exchangeable cations in cmol(+) kg^{-1}); BS: % base saturation

Index